Catalog of the Native Land and Freshwater Molluscs of the Hawaiian Islands

ROBERT H. COWIE
NEAL L. EVENHUIS
CARL C. CHRISTENSEN

BACKHUYS PUBLISHERS
Leiden
1995

Authors' address:

Department of Natural Sciences
Bishop Museum, P.O. Box 19000
Honolulu, Hawaii 96817-0916, USA

Contribution No. 94-012 to the Hawaii Biological Survey

Front cover: *Achatinella lila* Pilsbry, with newborn young. (Photograph William P. Mull)

CIP-DATA KONINKLIJKE BIBLIOTHEEK, DEN HAAG

Cowie, Robert H.

Catalog of the native land and freshwater molluscs of the Hawaiian Islands / Robert H. Cowie, Neal L. Evenhuis, Carl C. Christensen. - Leiden : Backhuys Publishers
With index, ref.
ISBN 90-73348-43-9
NUGI 823
Subject headings: snails ; Hawaii ; catalogues / mollusca ; Hawaii ; catalogues.

Copyright: Backhuys Publishers, Leiden, The Netherlands, 1995

All rights reserved. Nothing from this publication may be reproduced, stored in a computerized system or published in any form or in any manner, including electronic, mechanical, reprographic or photographic without prior written permission from the publishers, Backhuys Publishers, P.O. Box 321, 2300 AH Leiden, The Netherlands.

Insofar as photocopies from this publication are permitted by the Copyright Act 1912, Article 16B and Royal Netherlands Decree of 20 June 1974 (Staatsblad 351) as amended in Royal Netherlands Decree of 23 August 1985 (Staatsblad 471) and by Copyright Act 1912, Article 17, the legally defined copyright fee for any copies should be transferred to the Stichting Reprorecht (P.O. Box 882, 1180 AW Amstelveen, The Netherlands). For reproductions of parts of this publication in compilations such as anthologies or readers (Copyright Act 1912, Article 16), permission must be obtained from the publisher.

Printed in The Netherlands.

Dedicated to Charles Montague Cooke, Jr.
(1874–1948)
in honor of his unparalleled contributions to
Pacific malacology

FOREWORD

The loss of biological diversity has become a global crisis and our efforts to do something about it have been greatly hampered by lack of information. For example, we don't even know to an order of magnitude the number of different kinds of organisms that inhabit the earth. This information is crucial to the development of effective strategies for conservation of biological diversity and the preservation of fragile native ecosystems throughout the world that are vulnerable to damaging disturbance by human activities including the introduction of other alien species.

Bishop Museum has long recognized the value of information deriving from the classification and distribution of species. Beginning in the late 1800s, the Museum established strong programs to study and document the plants and animals of Hawaii. That effort, which continues today, has become the largest single source of primary information on Hawaiian organisms. During the past 15 years the Museum has been involved in a concerted effort to modernize its collection facilities and computerize collection records in order to make these information resources more readily available to researchers, resource managers and others, and has directed its research efforts in the natural sciences to the publication of comprehensive treatments of Hawaiian organisms. In recognition of the importance of these efforts, in 1992 the Hawaii State Legislature designated Bishop Museum as the Hawaii Biological Survey and charged it with the task of developing a comprehensive inventory of the plants and animals of the state. Work toward this goal is being conducted by the Museum in collaboration with numerous partners including the National Biological Survey and other federal, state, and local agencies.

The information compiled in these pages represents a major milestone in this program. It is the first comprehensive catalog of the native land and freshwater snails of Hawaii. It brings together in one place the important pioneering studies of Pilsbry, Cooke, Baker, Welch, Kondo, and others, and for the first time provides an exact enumeration of native land and freshwater snail species in the Hawaiian Archipelago. In particular this work represents a synthesis of the far-sighted studies originally undertaken by the late C. Montague Cooke, Jr. who, from 1902-1948, developed Bishop Museum's Pacific land snail collection into the world's largest, providing much of the basic taxonomic foundation on which this catalog is based.

The geographic isolation and rugged terrain of Hawaii have combined to provide ideal conditions for species formation. These features have resulted in the evolution of a spectacular assemblage of organisms from just a small number of founder species. The almost 800 snail species enumerated in this catalog probably evolved from only a few dozen founders. To put this in a larger perspective, this number of Hawaiian snails is comparable to the total number of land snail species in the entire fauna of North America north of Mexico.

Unfortunately, these same native Hawaiian ecosystems are tremendously fragile, and alien introductions, habitat modification, and urbanization have had seriously deleterious effects on the snail fauna. This rich assemblage of species is disappearing at an alarming rate. By some estimates, up to 75% of the species are now extinct and many of those surviving are endangered or threatened. Scientists and others are combining their efforts in an attempt to save the remaining fauna from possible extinction. This catalog will be a fundamental reference to systematists, evolutionary biologists, ecologists, conservation biologists and resource managers in providing them with essential information to assist in efforts to protect this highly endangered fauna.

——Allen Allison, Assistant Director
Research and Scholarly Studies, Bishop Museum

CONTENTS

The following contents table provides numbers of genera and species, by family, of native species, questionably native species (includes terrestrial species that are not known for certain to occur in the Hawaiian Islands and freshwater species that occur in the Hawaiian Islands but may be introduced), and non-native species described from Hawaiian material. Other non-native species that were described from non-Hawaiian material are not included in the catalog nor in this table.

					Page
Foreword					iii
Acknowledgments					1
Introduction					3
Explanatory Information					5

Systematic Catalog

Family	Habitat	Gen. (spp.) native	Gen. (spp.) ?native	Gen. (spp.) non-native	Page
Neritidae	freshwater	1 (2)	—	—	13
Hydrocenidae	terrestrial	1 (2)	—	—	16
Helicinidae	terrestrial	2 (14)	1 (5)	—	17
Hydrobiidae	freshwater	—	1 (1)	—	24
Thiaridae	freshwater	—	2 (5)	—	25
Ellobiidae	terrestrial	7 (10)	—	—	27
Lymnaeidae	freshwater	2 (5)[1]	—	—	31
Ancylidae	freshwater	—	1 (1)	—	34
Achatinellidae	terrestrial	12 (209)[1]	? (3)[2]	1 (1)	35
Amastridae	terrestrial	8 (325)	1 (1)	—	89
Pupillidae	terrestrial	5 (56)	1 (1)	2 (2)	130
Ferussaciidae	terrestrial	—	—	1 (1)	140
Subulinidae	terrestrial	—	—	3 (4)	141
Endodontidae	terrestrial	4 (33)	1 (1)	—	143
Punctidae	terrestrial	1 (1)	—	—	147
Succineidae	terrestrial	3 (42)	1 (3)	—	148
Helicarionidae	terrestrial	4 (60)	—	—	153
Zonitidae	terrestrial	4 (10)	1 (1)[3]	1 (1)	162
Milacidae	terrestrial	—	—	1 (1)	166
Limacidae	terrestrial	—	—	2 (4)	167
Incertae sedis		1 (1)	1 (1)	—	169
Totals		**55 (770)**	**12 (23)[2]**	**11 (14)**	

[1] Includes 1 taxon of uncertain generic placement.
[2] Two taxa of uncertain generic placement and a single species in *Tornatellides*.
[3] A single taxon of uncertain placement in the "zonitoid" families.

Checklist	170
Bibliography	205
Index of Listed Taxa	236

ACKNOWLEDGMENTS

For assisting us with obtaining literature we thank Peter Mordan, John Taylor and the other staff of the mollusc section of The Natural History Museum (London) [BMNH], Gary Rosenberg (Academy of Natural Sciences, Philadelphia [ANSP]), Ken Boss and Silvard Kool (Museum of Comparative Zoology, Harvard University [MCZ]), Alan Kabat (National Museum of Natural History, Washington, D.C. [USNM]) and Henk Mienis (Hebrew University of Jerusalem). The library staffs of the BMNH, USNM and Bishop Museum assisted us greatly in our search for often obscure literature.

For review or comments on all or part of the manuscript we thank Rüdiger Bieler (Field Museum of Natural History, Chicago), Ken Boss (MCZ), Phil Colman (Australian Museum, Sydney), Mike Hadfield (University of Hawaii, Honolulu), Alan Kabat (USNM), Fred Naggs (The Natural History Museum, London), Scott Miller (Bishop Museum, Honolulu), Winston Ponder (Australian Museum, Sydney) and Gary Rosenberg (Academy of Natural Sciences, Philadelphia).

In addition, Anthea Gentry (ICZN), Bob Hershler (USNM), Richard Preece (University of Cambridge), Barry Roth (California), Phil Tubbs (ICZN) and Catherine Unabia (University of Hawaii, Honolulu) assisted us on more specific points. Gary Rosenberg and Silvard Kool searched for information for us in the ANSP and MCZ collections, respectively.

The Cooke Foundation and the John D. and Catherine T. MacArthur Foundation are gratefully acknowledged for providing partial support for this project.

This publication constitutes Contribution No. 94-012 of the Hawaii Biological Survey

INTRODUCTION

This catalog lists all the species-group and genus-group names that have been applied to the native land and freshwater mollusc fauna of the Hawaiian Islands. Full citations are given for each name. The current status of each name is indicated. The type locality for each available species-group name is provided and the island(s) on which each valid taxon is known to occur are indicated. Following the main body of the catalog, a checklist is provided that simply lists all the names, for ease of reference. Background information on each major group (family or genus) and an introduction to other relevant literature are also provided, making the catalog a basic point of reference for studies of the Hawaiian land and freshwater mollusc fauna.

The native Hawaiian land snail fauna is probably the most diverse in the world for an area the size of the Hawaiian Islands. Not only does it manifest immense diversity, but all except 2 to 4 (plus perhaps a number of the Ellobiidae) of the 763 nomenclaturally valid species recognized in this catalog as definitely Hawaiian are endemic to the archipelago, most of them restricted to single islands. [An additional 16 species are listed as questionably belonging to, or questionably excluded from, the Hawaiian fauna, and a further 14 species are listed as having been described from Hawaiian material or being senior synonyms of species described from Hawaiian material.]

The native freshwater mollusc fauna (all gastropods) by contrast has very few species (seven). There are an additional 7 species, originally described from the Hawaiian Islands, that may be native but perhaps more likely are introduced, including 1 for which the nominate subspecies is extralimital but is represented in the Hawaiian Islands by another subspecies. The freshwater species do not show the same high level of endemism as do the land snails. Brackish water species are excluded from this catalog.

Unfortunately, a high proportion of the fauna has become extinct, and the remainder is seriously threatened. Habitat destruction and modification by early Polynesian settlers, combined with their introduction of pigs, dogs, the Polynesian Rat and non-native plants, began the process. Subsequent loss of habitat to agricultural and urban development followed the arrival of Europeans, and the increased rates of introduction of non-native plants and predators accelerated the process. Over-collecting by late 19th and early 20th century shell collectors no doubt also had an impact, but the introduction in the 1950s of the predatory snails *Gonaxis kibweziensis* (Smith), *G. quadrilateralis* (Preston) and especially *Euglandina rosea* (Férussac) in well-intentioned but ill-conceived attempts to control the introduced Giant African Snail, *Achatina fulica* Bowdich, led to a precipitous decline in the native fauna (Hadfield, 1986). Even the freshwater fauna is not immune from attack by *E. rosea* (Kinzie, 1992). Perhaps only 25% of the native species are still extant (Solem, 1990). Some groups appear to be more precariously poised than others. For instance, of the 325 species of Amastridae listed herein, probably fewer than 10 have been observed alive in the last 2 decades. Succineidae, on the other hand, are still frequently seen, sometimes in areas with abundant *E. rosea*.

The land snail fauna was highly complex, not only in species diversity, but also in intra-specific morphological diversity involving subtle differences in shell shape and sculpture, as well as extensive diversity of shell color and pattern (notably in the Achatinellinae). Often, particular forms would be confined to very precise localities, such as a single valley or ridge. Many of these forms were originally described as species. However, thorough revisionary work in the light of a modern species concept would probably reduce many to synonymy. In addition, 2 apparently similar forms might sometimes be described as different simply on the basis of their being from different localities. Not only has this led to an exaggerated number of described species, but it may well also have led to an over-estimate of the degree of single-island endemism. Nevertheless, there still remain (at least in museum collections) large numbers of undescribed species. Solem (1976) estimated that 199-205 species of Hawaiian "endodontoids" are represented in the Bishop Museum collections, but only 31 of them had been described prior to 1976. [Four additional endodontids have since been described.] The systematics of the Hawaiian fauna is therefore very insecure in places. This catalog is based purely on the latest revisions, and makes no major revisionary changes. In addition, it is also, therefore, not possible to give an accurate estimate of the real number of Hawaiian land snail species. It may be more or less than the number recognized as nomenclaturally valid in the present catalog, although Solem (1990) considered it very probable that the number would increase with further study.

The systematics of 3 major groups within the fauna have been dealt with comprehensively in the *Manual of Conchology* (Achatinellidae, Amastridae and Pupillidae). Other groups have been treated elsewhere. However, there has only been 1 attempt to compile a complete list of the entire fauna (Caum, 1928) and this list contains a significant number of errors and omissions. In addition, a number of major revisionary works have been published since Caum's list. The present catalog provides extensive information additional to Caum's work. Zimmerman (1948) and Solem (1990) summarized the number of species in each family.

As a natural system for evolutionary and ecological studies, the Hawaiian land snail fauna had immense potential (e.g., Gulick, 1905) that has not been realized. Despite the extinction of much of the fauna, the potential still exists. Museum collections are especially valuable in this regard, particularly as it is now possible to extract DNA from preserved specimens. Using modern techniques, including DNA analysis, it should still be possible to investigate the systematics and phylogenetic origins of the fauna, speciation within the fauna, aspects of population genetics, and so on. We hope that this catalog will provide a rigorous basis for future work.

EXPLANATORY INFORMATION

SCOPE

This catalog lists all published species-group names found in the literature, available and unavailable according to the *International Code of Zoological Nomenclature* (ICZN, 1985), that apply to the native land and freshwater molluscan taxa of the Hawaiian Islands, including the Northwestern Hawaiian Islands (see Map). These are all gastropods, there being no other classes (e.g., Bivalvia) represented in the native fauna. It does not include names of introduced taxa unless these were originally described from the Hawaiian Islands or are recognized in the literature as senior synonyms of taxa described from the Hawaiian Islands; in both cases this is clearly indicated. Brackish water species are not included; the supra-littoral Ellobiidae, as pulmonates, are included; but the Siphonariidae, although pulmonates, are intertidal and subtidal (Kay, 1979) and are therefore excluded as being marine. Kay (1979) included both brackish and freshwater species of Neritidae, the Ellobiidae, and the Siphonariidae in her treatment of the marine fauna and that work should be referred to for more complete coverage of these groups.

ADHERENCE TO THE *CODE*

Wherever possible, the catalog follows the requirements of the *International Code of Zoological Nomenclature* (the "*Code*"). A very small number of exceptions occur in the treatment of some names in accordance with common usage and in the interests of taxonomic stability. These exceptions are annotated.

We have not attempted revisionary work [e.g., decisions as to whether names originally published as "forms", "varieties", etc. should now be considered as subspecies or be synonymized in the light of *Code* Art. 45(c)]; we simply provide a nomenclatural catalog. Neither have lectotypes been designated where such are needed; this must await more detailed revisionary study. In some instances we have indicated new synonymies and new combinations. These have been introduced almost entirely out of nomenclatural necessity and in order to comply with the *Code*; and with a very small number of annotated exceptions, they do not represent taxonomic judgments on our part.

ARRANGEMENT AND TREATMENT OF TAXA

Family-level classifications of different authors for the land snails vary considerably (e.g., Boss, 1982; Tillier, 1989; Vaught, 1989; Zilch, 1959-1960). Without implying any phylogenetic opinion, we follow Vaught (1989) for the sequence of families herein. Subfamilies (where recognized) appear in alphabetical order within families, as do genera within families/subfamilies and subgenera within genera. No other supra-specific taxa (e.g., tribes, superfamilies) are used. All species-group names (valid and invalid, available and unavailable, distinguished by typographical treatment—see below) are listed alphabetically within genera/subgenera.

Assignment of subfamilies, genera and subgenera within families follow appropriate revisions, as indicated under each group, and with exceptions fully explained.

Genus-group synonyms (available and unavailable) are listed chronologically immediately under the genus-group heading. These synonymies are not exhaustive; we do not consider extralimital genus-group names (i.e., names that, as far as we know, have not been used in combination with species-group names in the Hawaiian fauna); and only the primary taxonomic literature has been covered from this perspective. Misidentifications and incorrect spellings are listed only if their omission might cause confusion. More complete and extralimital synonymies at the genus-group level can be found elsewhere (e.g., Vaught, 1989).

Taxa of uncertain placement are listed at the end of the most appropriate taxon (e.g., taxa of uncertain subgeneric placement within a certain genus are placed at the end of that genus). Two names of unknown taxonomic placement are listed at the end of the catalog.

TYPOGRAPHICAL TREATMENT OF NAMES

Family-group and genus-group headings are centered in boldface upper case type. Valid genus-group names are listed flush left in boldface upper case type. Valid, available, species-group names are listed in boldface and placed flush left; valid infraspecific names are in boldface flush left but are preceded by a "+" to facilitate distinguishing between specific and infraspecific taxa, as currently considered in the systematic literature. Each valid species-group name is followed, on the same line, by an indication in parentheses of island(s) from which the taxon is known (see below under Island Distribution). Synonyms for both genus- and species-group names are listed in italics flush left, upper case for genus-group names, lower case for species-group names. In the species-group, 2 junior homonyms (*seminulum* Boettger [Pupillidae] and *inconspicua* Ancey [Succineidae]) for which no synonyms are available as replacement names and for which we do not provide new names are listed in boldface italics flush left but with the islands they are recorded from given as for valid names. Unavailable names are listed in plain Roman type, flush left, upper case for genus-group names, lower case for species-group names.

Names of the small number of introduced taxa that have been originally described from the Hawaiian Islands are included, following the same typographical conventions, but are also underlined. If they have been synonymized with a non-Hawaiian taxon, the latter name is also included, also underlined, but, as a non-Hawaiian taxon, without the full bibliographic and locality information provided for Hawaiian taxa (see next section).

TAXONOMIC REFERENCES

The taxonomic reference for a genus-group name follows the name. The reference consists of author(s), date of publication and page number (and plate/figure number if these formed part of the original publication). For nomenclaturally available genus-group names, the reference is followed by information on the type species (see below). Type species information is not given for unavailable genus-group names. For genus-group names that have changed status (e.g., from subgenus to genus, section to genus, etc.), the original status is placed in parentheses after the page number, e.g., "**LYROPUPA** Pilsbry, 1900: 432 (as *Nesopupa* sect.)".

For species-group names, on the line following the name and indented, the name is given in its original generic combination (including subgenus if in the original description, and using the original spelling , even if incorrect) and with its original status indicated (e.g., subspecies, "var.", "color-form", etc., as necessary). The name is followed by its author(s), date of publication, page number and plate/figure number(s). If the page was not numbered in the original publication, the correct page number is given in brackets. The author and date act as a reference to the work as listed in the Bibliography section. If an author published more than 1 work in the same year, a suffix (a, b, c, etc.), indicating chronological order of publication, is attached to the date in both the catalog text and the Bibliography. If the current status of a species-group name differs from that of the original proposal of the name, annotation, including reference to publications in the Bibliography, is given in a Remarks section, which follows on the next line.

Authors cited are those responsible for the names listed, in accordance with the *Code*. Authorship by more than 1 author is listed as in the original work. If authorship of the name differs from that of the work, the citation takes the form, e.g., "Smith *in* Gulick & Smith, 1873", so the work can be located in the Bibliography (in this case, under "Gulick & Smith, 1873"). Authors' names containing the articles "de", "von", "van" are cited and alphabetized in the Bibliography by the main name, e.g., "**PLEUROPOMA** Möllendorff, 1893" in the main body of the catalog, and "Möllendorff, O.F. von. 1893" in the Bibliography.

The date given for a work cited in the catalog is the date of publication, as determined by extensive bibliographic research. If the date printed in the original work is incorrect, the correct date is placed in square brackets in the Bibliography, in accordance with Recommendation 22A(5) of the *Code*, but the brackets are omitted in the main body of the catalog.

When an author published the same name as new for the same taxon in more than 1 place, the later citation(s) is(are) given immediately following the page number (or plate and figure number(s), if present) in square brackets; if spelling differed in the later work(s) this is indicated in parentheses; e.g. "*Achatinella gouldi* Newcomb, 1853: 21 [1854a: 4, pl. 22, fig. 1 (as "*gouldii*"); 1854b: 129, pl. 22, fig. 1 (as "*gouldii*")]". An exception to this (*Striatura*) necessitates a double entry because of the designation of different type species for the 2 proposals of the genus name.

Page numbers cited are those on which the name first appeared in the original work. Occasionally, such as cases where the name first appeared in a list or key, with the actual description beginning on a subsequent page, we cite the page numbers of both the list, key, etc., and of the actual description.

If a taxon is mentioned but is not Hawaiian, e.g., a non-Hawaiian type species of a genus-group name, the name, author and date are followed by "not Hawaiian" in square brackets, e.g. "*Lamellina serrata* Pease, 1861 [not Hawaiian]".

TYPE SPECIES

Type species of available genus-group names are given with their correct original generic combination, authorship, date, and spelling, even if this differs from their citation in the fixation of the species as the type (how it differs being given in parentheses). The senior synonym, if any, of the type species is given in square brackets, e.g., "Type species: *Ellobium midae* Röding, 1798 (as "*Auricula*") [= *Voluta aurismidae* Linnaeus, 1767]".

Method of fixation of type species follows the *Code* Art. 68 and 69, using the terms original designation, indication (use of "*typica*"), monotypy, absolute tautonymy, subsequent designation, and subsequent monotypy.

HOMONYMS AND REPLACEMENT NAMES

Homonymy of species-group names is indicated in the Remarks section under the name. In most cases, the junior homonym has already been synonymized with another, earlier name, or a replacement name has been, or is here, provided from synonymy. In 2 cases (*seminulum* Boettger [Pupillidae] and *inconspicua* Ancey [Succineidae]) in which a replacement name appears necessary but there are no synonyms, we have refrained from proposing a new name, pending further research.

Homonymy of genus-group names is indicated in square brackets following the type species fixation, e.g., "*LAMELLINA* Pease, 1861: 439. Type species: *Lamellina serrata* Pease, 1861 [not Hawaiian], by monotypy. [Preoccupied, Bory de Saint-Vincent, 1826]". Replacement names are listed with the preoccupied name they replace in square brackets following the type species fixation, e.g., "**KAUAIA** Sykes, 1900: 355. Type species: *Achatinella kauaiensis* Newcomb, 1860, automatic. [n.n. for *Carinella* Pfeiffer, 1875]".

EMENDATIONS AND UNAVAILABLE NAMES

Emendations (none justified in the present catalog) are available names according to the *Code*. Both genus-group and species-group emendations are listed in the catalog with full reference (author(s), date, page) and an indication of the name being emended, e.g., "*Achatina accincta* Gould, 1852: 88. Unjustified emendation of *accineta* Mighels, 1845". Since, in the present catalog, no emendation is justified, all are junior synonyms of the name being emended and are therefore printed in italics. No other information is given except for explanatory annotations in the Remarks section, if necessary.

In cases of incorrect original spellings (i.e., if more than 1 spelling appears for the same taxon in the original publication), we have determined a correct original spelling following the First Reviser Principle (*Code* Art. 24). Incorrect original spellings are listed with full citation (author(s), date, page), followed by an indication of the name for which it is an incorrect spelling. Genus-group names are listed as, e.g., "STURYANELLA: Pilsbry & Cooke, 1934: 54. Incorrect original spelling of *Sturanyella*." Species-group names are listed as, e.g., "meinickei.", followed on the next line by "*Achatinella lehuiensis meinickei* Pilsbry & Cooke, 1921: 109. Incorrect original spelling of *meineckei* Pilsbry & Cooke." No other information is provided except for explanatory details in the Remarks section, if necessary. Incorrect subsequent spellings are not listed but may be mentioned in annotations, for clarity.

Other unavailable names are listed with full citation and an indication of why the name is unavailable (e.g., "*Nom. nud.*").

MISIDENTIFICATIONS

Misidentifications are excluded from the catalog unless inclusion is deemed necessary for clarity (*Sphyradium*, *Tornatellinops*). The names of misidentified taxa are italicized but separated from authorship by a colon, e.g., "*SPHYRADIUM*: authors, not Charpentier, 1837, misidentification".

GENDER ENDINGS OF SPECIES-GROUP NAMES

Names in synonymy and unavailable names are cited in their original orthography, but valid names have been changed, if necessary, so that their ending agrees in gender with the genus with which they are combined in the catalog (*Code* Art. 31). If we were unable to determine adjectival status for a name, we treated it as a noun (Art. 31(i)) and retained the original orthography; it is possible that some names remain incorrect. For example, *micromphala* (*Tornatellides*) is derived from the Greek adjective "mikros", meaning small, and the Greek noun "omphalos", meaning umbilicus. When combined and Latinized, *micromphala* can be used as either an adjective, i.e., with a small umbilicus, or a noun, i.e., one with a small umbilicus (see Brown, 1956: 51). In this instance it is not possible to determine the intended usage of the original author and so we treat *micromphala* as a noun in apposition and therefore do not change its gender now that it is in combination with the masculine genus name, *Tornatellides*.

NOMENCLATURAL CHANGES

Changes in generic combination, synonymy and status, newly presented in this catalog, are indicated in boldface, respectively, by "**N. comb.**" following the type locality information, "**N. syn.**" in the Remarks section immediately following the explanation of the change (species-group names) or immediately following the type species information (genus-group names), and "**n. stat.**" following the taxon name and author (only 1 instance: Pacificellinae). With rare exceptions, the above changes are purely nomenclatural. Specific reasons for changes are annotated, and the very few instances in which we have had to make a taxonomic judgment are fully explained.

MISCELLANEOUS ANNOTATIONS

Under each family heading, explanatory and other useful information is given. Similarly, if deemed necessary or useful, such information is provided for genus-group taxa immediately under the genus-group synonymy. Annotations other than those indicated in the above paragraphs are placed in square brackets immediately following the item to be clarified or, in the case of species-group names if the annotations are more extensive, placed in a separate Remarks section on the line after the citation and type species information. The Remarks section is used, in particular, to indicate the current status of the name (e.g., synonym, subspecies) in the light of the most recent treatment, if different from its originally published status.

TYPE LOCALITIES

Immediately following the author and citation for each available species-group name, type-locality information is given, as follows: the island on which the type locality is to be found, followed by a colon, and then (in quotes) the locality itself as originally published. If the locality in the original publication was simply the name of an island, with no additional locality details, then the name of the island is not repeated but is followed by "no additional details" in square brackets, e.g., "Maui: [no additional details]". Similarly, if the locality was given originally as "Hawaiian Islands", "Sandwich Islands", "îles Hawaï", "îles Sandwich", "Sandwichsinseln", "in Insulis Sandwich", or any other indication that the locality was no more restricted than to the whole archipelago, we simply state: "Hawaiian Islands: [no additional details]". In all cases in which the term "Sandwich" was used, we have verified that it did not refer to the islands in Vanuatu or the Bismarck Archipelago, which were formerly known by that name (see Motteler, 1986). If the type locality is not restricted to the Hawaiian Archipelago, we provide the original published locality in quotes. Sometimes more than 1 locality was given in the original publication. Unless the type locality has been clarified subsequently, we list all the original localities verbatim following the above conventions. In many instances a subsequent author has attempted to restrict or designate the type locality, but unless a lectotype was designated, this has no nomenclatural status. If a valid lectotype designation has been made, only the lectotype locality is given. If there is some question as to the correctness of the originally published type locality, this is indicated in square brackets immediately following the originally published type locality information, with reference to the publication in which the possible error was indicated, e.g., "Oahu: "Kolau [= Koolau] poko" [?error = "Waialee" (Welch, 1954: 98)]". In a few cases no type locality appears to have been given; we indicate this by stating: "Type locality not given". Type locality information is not provided for unavailable names.

ISLAND DISTRIBUTION

Each valid name is followed, on the same line, by an indication in parentheses of the island(s) from which the taxon is known. The names of the Northwestern Hawaiian Islands are given in full; those of the main islands are abbreviated, as indicated below. If there is some question about the taxon's presence on a particular island the letter/name pertaining to that island is preceded by a "?". If it is not clear from which island(s) it is known, we simply state "Hawaiian Islands". If there is a question whether it occurs in the Hawaiian Islands this is indicated by "?not Hawaiian" or "?Hawaiian". If it is not known whether the taxon occurs in the Hawaiian Islands or not, this is indicated with "unknown". Locality information is derived largely from the *Manual of Conchology*, Caum (1928), the original descriptions, and the most recent revisions; if other sources have been used, they are indicated. If a particular taxon has been recorded from only certain islands, but has been synonymized with 1 or more taxa recorded from other islands, the islands given for the valid name are the combined islands of all the synonymous taxa. If a name represents a taxon introduced to the Hawaiian Islands this is indicated immediately following the list of islands. If we are uncertain whether the taxon is introduced we indicate this by a "?", i.e., "?introduced", but we also underline the name (see above). Distributions of these introduced taxa, as indicated herein, may well be inaccurate; these species are readily spread from island to island by accident and current distributions may well be wider than those indicated, which are derived solely from the literature examined.

ABBREVIATIONS

The following abbreviations are used throughout the catalog:

ISLANDS:
H = Hawaii
K = Kauai
Kah = Kahoolawe
L = Lanai
M = Maui

EXPLANTORY INFORMATION

ABBREVIATIONS (continued)
Mo = Molokai
N = Niihau
O = Oahu

OTHER ABBREVIATIONS:
ANSP	= Academy of Natural Sciences, Philadelphia
Art.	= Article(s) (of the *Code*)
BMNH	= The Natural History Museum, London
BPBM	= Bernice P. Bishop Museum, Honolulu
Code	= International Code of Zoological Nomenclature (ICZN, 1985)
fig(s).	= figure(s)
gen.	= genus or genera
ICZN	= International Commission on Zoological Nomenclature
in litt.	= *in litteris*
MCZ	= Museum of Comparative Zoology, Harvard University, Cambridge
N. comb.	= New combination
n.n.	= new name (replacement name)
n. stat.	= new status
N. syn.	= New synonymy
Nom. nud.	= *Nomen nudum*
p.	= page(s)
pl(s).	= plate(s)
q.v.	= which see
sect.	= section
s.l.	= *sensu lato*
s. str.	= *sensu stricto*
sp(p).	= species (singular and plural)
subg.	= subgenus
subsect.	= subsection
USNM	= United States National Museum of Natural History, Washington, DC
var.	= variety

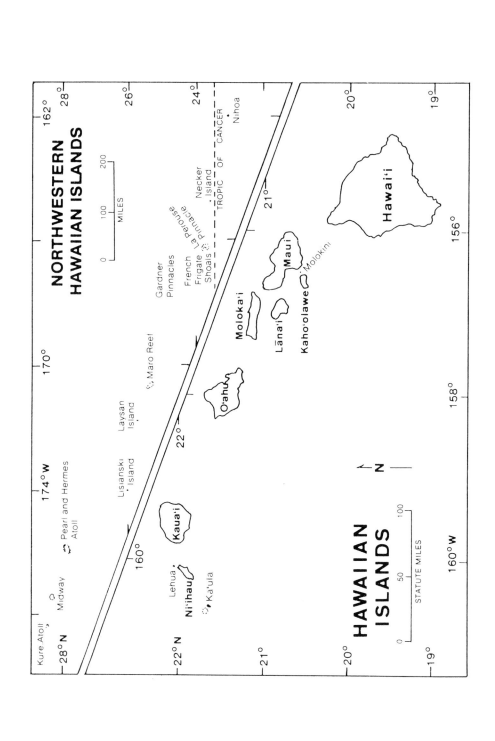

SYSTEMATIC CATALOG

Family NERITIDAE

Neritidae are found in marine, brackish and freshwater habitats. The most recent treatment of the family in the Hawaiian Islands was by Kay (1979). We include here only the 2 taxa Kay indicated as occurring in entirely freshwater habitats, excluding some of the names listed by Caum (1928: 5). *Theodoxus cariosus* (Wood, 1928), although diadromous, does not inhabit completely fresh water (Kay, 1979; Maciolek, 1978) and is excluded. Also excluded is *Neritilia hawaiiensis* Kay, 1979. Although it inhabits ponds with no direct connection to the ocean, these so-called "anchialine" ponds are brackish, and *N. hawaiiensis* cannot therefore be considered a freshwater species. *Neritina vespertina* Sowerby, 1849, may also not inhabit completely fresh water (Maciolek, 1978) but we include it. These inclusions/exclusions are necessarily somewhat arbitrary.

Neritina spinosa Sowerby, 1825 (= *Clithon spinosus* of more recent authors) was listed by Caum (1928: 5), having been indicated by Récluz (1850: 147) and Garrett (according to Starmühlner, 1976: 507) as from the Hawaiian Islands. Caum's listing is probably in error, as it was not dealt with by Kay (1979) nor recorded from the Hawaiian Islands by Haynes (1988). It is found near the mouths of rivers (Starmühlner, 1976) and so may not be considered a truly freshwater species. We exclude it from this catalog of the Hawaiian fauna.

Thus, the Hawaiian truly freshwater neritid fauna is considered here to comprise just 2 species and no infraspecific taxa.

Subfamily NERITINAE

Genus NERITINA Lamarck

NERITINA Lamarck, 1816: 11, pl. 455. Type species: *Nerita pulligera* Linnaeus, 1767 (as "*Neritina*") [not Hawaiian], by subsequent designation of Children (1823: 247) [see also ICZN (1931: 23; Opinion 119), ICZN (1957a: 166, 170, 187; Direction 72)].

Smith (1992: 64) indicated "Rafinesque, C.S. (1815)" as the author of *Neritina*, but cited Rafinesque's 1820 "*Annals of Nature* [etc.]". "Neritina" in Rafinesque's 1815 "*Analyse de la Nature* [etc.]" is a *nom. nud.* (ICZN, 1957a: 170, 187; Direction 72). In addition, Smith (1992: 64) indicated "Stiles (1931)" [= ICZN (1931; Opinion 119)] as having designated the type species.

Kay (1979) did not indicate subgeneric placements. We follow the subgeneric placements of Baker (1923).

Subgenus NERIPTERON Lesson

NERIPTERON Lesson, 1831b: 384. Type species: *Neritina taitensis* Lesson, 1831 (as "1830") [not Hawaiian], by subsequent designation of Baker (1923: 143).

We follow Kay's (1979) treatment of the species-group names, in which, by indicating *vespertina* Sowerby as endemic to the Hawaiian Islands, she implied that it is not synonymous with *taitensis* Lesson, as both Tryon (1888: 73 [as "*tahitensis*"]) and Caum (1928: 5 [as "*tahitensis*"]) had considered. *N. taitensis* was described from Tahiti and is here considered not Hawaiian. Consequently, *taitensis* and a number of putative synonyms in Tryon (1888b) and Caum (1928) (i.e., *auriculata* "Sowerby" [?not Lamarck, 1816], *lamarckii* Deshayes, 1838, *marginata* Rousseau, 1854 [as "Hombron & Jacquinot"]) are excluded from the present catalog.

Kay (1979) considered "*sandwichensis* Reeve" a synonym of *vespertina*. Caum (1928) listed "*sandwichensis* Reeve" in the synonymy of *taitensis* but also included *vespertina*. Kay (1979) considered *sandwichensis* Deshayes a synonym of *Theodoxus cariosus* (Wood), a taxon we have excluded from this catalog as not inhabiting entirely fresh water. Reeve (1855c: pl. 17) attributed *sandwichensis* to Deshayes; he was not proposing it as a new species. Therefore "*sandwichensis* Reeve" is simply a misidentification and is not included in this catalog.

solidissima.
   Neritina solidissima Sowerby, 1849: 541, pl. 116, fig. 273. Type locality not given.
      Remarks. Synonym of *vespertina* Sowerby, *teste* Kay (1979: 67). Sowerby cited "*Pro. Zool. Soc.*, 1849" but there appear to be no papers by Sowerby on this species in any of the issues of this journal for 1848–1851.
vespertina.
   Neritina vespertina Jay, 1839: 66. *Nom. nud.*
      Remarks. Probably originally a manuscript name of Nuttall.
vespertina. (Hawaiian Islands)
   Neritina vespertina Sowerby, 1849: 509, pl. 113, figs. 131, 132. Hawaiian Islands: [no additional details].
      Remarks. Kay (1979: 67) placed *vespertina* Sowerby in the genus *Theodoxus* Montfort; on the advice of C.R. Unabia (pers. comm.) we place it in *Neritina* Lamarck (cf. Haynes, 1988: 94). Originally a manuscript name of Nuttall.

Subgenus NERITONA Martens

NERITONA Martens, 1869: 22. Type species: *Neritina labiosa* Sowerby, 1841 [not Hawaiian], by monotypy.

gigas.
   Neritina (*Neripteron*) *gigas* Lesson, 1842: 187. Hawaiian Islands: [no additional details].
      Remarks. Synonym of *granosa* Sowerby, *teste* Kay (1979: 65)
granosa. (Hawaiian Islands)
   Neritina granosa Sowerby, 1825: 45 [name only], xi [description]. "one of the South Sea Islands".
papillosa.
   Neritina papillosa Jay, 1839: [117]. *Nom. nud.*
      Remarks. Listed in synonymy with *granosa* Sowerby by Jay.

Incertae Sedis in Genus NERITINA Lamarck

atra.
Neritina atra Jay, 1839: 65. *Nom. nud.*
Remarks. Locality given as "Sandwich Islands". We do not know whether this name was intended for a marine, brackish or freshwater taxon, and although Jay attributed the name to Nuttall, it is possible that it was intended as *atra* Gray, 1831 [not Hawaiian] or *atra* Lesson, 1831 [not Hawaiian].

Family HYDROCENIDAE

Operculate groups have radiated widely in terrestrial habitats in the Pacific, with the families Helicinidae, Diplommatinidae and Assimineidae being especially diverse. Only the Helicinidae (q.v.) and the much less diverse Hydrocenidae are represented in the Hawaiian fauna.

Pilsbry (1928) described 3 species of Hydrocenidae, all from the same locality. All 3 were listed by Caum (1928: 5). Zimmerman (1948: 98), based on the views of C.M. Cooke, Jr., suggested that these 3 taxa were all forms of a single species, but did not formally synonymize them. Thompson & Huck (1985), the most recent authors to discuss the group in Hawaii, also seemed to think that they were probably all 1 species, but only synonymized 2 of them. Solem (1990: 29) considered the Hawaiian fauna to consist of just 1 species. Nomenclaturally, there remain 2 valid species-group names.

Genus GEORISSA Blanford

GEORISSA Blanford, 1864: 463. Type species: *Hydrocena pyxis* Benson, 1856 [not Hawaiian], by original designation.

cookei. (K)
> *Georissa cookei* Pilsbry, 1928: 3, fig. 1a. Kauai: "ridge between Hanalei and Wailua . . . on the Pole Line Trail between Kualapa and the summit camp, elevation about 2,000 feet".

kauaiensis. (K)
> *Georissa kauaiensis* Pilsbry, 1928: 4, fig. 1c. Kauai: "ridge between Hanalei and Wailua . . . on the Pole Line Trail between Kualapa and the summit camp, elevation about 2,000 feet".

neali.
> *Georissa neali* Pilsbry, 1928: 4, fig. 1b. Kauai: "ridge between Hanalei and Wailua . . . on the Pole Line Trail between Kualapa and the summit camp, elevation about 2,000 feet".
> *Remarks.* Synonym of *cookei* Pilsbry, *teste* Thompson & Huck (1985: 81). We act here as first revisers and select *cookei* Pilsbry as the senior synonym, following *Code* Recommendation 24A.

Family HELICINIDAE

Among the terrestrial operculate snails, only the Helicinidae constitute a major part of the Hawaiian fauna. The Hydrocenidae (above) are represented in the Hawaiian Islands by only 2 species (which may be conspecific). Together, these 2 families make up the Hawaiian terrestrial operculate snail fauna.

Much confusion, especially among the genus-group names, has been generated, not only by the inadequacies of Wagner's (1905, 1907–1911) monographs (see Baker, 1922: 29; Solem, 1959: 166–167), but also by the acceptance by various authors (Neal, 1934: 38; Pilsbry & Cooke, 1934: 54; Solem, 1959: 168) of Baker's (1922: 43) invalid designation of *Helicina laciniosa* Mighels, 1845 as the type species of *Sturanya* (see under the genus *Pleuropoma*, below). Neal (1934) was the last to revise the Hawaiian Helicinidae. We follow the concepts of her revision but take into account 2 subsequent changes made by Johnson (1949: 226, 228).

Caum (1928) listed 18 species (including 4 he considered erroneously reported from the Hawaiian islands) and 6 subspecies. Zimmerman's synopsis (1948: 98) indicated 7 described species and 14 "varieties" of *Orobophana* and 9 species and 29 "varieties" of *Pleuropoma*, no doubt following Neal (1934). Solem (1990: 29) also indicated 16 species and 43 "varieties", probably simply following Zimmerman and Neal without taking account of Johnson (1949) who reduced 2 species to synonymy.

The Hawaiian fauna consists of 14 species (6 in *Orobophana* and 8 in *Pleuropoma*) and 44 infraspecific taxa (14 in *Orobophana* and 30 in *Pleuropoma*), plus 5 taxa that may not be Hawaiian.

Genus OROBOPHANA Wagner

OROBOPHANA Wagner, 1905: 415. Type species: *Helicina uberta* Gould, 1847, by subsequent designation of Baker (1922: 43).

baldwini. (K)
>*Helicina baldwini* Ancey, 1904a: 126, pl. 7, fig. 24. Kauai: "Kipu".

+beta. (O)
>*Helicina laciniosa* var. *beta* Pilsbry & Cooke, 1908: 202, fig. 4. Oahu: "Kapalama".
>>*Remarks.* Variety of *uberta* Gould, *teste* Neal (1934: 32).

+borealis. (O)
>*Orobophana uberta* var. *borealis* Neal, 1934: 31, figs. 31, 32. Oahu: "Punaluu".

+bryani. (O)
>*Orobophana uberta* var. *bryani* Neal, 1934: 24, fig. 21. Oahu: "Waianae Mountains, Kolekole Pass".

cookei. (K)
>*Orobophana cookei* Neal, 1934: 14, figs. 4, 5. Kauai: "Kalalau, on cliff on west side of valley, altitude about 200 feet".

+exanima. (O)
    Orobophana uberta var. *exanima* Neal, 1934: 37, fig. 43. Oahu: "Kahuku, coral bluff half a mile west of town".

+hibrida. (O)
    Orobophana uberta var. *hibrida* Neal, 1934: 35, figs. 40, 41. Oahu: "Kawaiiki".

juddii. (K)
    Helicina juddii Pilsbry & Cooke, 1908: 208, fig. 13. Kauai: "Koloa beach".

+lihueensis. (K)
    Orobophana baldwini var. *lihueensis* Neal, 1934: 18, figs. 12–14. Kauai: "Lihue district, back of Grove Farm reservoir".

+lymaniana. (O)
    Helicina lymaniana Pilsbry & Cooke, 1908: 208, fig. 12. Oahu: "Waialua".
      Remarks. Variety of *uberta* Gould, *teste* Neal (1934: 33).

+magdalenae. (O)
    Helicina magdalenae Ancey, 1890: 342. Oahu: "Tantalus".
      Remarks. Variety of *uberta* Gould, *teste* Neal (1934: 27).

+makuaensis. (O)
    Orobophana uberta var. *makuaensis* Neal, 1934: 27, fig. 29. Oahu: "Waianae Mountains, Makua Valley, east side".

meineckei. (K)
    Orobophana meineckei Neal, 1934: 18, figs. 15–17. Kauai: "Hanakoa Valley".

+nuuanuensis. (O)
    Helicina nuuanuensis Pilsbry & Cooke, 1908: 206, fig. 10. Oahu: "Nuuanu, upper part".
      Remarks. Variety of *uberta* Gould, *teste* Neal (1934: 34).

+percitrea. (O)
    Orobophana uberta var. *percitrea* Neal, 1934: 37, fig. 42. Oahu: "top of ridge between Halawa and Aiea Valleys".

+praemagna. (K)
    Orobophana stokesii var. *praemagna* Neal, 1934: 16, figs. 8, 9. Kauai: "south of Wailua River, half a mile from ocean, altitude 5 to 10 feet".

rhodostoma.
    Helicina constricta var. *rhodostoma* Pfeiffer, 1850c: 22. Type locality not given.
      Remarks. Primary junior homonym of *rhodostoma* Gray, 1824 [not Hawaiian]. Synonym of *magdalenae* Ancey, *teste* Neal (1934: 27). Originally a manuscript name of Mighels (Pfeiffer, 1850c: 22).

stokesii. (K)
    Orobophana stokesii Neal, 1934: 15, figs. 6, 7. Kauai: "Koloa, Makahuena Point".

+subtenuis. (O)
    Orobophana uberta var. *subtenuis* Neal, 1934: 31, figs. 33, 34. Oahu: "Mount Tantalus, north inner side of crater".

uberta. (O)
    Helicina uberta Gould, 1847b: 202. Maui: [no additional details]; Oahu: "Oahu Mountains".
      Remarks. Neal (1934: 22–23) considered the type locality to be the Waianae Mountains of Oahu.

+wilderi. (O)
    Orobophana uberta var. *wilderi* Neal, 1934: 26, figs. 26, 27. Oahu: "Waianae Mountains, Mokuleia".

Genus PLEUROPOMA Möllendorff

PLEUROPOMA Möllendorff, 1893: 140 (as *Helicina* sect.). Type species: *Helicina dichroa* Möllendorff, 1893 [not Hawaiian], by original designation.
STURANYA Wagner, 1905: 383. Type species: *Helicina plicatilis* Mousson, 1865, by subsequent designation of Kobelt (1907: 234).
STURANYELLA Pilsbry & Cooke, 1934: 54. Type species: *Helicina plicatilis* Mousson, 1865 [not Hawaiian], by original designation.
STURYANELLA: Pilsbry & Cooke, 1934: 54. Incorrect original spelling of *Sturanyella*.

Neal (1934: 38), Pilsbry & Cooke (1934: 54) and Solem (1959: 168) incorrectly considered the type species of *Sturanya* to be *Helicina laciniosa* Mighels, 1845, by subsequent designation of Baker (1922: 43). This has led to much confusion as to the correct status of *Sturanya*. Whether it is best retained as a valid genus or placed in synonymy with *Pleuropoma* can only be decided by further study, particularly of the type species. We follow Neal (1934: 38) in placing it as a synonym of *Pleuropoma*.

Subgenus APHANOCONIA Wagner

APHANOCONIA Wagner, 1905: 388. Type species: *Helicina verecunda* Gould, 1859 [not Hawaiian], by subsequent designation of Gude (1921: 366).
SPHAEROCONIA Wagner, 1909b: 189. Type species: *Helicina verecunda* Gould, 1859 [not Hawaiian], by subsequent designation of Baker (1922: 43).

Knight *et al.* (1960: 286) indicated the subsequent designation of the type species of *Aphanoconia* as by "Gude, 1914", but we have been unable to locate this work. It may be a simple error since the second and third parts of Gude's treatment of the molluscs in the *Fauna of British India*, with the type of *Aphanoconia* being designated in the third part, were published in 1914 and 1921, respectively. Wenz (1938b: 443), apparently incorrectly, indicated *Helicina* (*Sulphurina*) *sphaeroconus* Möllendorff, 1895 as the type of *Sphaeroconia*.

The genus-group names *Aphanoconia* and *Sphaeroconia* are objective synonyms; the former has priority. Neal (1934: 83) treated *Sphaeroconia* as a valid subgenus, although she included *Aphanoconia* as a synonym of *Pleuropoma*, apparently not realizing that their type species are the same. We follow Neal's concept of the subgenus, although it takes the name *Aphanoconia*. Knight *et al.* (1960: 286) and Vaught (1989: 14) also treated *Aphanoconia* as a subgenus of *Pleuropoma*, while Pilsbry & Cooke (1934: 54) and Solem (1959: 168; 1989: 465) placed it as a synonym of *Pleuropoma*.

hawaiiensis.
    Helicina hawaiiensis Pilsbry & Cooke, 1908: 204, fig. 6. Oahu [?error = Kauai (Neal, 1934: 85)]: [no additional details].
        Remarks. Johnson (1949: 228), quoting apparently unpublished statements of C.M. Cooke, Jr., indicated that *hawaiiensis* is a synonym of *rotelloidea* Mighels "by the identification of the type".

kauaiensis. (K)
    Helicina kauaiensis Pilsbry & Cooke, 1908: 205, fig. 8. Kauai: "upper part of Milolii".

+knudseni. (K)
    Helicina knudseni Pilsbry & Cooke, 1908: 204, fig. 7. Kauai: "Olokele".
        Remarks. Variety of *hawaiiensis* Pilsbry & Cooke, *teste* Neal (1934: 86).

+makalii. (K)
: *Pleuropoma (Sphaeroconia) hawaiiensis* var. *makalii* Neal, 1934: 90, fig. 106. Kauai: "Wailua, north fork of Wailua River".

+mauiensis. (M)
: *Pleuropoma (Sphaeroconia) rotelloidea* var. *mauiensis* Neal, 1934: 96, fig. 116. Maui: "West Maui, Maunahooma".

niihauensis. (N)
: *Pleuropoma (Sphaeroconia) niihauensis* Neal, 1934: 91, fig. 111. Niihau: "southwestern shore, at Kiekie".

+orientalis. (K)
: *Pleuropoma (Sphaeroconia) kauaiensis* var. *orientalis* Neal, 1934: 91, fig. 110. Kauai: "Nonou mountains, west side".

rotelloidea. (K, O)
: *Helicina rotelloidea* Mighels, 1845: 19. Oahu: [no additional details].
: Remarks. Johnson (1949: 228) designated a lectotype (MCZ 156499) and quoted C.M. Cooke, Jr. as believing that the "type did not come from Oahu . . . but from Kauai".

+sola. (Mo)
: *Pleuropoma (Sphaeroconia) rotelloidea* var. *sola* Neal, 1934: 95, figs. 114, 115. Molokai: "Kaunakakai district, Kalihi Valley".

sulculosa. (H)
: *Helicina sulculosa* Ancey, 1904a: 127, pl. 7, fig. 25. Hawaii: "Olaa".

Subgenus PLEUROPOMA Möllendorff

PLEUROPOMA Möllendorff, 1893: 140 (as *Helicina* sect.). Type species: *Helicina dichroa* Möllendorff, 1893 [not Hawaiian], by original designation.

STURANYA Wagner, 1905: 383. Type species: *Helicina plicatilis* Mousson, 1865, by subsequent designation of Kobelt (1907: 234).

STURANYELLA Pilsbry & Cooke, 1934: 54. Type species: *Helicina plicatilis* Mousson, 1865 [not Hawaiian], by original designation.

STURYANELLA: Pilsbry & Cooke, 1934: 54. Incorrect original spelling of *Sturanyella*.

See comments under genus *Pleuropoma*.

+alpha. (O)
: *Helicina laciniosa* var. *alpha* Pilsbry & Cooke, 1908: 203, fig. 5. Oahu: "Mt. Tantalus".

berniceia.
: *Helicina berniceia* Pilsbry & Cooke, 1908: 207, fig. 11. Kauai: "Limahuli".
: Remarks. Johnson (1949: 226), quoting apparently unpublished statements of C.M. Cooke, Jr., considered *berniceia* a synonym of *laciniosa* Mighels.

+bronniana. (Hawaiian Islands)
: *Helicina bronniana* Philippi, 1847: 124. Hawaiian Islands: [no additional details].
: Remarks. Neal (1934: 98–99) indicated considerable uncertainty regarding the correct placement of this taxon. It has been treated as a synonym of *rotelloidea* Mighels (e.g., by Sykes, 1900: 397) and a "variety" of *laciniosa* Mighels (by Pilsbry & Cooke, 1908: 5). Neal suggested it might be "more probably [related] to *Orobophana uberta* variety *magdalenae*", but relegated it to "doubtful and excluded species". We place it here based on Pilsbry & Cooke (1908: 5), this being the most recent definitive statement.

+canyonensis. (K)
: *Pleuropoma laciniosa* var. *canyonensis* Neal, 1934: 68, fig. 71. Kauai: "Olokele Canyon".

HELICINIDAE [21]

+**delta**. (N, K)
    Helicina laciniosa var. *delta* Pilsbry & Cooke, 1908: 201, fig. 2. Kauai: "Ekaula, below Puukapele".

dissotropis.
    Helicina dissotropis Ancey, 1904a: 127, pl. 7, figs. 22, 23. Oahu: [no additional details].
        Remarks. Synonym of *sandwichiensis* Souleyet, *teste* Neal (1934: 50).

+**ferruginea**. (O)
    Pleuropoma laciniosa var. *ferruginea* Neal, 1934: 46, figs. 47, 48. Oahu: "Waianae Mountains, Leilehua".

+**gamma**. (O)
    Helicina laciniosa var. *gamma* Pilsbry & Cooke, 1908: 202, fig. 3. Oahu: "Ewa".
        Remarks. Pilsbry & Cooke (1908: 202) gave both Ewa and Wahiawa as localities but the label associated with the "type" (BPBM 14902) says "Ewa".

+**gemina**. (O)
    Pleuropoma oahuensis var. *gemina* Neal, 1934: 81, figs. 97–99. Oahu: "Waianae Mountains, Mokuleia, ridge west of Makaleha Valley".

+**globuloidea**. (K)
    Pleuropoma laciniosa var. *globuloidea* Neal, 1934: 67, figs. 67, 68. Kauai: "Nonou mountains, southwest side".

+**honokowaiensis**. (M)
    Pleuropoma laciniosa var. *honokowaiensis* Neal, 1934: 73, fig. 76. Maui: "West Maui, Honokowai".

+**kaaensis**. (L)
    Pleuropoma laciniosa var. *kaaensis* Neal, 1934: 76, fig. 85. Lanai: "western end".

+**kahoolawensis**. (Kah)
    Pleuropoma laciniosa var. *kahoolawensis* Neal, 1934: 76, fig. 86. Kahoolawe: "Hakioawa".

+**kiekieensis**. (N)
    Pleuropoma laciniosa var. *kiekieensis* Neal, 1934: 69, fig. 72. Niihau: "southwestern shore, at Kiekie".

+**konaensis**. (H)
    Pleuropoma laciniosa var. *konaensis* Neal, 1934: 77, fig. 88. Hawaii: "Puuwaawaa, Mawai, near Puu Henahena".

+**kulaensis**. (M)
    Pleuropoma laciniosa var. *kulaensis* Neal, 1934: 74, fig. 78. Maui: "East Maui, Keokea".

laciniosa. (K)
    Helicina laciniosa Mighels, 1845: 19. Oahu: [?error = Kauai (Johnson, 1949: 226); no additional details].

+**laula**. (O)
    Pleuropoma laciniosa var. *laula* Neal, 1934: 48, figs. 51, 52. Oahu: "Waianae Mountains, Makua, southern slope".

+**matutina**. (K)
    Pleuropoma laciniosa var. *matutina* Neal, 1934: 68, figs. 69, 70. Kauai: "Koloa, Mahaulepu, dune near Kapunakea Pond".

+**molokaiensis**. (Mo)
    Pleuropoma laciniosa var. *molokaiensis* Neal, 1934: 69, fig. 73. Molokai: "east-central mountains, Waikolu Valley, above Kaluahauoni".

+**moomomiensis**. (Mo)
    Pleuropoma laciniosa var. *moomomiensis* Neal, 1934: 72, fig. 75. Molokai: "Moomomi, sand dunes".

nonouensis. (K)
> *Pleuropoma nonouensis* Neal, 1934: 79, figs. 92–94. Kauai: "Nonou Mountains, northeast side".

oahuensis. (O)
> *Helicina oahuensis* Pilsbry & Cooke, 1908: 199, fig. 1. Oahu: "back of Leilehua Ranch-house, Waianae Mts."

+perparva. (O)
> *Pleuropoma laciniosa* var. *perparva* Neal, 1934: 64, figs. 60, 61. Oahu: "Kawailoa (Kailua)".

+piliformis. (L)
> *Pleuropoma laciniosa* var. *piliformis* Neal, 1934: 74, figs. 80, 81. Lanai: "Mahana District".

+praeparva. (K)
> *Pleuropoma laciniosa* var. *praeparva* Neal, 1934: 67, figs. 64–66. Kauai: "Wailua River, near north fork".

+pusilla. (O)
> *Pleuropoma laciniosa* var. *pusilla* Neal, 1934: 49, fig. 53. Oahu: "Waianae Mountains, Palehuaiki".

+sandwichiensis. (O)
> *Helicina sandwichiensis* Souleyet, 1852: 529, pl. 30, figs. 1–5. Hawaiian Islands: [no additional details].
>> *Remarks*. Variety of *laciniosa* Mighels, *teste* Neal (1934: 50). Possibly a junior synonym of *fulgora* Gould (see questionably included Helicinidae in the Hawaiian fauna). See Neal (1934: 51–52) for details.

+signata. (O)
> *Pleuropoma laciniosa* var. *signata* Neal, 1934: 63, fig. 59. Oahu: "Malaekahana".

+spaldingi. (O)
> *Pleuropoma laciniosa* var. *spaldingi* Neal, 1934: 48, figs. 49, 50. Oahu: "Waianae Mountains, Puu Kaala, near spring, elevation about 3,000 feet".

subsculpta. (O)
> *Pleuropoma subsculpta* Neal, 1934: 82, figs. 100, 101. Oahu: "Waianae Mountains, Makua".

Questionably Included HELICINIDAE in the Hawaiian Fauna

Neal (1934) included the species listed below in a section of "Doubtful and excluded species", all in the genus *Helicina*. In this section, Neal also included *bronniana* Philippi (dealt with here under *Pleuropoma*) and *pisum* Rousseau (as "Hombron & Jacquinot") (a primary junior homonym of *pisum* Philippi, but not Hawaiian). Revisionary study would, no doubt, refer at least some of the species here listed in *Helicina* to other genera.

Genus HELICINA Lamarck

HELICINA Lamarck, 1799: 76. Type species: *Helicina neritella* Lamarck, 1801 [not Hawaiian], by subsequent monotypy in Lamarck (1801: 94).

Children (1823: 239) designated "*Helicina neritella* Lamarck, 1799" as the type species, but Lamarck's description of *Helicina* included no species. Lamarck (1801: 94) was the first to include any species (*neritella* only) in *Helicina*.

antoni. (?not Hawaiian)
> *Helicina antoni* Pfeiffer, 1849a: 88. Type locality not given.
>> *Remarks.* Probably not a Hawaiian species (Neal, 1934: 98).

constricta. (?not Hawaiian)
> *Helicina constricta* Pfeiffer, 1849b: 120. Tahiti and Hawaiian Islands ["Otaheite and the Sandwich Islands"]: [no additional details].
>> *Remarks.* Also published by Pfeiffer (1850c: 22), citing the original description. Neal (1934: 29) outlined the history of this name. It appears that it may refer to 2 taxa: a "typical form" (i.e. *constricta*) from Tahiti, the identity of which "has not been determined", and a "variety" (now called *magdalenae* Ancey) from Oahu. Neal also suggested that "possibly the reference to Tahiti is wrong", and that both the "typical form" and the "variety" may have come from Oahu, and that, if this were the case, the 2 would be considered the same, i.e. a variety of *Orobophana uberta* (Gould). The earlier name, *constricta* Pfeiffer, would then have priority over *magdalenae* Ancey.

crassilabris. (?not Hawaiian)
> *Helicina crassilabris* Philippi, 1847: 125. Hawaiian Islands: [no additional details].
>> *Remarks.* Probably not a Hawaiian species (Neal, 1934: 99).

fulgora. (?not Hawaiian)
> *Helicina fulgora* Gould, 1847b: 201. Samoa: "Upolu and Manua".
>> *Remarks.* Possibly a senior synonym of *sandwichiensis* Souleyet; mention of Hawaii as locality by Wagner apparently erroneous (Neal, 1934: 51–52).

pisum. (?not Hawaiian)
> *Helicina pisum* Philippi, 1847: 124. "*Insulae Sandwich*" [?= Vate or Sandwich Island or Savage Island (Sykes, 1900: 397; Neal, 1934: 100)]: [no additional details].
>> *Remarks.* Probably not a Hawaiian species (Neal, 1934: 99–100).

Family HYDROBIIDAE

The single species was described in the genus *Paludina* by Mighels (1845: 22) and treated subsequently in this genus by Küster (1852b: 34). Caum (1928: 7) referred it to *Paludestrina* and listed it in the Viviparidae. *Paludestrina* d'Orbigny, 1841 is a junior objective synonym of *Hydrobia* Hartmann, 1821, with type species *Cyclostoma acutum* Draparnaud [not Hawaiian] (see Pilsbry & Baker, 1958: 116). However, Mighels' material seems lost (Johnson, 1949) and we feel unable to place this taxon correctly.

Hydrobiids had not subsequently been recorded from the Hawaiian Islands in the literature until very recent reports from archaeological investigations (Athens *et al.*, 1994; Athens & Ward, 1993). However, a number of species (none adequately identified) are represented in the collections of Bishop Museum (Honolulu). All these taxa, including Mighels' species (*porrecta*), may be referable to the genus *Tryonia* Stimpson, and may be introduced.

Incertae sedis in family HYDROBIIDAE

porrecta. (O; ?introduced)
Paludina porrecta Mighels, 1845: 22. Oahu: [no additional details].

Family THIARIDAE

The names here included in the native Hawaiian fauna are based on Caum (1928: 7) with island occurrences derived from that work and from the original literature. It is possible that some or all of the 5 taxa (4 species and 1 subspecies of an extralimital nominate species) are introduced, although we are not aware of any published statements definitively indicating this. There are a number of other extralimital taxa, not included here, that appear to have been introduced to the Hawaiian Islands, although their identity has not been definitively clarified, and there may be confusion between these and the purportedly native taxa.

Both specific and generic limits in the Thiaridae appear often to be poorly understood. We treat *Tarebia* as a genus, not as a subgenus of *Thiara*, following Morrison (1952: 8, 1954: 379) and Vaught (1989: 28) but contrary to Pace (1973: 51–67). We include *mauiensis* Lea in *Tarebia*. As far as we are aware, the remainder of the species described from the Hawaiian Islands have never been referred to particular subgenera. We place them in the genus *Thiara* with subgeneric placement uncertain.

Genus TAREBIA Adams & Adams

TAREBIA Adams & Adams, 1854b: 304 (as *Vibex* subg.). Type species: *Melania celebensis* Quoy & Gaimard, 1834 [= *Melania granifera* Lamarck, 1816] [not Hawaiian], by subsequent designation of Abbott (1948: 290).

The history of the type species designation for *Tarebia* was outlined by Morrison (1954: 379; see also Pace, 1973: 64), who designated *Melania semigranosa* von dem Busch, 1842 [= *Melania granifera* Lamarck, 1816] [not Hawaiian] as the type. Apparently, Morrison was not aware of the earlier valid designation of *Melania celebensis* Quoy & Gaimard, 1834, by Abbott (1948: 290).

+**mauiensis**. (Mo, M; ?introduced)
    Melania mauiensis Lea, 1856: 145. Maui: [no additional details].
        Remarks. Subspecies or "race" of *granifera* Lamarck [not Hawaiian], *teste* Abbott (1952: 72, 112). Caum (1928: 7) treated *mauiensis* as a senior synonym of *granifera* Lamarck [not Hawaiian].

tahitensis.
    Melania tahitensis Brot, 1877b: 323. *Nom. nud.*
        Remarks. Probably originally a Pease manuscript name (Ancey, 1904a: 127), listed in the synonymy of *mauiensis* Lea by Brot. Synonym of *mauiensis* Lea, *teste* Caum (1928: 7).

Genus THIARA Röding

THIARA Röding, 1798: 109. Type species: *Helix amarula* Linnaeus, 1758 [not Hawaiian], by subsequent designation of Brot (1874: 7; as "*Tiara* Bolten").

MELANIA Lamarck, 1799: 75. Type species: *Helix amarula* Linnaeus, 1758 [not Hawaiian], by monotypy.

Gray (1847: 152) incorrectly designated *Melania hollandri* Férussac (as "*holandri*") as the type species of *Thiara*; *hollandri* was not among the originally included species. Wenz (1939: 712) and Smith (1992: 77, as "*Thiara* Bolten"), explicitly, and Abbott (1948: 288), implicitly, indicated that *amarula* was the type species of *Thiara* by monotypy. However, these listings are incorrect as there were 2 originally included valid species. Morrison (1954: 378) outlined the history of the type species designation for *Thiara*. We are uncertain as to the correct current subgeneric placement of the species included here in the genus *Thiara*.

baldwini. (M; ?introduced)
> *Melania baldwini* Ancey, 1899: 273, pl. 12, fig. 6. Maui: "Lahaina".

contigua.
> *Melania contigua* Pease, 1870c: 7. Kauai: [no additional details].
>> Remarks. Synonym of *indefinita* Lea & Lea, *teste* Caum (1928: 7).

indefinita. (K, O, Mo, M; ?introduced)
> *Melania indefinita* Lea & Lea, 1851: 187. Philippines: "Naga, Luzon".

kauaiensis. (K; ?introduced)
> *Melania kauaiensis* Pease, 1870c: 7, pl. 3, fig. 6. Kauai: [no additional details].

newcombii.
> *Melania newcombii* Lea, 1856: 145. Oahu: [no additional details].
>> Remarks. Synonym of *indefinita* Lea & Lea, *teste* Caum (1928: 7).

oahuensis.
> *Melania oahuensis* Brot, 1872: 43, pl. 3, fig. 2. Oahu: [no additional details]; Molokai: [no additional details].
>> Remarks. Synonym of *indefinita* Lea & Lea, *teste* Caum (1928: 7). Probably originally a manuscript name of Pease.

paulla.
> *Melania paulla* Brot, 1872: 43. *Nom. nud.*
>> Remarks. Synonym of *indefinita* Lea & Lea, *teste* Caum (1928: 7). Originally a manuscript name of Dunker, which Brot simply considered as referring to juvenile *oahuensis* Brot ["n'est que le jeune âge de *M. oahuensis*" (Brot, 1872: 43)].

verrauiana. (Hawaiian Islands; ?introduced)
> *Melania verrauiana* Lea, 1856: 144. Hawaiian Islands: [no additional details].

Family ELLOBIIDAE

The Ellobiidae are pulmonates, but their placement within the Pulmonata has differed in various higher classifications (see Bieler, 1992; Boss, 1982; Zilch, 1959a). The most recent treatment of the Hawaiian Ellobiidae was by Kay (1979, as "Melampidae"), who placed them in the Basommatophora. They are supralittoral in habitat. In the Hawaiian Islands they are also found around brackish, anchialine ponds in lava flows (Kay, 1979).

We follow Kay (1979: 490–93) for the treatment of species-group names. We also include *Auriculastra elongata* (Küster, 1844), *Melampus lucidus* Pease, 1869, and *Ellobium fuscum* (Küster, 1844), which were listed by Caum (1928: 8) but not dealt with by Kay (1979), giving a total of ten nomenclaturally valid species and no infraspecific taxa. Treatment of genus-group and family-group names differs somewhat among authors; we follow Zilch (1959a). Island distributions derive from the literature we have seen [and in 1 case from the personal observations of 2 of us (RHC, CCC)]; probably most species are more widely distributed.

Subfamily CASSIDULINAE

Genus ALLOCHROA Ancey

ALLOCHROA Ancey, 1887: 288. Type species: *Auricula bronnii* Philippi, 1846 (as "*Melampus bronni*"), by original designation.

bronnii. (Hawaiian Islands)
    Auricula bronnii Philippi, 1846: 98. Hawaiian Islands: [no additional details].
brownii.
    Ellobium brownii Adams & Adams, 1855f: 237. *Nom. nud.*
        Remarks. Synonym of *bronnii* Philippi, *teste* Kay (1979: 491).
conica.
    Laimodonta conica Pease, 1863: 242. "Pacific Inlands [sic]" (in publication title): [no additional details].
        Remarks. Synonym of *bronnii* Philippi, *teste* Caum (1928: 8).
sandwichiensis.
    Auricula sandwichiensis Souleyet, 1852: 524, pl. 29, figs. 29–32. Hawaiian Islands: [no additional details].
        Remarks. Synonym of *bronnii* Philippi, *teste* Kay (1979: 491).

Genus AURICULASTRA Martens

AURICULASTRA Martens *in* Möbius, Richters & Martens, 1880: 207 (as *Marinula* subg.). Type species: *Auricula elongata* Küster, 1844, by monotypy.

Zilch (1959a: 76) and Smith (1992: 212), apparently incorrectly, indicated *Auricula subula* Quoy & Gaimard, 1832 [not Hawaiian] as the type species, the latter author stating that the type fixation was by original designation.

elongata. (Hawaiian Islands)
> *Auricula elongata* Küster, 1845: 53, pl. 8, figs. 6–8. Hawaiian Islands: [no additional details].
> *Remarks.* Originally a manuscript name of Parreyss. Listed by Caum (1928: 8) but not dealt with by Kay (1979).

Subfamily ELLOBIINAE

Genus ELLOBIUM Röding

ELLOBIUM Röding, 1798: 105. Type species: *Ellobium midae* Röding, 1798 [= *Voluta aurismidae* Linnaeus, 1767] [not Hawaiian], by subsequent designation of Gray (1847: 179).

AURICULA Lamarck, 1799: 76. Type species: *Ellobium midae* Röding, 1798 (as "*Auricula*") [= *Voluta aurismidae* Linnaeus, 1767] [not Hawaiian], by subsequent designation of Children (1823: 241).

Caum (1928: 8) listed *owaihiensis* Chamisso, 1829 in *Auricula*, as well as including it in the synonymy of *Auriculella auricula* (Férussac) (Achatinellidae). We place it with *Auriculella*.

fuscum. (Hawaiian Islands)
> *Auricula fusca* Küster, 1844: 38, pl. 5, figs. 18–20. Hawaiian Islands: [no additional details]. **N. comb.**
> *Remarks.* The new combination is introduced purely as a matter of nomenclatural necessity, since *Auricula* Lamarck is a junior objective synonym of *Ellobium* Röding. According to G. Rosenberg (*in litt.* to RHC, 28 September 1993), this species most closely resembles *Melampus castaneus* among the Hawaiian Ellobiidae recognized by Kay (1979), but we refrain from formally synonymizing it, pending future revisionary study. Originally a manuscript name of Philippi. Listed by Caum (1928: 8) but not dealt with by Kay (1979).

Subfamily MELAMPODINAE

Genus MELAMPUS Montfort

MELAMPUS Montfort, 1810: 318. Type species: *Bulimus coniformis* Bruguière, 1789 [= *Voluta coffeus* Linnaeus, *teste* Zilch (1959a: 65)] [not Hawaiian], by original designation.

PIRA Adams & Adams, 1855f: 244 (as *Tralia* subg.). Type species not designated.

Kay (1979: 492) retained *Pira* Adams & Adams as a valid genus for *sculptus* Pfeiffer. We follow Thiele (1931: 467), Zilch (1959a: 66) and Vaught (1989: 75) and include *Pira* as a synonym of *Melampus*. We have been unable to find a type species designation for *Pira*.

castaneus. (Hawaiian Islands)
> *Voluta castanea* Megerle von Mühlfeld, 1816: 4, pl. 1, fig. 2. "Ostindien".

fricki.
> *Melampus fricki* Pfeiffer, 1859b: 29. Hawaiian Islands: [no additional details].
> *Remarks*. Synonym of *sculptus* Pfeiffer, *teste* Kay (1979: 492).

lucidus. (O)
> *Melampus lucidus* Pease, 1869b: 75. Oahu: [no additional details].
> *Remarks*. Listed by Caum (1928: 8) but not dealt with by Kay (1979). Smith (1992: 217) considered the type locality unknown.

parvulus.
> *Melampus parvulus* Pfeiffer, 1854c: 147. *Nom. nud.*

parvulus. (O)
> *Melampus parvulus* Pfeiffer, 1856f: 24. Oahu: [no additional details].

pseudocommodus.
> *Melampus pseudocommodus* Schmeltz, 1869: 135. *Nom. nud.*
> *Remarks*. Treated by Schmeltz as a synonym of *semiplicatum* Adams & Adams (as "*semiplicatus* Pease").

sculptus. (Hawaiian Islands)
> *Melampus sculptus* Pfeiffer, 1859b: 29. Admiralty Islands.
> *Remarks*. Referred to *Pira* by Kay (1979: 492).

semiplicatum.
> *Ellobium semiplicatum* Adams & Adams, 1854l: 8. Singapore.
> *Remarks*. Synonym of *sculptus* Pfeiffer, *teste* Kay (1979: 492).

Subfamily PEDIPEDINAE

Genus LAEMODONTA Philippi

LIRATOR Beck, 1838: 108. *Nom. nud.*
LAEMODONTA Philippi, 1846: 98. Type species: *Auricula striata* Philippi, 1846 [= *Pedipes octanfracta* Jonas, 1845], by monotypy.
PLECOTREMA Adams & Adams, 1854j: 120. Type species: *Plecotrema typica* Adams & Adams, 1854, by indication (*Code* Art. 68(c)).
ENTERODONTA Sykes, 1894: 73. Type species: *Auricula striata* Philippi, 1846 [= *Pedipes octanfracta* Jonas, 1845], automatic. [Unnecessary n.n. for *Laemodonta* Philippi, 1846 (as *Laimodonta* Adams & Adams, 1855)].

Zilch (1959a: 75) listed "*Laimodonta* H. &. A. Adams" and "*Enterodonta* Sykes" in synonymy with *Allochroa* Ancey. We follow Vaught (1989: 75) in treating them under *Laemodonta* Philippi. "*Laimodonta*" (Bronn, 1847: 4; Adams & Adams, 1855a: 31, 34) is a misspelling of *Laemodonta* Philippi.

Caum (1928: 8) included "*Plecotrema labrellum* Adams" in the synonymy of *striata* Philippi. However, *labrella* was originally described by Deshayes (1830: 92) from the île de France. Adams & Adams (1854j: 122) listed *labrella* Deshayes without giving a locality. The name appears not to have been published by C.B. Adams (see Clench & Turner, 1950; Turner, 1956a,b; Johnson & Boss, 1972; Jacobson & Boss, 1973). It is not clear whether Caum was referring to *labrella* Deshayes or to something else. We exclude *labrella* from this catalog.

clausa.
> *Plectrotrema clausa* Adams & Adams, 1854j: 121. Hawaiian Islands: [no additional details].
> *Remarks*. Synonym of *octanfracta* Jonas, *teste* Kay (1979: 491).

inaequalis.
> *Plecotrema inaequalis* Adams & Adams 1854j: 122. *Nom. nud.*
> *Remarks.* Considered a synonym of *striata* Philippi by Caum (1928: 8); *striata* subsequently synonymized with *octanfracta* Jonas by Kay (1979: 491). Synonym of *octanfracta* Jonas. Adams & Adams (1854j: 122) indicated that this name had been published, in combination with the genus *Pedipes*, in C.B. Adams' *Contributions to Conchology*. However, Clench & Turner (1950: 292) considered it a manuscript name never published by C.B. Adams, and we have been unable to find it in the *Contributions to Conchology*.

multisulcatus.
> *Lirator multisulcatus* Beck, 1838: 108. *Nom. nud.*
> *Remarks.* Considered a synonym of *striata* Philippi by Caum (1928: 8); *striata* subsequently synonymized with *octanfracta* Jonas by Kay (1979: 491). Synonym of *octanfracta* Jonas.

octanfracta. (Hawaiian Islands)
> *Pedipes octanfracta* Jonas, 1845: 169. ?Hawaiian Islands ("*Habitat ad insulas Sandwicensis?*"): [no additional details].

striata.
> *Auricula striata* Philippi, 1846: 98. Hawaiian Islands: [no additional details].
> *Remarks.* Synonym of *octanfracta* Jonas, *teste* Kay (1979: 491). Perhaps originally a manuscript name of "Adams" (see Philippi, 1846: 98).

Genus PEDIPES Férussac

PEDIPES Férussac, 1821e: 113. Type species: *Bulimus pedipes* Bruguière, 1789 [not Hawaiian], by absolute tautonymy.

Authorship of the genus was incorrectly attributed to Bruguière by Zilch (1959a: 68) and Vaught (1989: 75).

sandwicensis. (Hawaiian Islands)
> *Pedipes sandwicensis* Pease, 1860: 146. Hawaiian Islands ("Sandwich Islands" in publication title): [no additional details].

Subfamily PYTHIINAE

Genus BLAUNERIA Shuttleworth

BLAUNERIA Shuttleworth, 1854: 56. Type species: *Tornatellina cubensis* Pfeiffer, 1842 [= *Voluta heteroclita* Montagu, 1808, *teste* Zilch (1959a: 74)] [not Hawaiian], by monotypy.

gracilis. (O, Mo)
> *Blauneria gracilis* Pease, 1860: 145. Hawaiian Islands ("Sandwich Islands" in publication title): [no additional details].
> *Remarks.* A lectotype was selected by Kay (1965b: 26) (BMNH 1962770), but no records of new collections have been published since the original description, *teste* Kay (1979: 493). However, 2 specimens, collected from archaeological sites, 1 on Molokai in 1982 (identified by CCC) and the other on Oahu in 1993 (identified by RHC) are now held in the Bishop Museum.

Family LYMNAEIDAE

Hubendick (1952) considered all the Hawaiian species endemic to the archipelago, although not to particular islands, as is the case with most of the terrestrial species. The Hawaiian species occur in an unusually wide range of habitats for most lymnaeids: from streams and ponds to the surfaces of wet rocks and cliff faces, as well as anthropogenic habitats such as taro patches (Hubendick, 1952).

The present treatment is based on the revisions of Hubendick (1952) and Morrison (1969), but with generic and subgeneric status following Vaught (1989: 76). The systematics of the Hawaiian taxa seems very muddled but these most recent revisions leave a total of 5 valid species, 2 each in the genera *Erinna* and *Lymnaea*, and a single species, of which the placement and status are uncertain. Morrison (1969: 32) indicated confusion of native and introduced species by Hubendick (1952). Probably there are other introduced Lymnaeidae (and possibly Physidae) that have been confused with native species. The fauna requires further careful study.

Genus ERINNA Adams & Adams

ERINNA Adams & Adams *in* Adams, 1855: 120. Type species: *Erinna newcombi* Adams & Adams, 1855, by original designation.

aulacospira. (K, Mo, M, H)
>*Limnea aulacospira* Ancey, 1889c: 290. Maui: [no additional details].

hawaiensis.
>*Limnaea hawaiensis* Pilsbry, 1904: 790, 2 unnumbered figs. Hawaii: "in small streams in the mountains on the Hilo side".
>>*Remarks.* Synonym of *aulacospira* Ancey, *teste* Hubendick (1952: 315).

newcombi. (K)
>*Erinna newcombi* Adams & Adams *in* Adams, 1855: 120. Kauai: "Heneta River".

Genus LYMNAEA Lamarck

LYMNAEA Lamarck, 1799: 75. Type species: *Helix stagnalis* Linnaeus, 1758 [not Hawaiian], by subsequent designation of Fleming (1818: 312) [see ICZN (1957b: 321; Opinion 495); ICZN gave the page as 313].

Subgenus PSEUDISIDORA Thiele, 1931

PSEUDISIDORA Thiele, 1931: 476. Type species: *Lymnaea rubella* Lea, 1841, by monotypy [see Boss & Bieler, 1991: 4, 50].

The type species, *Lymnaea rubella* Lea, was first described in 1841 in the *Proceedings of the American Philosophical Society*, not in 1844 (in the *Transactions of the American Philosophical Society*) as stated by Boss & Bieler (1991: 50).

Many taxa were synonymized with *reticulata* Gould by Hubendick (1952). Subsequently, Morrison (1969) synonymized *reticulata* Gould with the earlier *producta* Mighels without explicitly indicating that the taxa previously synonymized with *reticulata* by Hubendick became synonyms of *producta*. Since we are following the latest revision (i.e., Morrison, 1969), the taxa synonymized with *reticulata* Gould by Hubendick (1952) now become new junior synonyms of *producta* Mighels.

affinis.
> *Lymnaea affinis* Souleyet, 1852: 528, pl. 29, figs. 42–44. Oahu ("dans les mêmes lieux que la précédente", i.e., *L. oahouensis*): [no additional details].
> *Remarks.* Synonym of *reticulata* Gould, *teste* Hubendick (1952: 321). Although Morrison (1969: 32) did not include *affinis* in the synonymy of either *reticulata* or *producta*, he did state that he had not seen specimens of it. Morrison's statement is very unclear and we therefore follow Hubendick's treatment of *affinis* as a synonym of *reticulata* and hence of *producta*. **N. syn.**

ambigua.
> *Limnaea ambigua* Pease, 1870c: 6, pl. 3, fig. 5. Hawaiian Islands (in publication title): [no additional details].
> *Remarks.* Synonym of *reticulata* Gould, *teste* Hubendick (1952: 321). Synonym of *producta* Mighels. **N. syn.**

binominis.
> *Limnaea binominis* Sykes, 1900: 391. Oahu: "Mts. near Honolulu".
> *Remarks.* Proposed as a new name for *sandwichensis* Clessin, 1886, which is preoccupied by *sandwicensis* Philippi, 1845. Synonym of *reticulata* Gould, *teste* Hubendick (1952: 321). Synonym of *producta* Mighels. **N. syn.**

compacta.
> *Limnaea compacta* Pease, 1870c: 6, pl. 3, fig. 4. Oahu: [no additional details].
> *Remarks.* Synonym of *reticulata* Gould, *teste* Hubendick (1952: 321). Synonym of *producta* Mighels. **N. syn.**

flavida.
> *Physa flavida* Clessin *in* Küster, Dunker & Clessin, 1886b: 364, pl. 51, fig. 9. Hawaiian Islands: [no additional details].
> *Remarks.* Synonym of *reticulata* Gould, *teste* Hubendick (1952: 321). Synonym of *producta* Mighels. **N. syn.**

hartmanni.
> *Physa hartmanni* Clessin *in* Küster, Dunker & Clessin, 1886b: 371, pl. 54, fig. 9. Hawaii (as "Hawai"): [no additional details].
> *Remarks.* Synonym of *reticulata* Gould, *teste* Hubendick (1952: 321). Synonym of *producta* Mighels. **N. syn.** Hubendick (1952) only recorded *reticulata* from Oahu, Kauai and Niihau, so the locality given by Clessin (*in* Küster, Dunker & Clessin, 1886b: 371) may mean the Hawaiian Islands rather than the Island of Hawaii itself.

moreletiana.
> *Physa moreletiana* Clessin *in* Küster, Dunker & Clessin, 1886a: 341, pl. 48, fig. 3. Hawaiian Islands: [no additional details].
> *Remarks.* Synonym of *reticulata* Gould, *teste* Hubendick (1952: 321). Synonym of *producta* Mighels. **N. syn.**

naticoides.
> *Physa naticoides* Clessin *in* Küster, Dunker & Clessin, 1886a: 341, pl. 48, fig. 5. Hawaiian Islands: [no additional details].
> *Remarks.* Synonym of *reticulata* Gould, *teste* Hubendick (1952: 321) and of *affinis* Souleyet, *teste* Morrison (1969: 32). Both these and *naticoides* here considered synonyms of *producta* Mighels. **N. syn.**

oahouensis.
> *Lymnaea oahouensis* Souleyet, 1852: 527, pl. 29, figs. 38–41. Oahu: [no additional details].
> *Remarks.* Synonym of *rubella* Lea, *teste* Hubendick (1952: 312) and Morrison (1969: 31).

peasei.
>*Physa peasei* Clessin *in* Küster, Dunker & Clessin, 1886a: 339, pl. 47, fig. 8. Hawaiian Islands: [no additional details].
>>*Remarks.* Considered a synonym of *reticulata* Gould by Hubendick (1952: 321) and of *affinis* Souleyet by Morrison (1969: 32). Both these and *peasei* here considered synonyms of *producta* Mighels. **N. syn.**

producta. (N, K, O, ?H)
>*Physa producta* Mighels, 1845: 21. Oahu: [no additional details].
>>*Remarks.* Hubendick (1952: 321) questioningly included *producta* in the synonymy of *reticulata* Gould. Pease (1870c: 5) and Caum (1928: 9) indicated that the types were lost and other specimens are not known. Morrison (1969: 31) considered it a valid species.

reticulata.
>*Physa reticulata* Gould, 1847d: 214. Hawaiian Islands: [no additional details].
>>*Remarks.* Synonym of *producta* Mighels, *teste* Morrison (1969: 31).

rubella. (N, K, O, Mo, M, H)
>*Lymnea rubella* Lea, 1841: 33 [1844: 12]. Oahu: [no additional details]

sandwicensis.
>*Limnaeus sandwicensis* Philippi, 1845: 63. Oahu: [no additional details].
>>*Remarks.* Synonym of *rubella* Lea, *teste* Hubendick (1952: 312) and Morrison (1969: 31).

sandwichensis.
>*Physa sandwichensis* Clessin *in* Küster, Dunker & Clessin, 1886a: 342, pl. 48, fig 7. Hawaiian Islands: [no additional details].
>>*Remarks.* Secondary junior homonym of *sandwicensis* Philippi (*Code* Art. 58). Replaced by *binominis* Sykes. Synonym of *reticulata* Gould, *teste* Hubendick (1952: 321). Synonym of *producta* Mighels. **N. syn.**

sinistrorsus.
>*Limnaeus volutatus* var. *sinistrorsus* Martens, 1866: 208. Hawaiian Islands: [no additional details].
>>*Remarks.* Synonym of *reticulata* Gould, *teste* Hubendick (1952: 321). Synonym of *producta* Mighels. **N. syn.**

turgidula.
>*Limnaea turgidula* Pease, 1870c: 5, pl. 3, fig. 3. Oahu: [no additional details].
>>*Remarks.* Synonym of *reticulata* Gould, *teste* Hubendick (1952: 321). Synonym of *producta* Mighels. **N. syn.**

volutata.
>*Limnea volutata* Gould, 1847c: 211. Oahu: [no additional details].
>>*Remarks.* Synonym of *rubella* Lea, *teste* Morrison (1969: 31). Morrison (1969: 32) considered Hubendick (1952) to have misidentified the introduced species *Galba viridis* (Quoy & Gaimard, 1832) as *L. volutata* Gould.

Incertae Sedis in LYMNAEIDAE

umbilicata. (O)
>*Physa umbilicata* Mighels, 1845: 21. Oahu: [no additional details].
>>*Remarks.* Hubendick (1952: 321) tentatively included *umbilicata* in the synonymy of *reticulata* Gould. Caum (1928: 9) indicated that the types were lost and other specimens are not known. Morrison (1969: 32) simply stated that he had not seen specimens of *umbilicata*, without further comment.

Family ANCYLIDAE

The Ancylidae (fresh water limpets) are represented by a single species in the Hawaiian Islands. It is possible that this species is introduced (at least not endemic—see Hubendick, 1967: 26), although we are not aware of any published statement definitively indicating this.

Genus FERRISSIA Walker

FERRISSIA Walker, 1903: 15 (as *Ancylus* sect.). Type species: *Ancylus rivularis* Say, 1817 [not Hawaiian], by original designation.

Subgenus PETTANCYLUS Iredale

PETTANCYLUS Iredale, 1943: 228 (as genus). Type species: *Ancylus tasmanicus* Tenison-Woods, 1876 [not Hawaiian], by original designation.

sharpi. (K, O, H; ?introduced)
> *Ancylus sharpi* Sykes, 1900: 394, pl. 12, figs, 14, 14a. Oahu: "on pali, head of Nuuanu Valley".
>> *Remarks.* See Hubendick (1967: 26) for a recent discussion.

Family ACHATINELLIDAE

The family Achatinellidae is 1 of 4 families endemic to the Pacific basin (if the very few species in Australasia, Southeast Asia, and islands of the Indian Ocean are considered artificial introductions), the others being the Amastridae, Partulidae (which are not Hawaiian), and Endodontidae (see Cowie, 1992; Tillier, 1989). The evolutionary origins of the Achatinellidae are not entirely clear, although Tillier (1989) grouped them with the Valloniidae, Pupillidae, and Pyramidulidae. Fossils, possibly referable to the Achatinellidae, have been found in the Late Palaeozoic of Europe and North America and in the Cretaceous of North America (Solem, 1979, 1981; Solem & Yochelson, 1979; see also Cowie, 1992). Achatinellidae are widespread in the Pacific (Cooke & Kondo, 1961) but only in the Hawaiian Islands have they evolved into relatively larger forms. These larger forms constitute the endemic Hawaiian Islands subfamily Achatinellinae. The spectacular array of color and pattern variation, exhibited in particular by the arboreal achatinelline genera *Achatinella* (endemic to the island of Oahu) and *Partulina*, were of significance in the early development of evolutionary thought (see Carson, 1987; Gulick, 1905; Wright, 1978) but the basic biology of most Achatinellidae remains poorly known. A notable exception is the recent work on the growth, demographics and population dynamics of a small number of species of Achatinellinae, accomplished with a view, in part, to providing a basis for conservation (Hadfield, 1986; Hadfield & Miller, 1989; Hadfield *et al.*, 1993; Hadfield & Mountain, 1981). Many of the Achatinellinae are already extinct. The remainder are seriously endangered, as are many other Pacific land snails, because of over-collecting, habitat destruction and the introduction of alien predators (Cowie, 1992; Hadfield, 1986; Hadfield *et al.*, 1993).

The assignment of genera to subfamilies herein follows the latest (partial) revision of Cooke & Kondo (1961). We do not use tribes in this arrangement. All Achatinellinae were referred to the single tribe Achatinellini by Cooke & Kondo (1961). Referral of the various genera to tribes in the other subfamilies by Cooke & Kondo (1961) is indicated in the comments under each genus.

In the *Manual of Conchology*, Pilsbry & Cooke (1914–1916) recognized 107 species in the "tornatellinid" subfamilies (= Achatinellidae excluding Achatinellinae) (excluding *Tornatellina vitrea* Dohrn, which they dealt with (p. 203) in *Tornatellides*, but "discarded"). Caum (1928) listed only 106 (missing *Auriculella chamissoi* (Pfeiffer) and also excluding *T. vitrea*). Cooke & Kondo (1961) added 3 new species (*Tornatellides* (*T.*) *oswaldi*, *Tornatellides* (*Aedituans*) *neckeri* (a new subgenus), *Philopoa singularis* (a new genus)), removed all *Tornatellina* spp. to *Tornatellinops* (now in *Pacificella*) or *Lamellidea*, transferred *thaanumi* Cooke & Pilsbry from *Tornatellides* to *Tornatellaria*, and moved *stokesi* Pilsbry & Cooke from *Tornatellaria* to *Tornatellides*. These changes, and including *vitrea* (listed here as *Tornatellides vitreus*), currently give a total of 111 species in the following genera: *Auriculella* (31 spp.), *Elasmias* (3), *Gulickia* (1), *Lamellidea* (8, including 1 introduced species), *Pacificella* (2), *Philopoa* (1), *Tornatellaria* (16), *Tornatellides* (49). In addition, there are 17 infraspecific taxa, in *Auriculella* (5), *Elasmias* (1), *Lamellidea* (2), *Pacificella* (1),

Tornatellaria (2) and *Tornatellides* (6). Solem (1990: 29) indicated 106 species and 11 subspecies.

Treatment of the Achatinellinae in the *Manual of Conchology* (Pilsbry & Cooke, 1912–1914) was particularly complex. However, it is clear that taxa considered therein as species were treated in distinct, numbered sections, while taxa considered as subspecies were treated in subsections designated by lower case letters following the main numbered section for the nominate taxon. Other taxa ("forms", "mutants", etc.) were often discussed as if they were distinct, but were usually included in the synonymy of the appropriate species or subspecies and not treated in a numbered section or lettered subsection; their treatment seems to have been as synonyms. In most cases we have followed this interpretation from the perspective of ascertaining specific, subspecific or synonymic status. In instances in which our interpretation differs from this or in which the treatment in the *Manual* is unclear, we have provided annotations. This problem is especially notable in *Achatinella* subg. *Achatinellastrum*, but occurs in other genera/subgenera and in the family Amastridae (Hyatt & Pilsbry, 1910–1911).

We here recognize 99 valid species of Achatinellinae, with details discussed under each genus. In addition, 3 taxa, listed by Pilsbry & Cooke (1914a: 366) among "Unrecognized or undescribed Achatinellidae, etc.", are listed here as "*Incertae Sedis* in Achatinellidae". This gives an overall total of 213 nomenclaturally valid species of Achatinellidae. One of these (*oblonga* Pease, listed under *Lamellidea*) is considered as not being native to the Hawaiian Islands.

STATUS OF NEWCOMB'S NAMES

Confusion has arisen regarding the authorship of certain names of Achatinellidae and Amastridae, generally attributed to Wesley Newcomb (see Clarke, 1958). According to Waterhouse (1937), the part of the *Proceedings of the Zoological Society of London* containing Newcomb's (1854b) paper "Descriptions of seventy-nine new species of *Achatinella* . . ..", which includes descriptions of both Achatinellinae and Amastridae, was published on 14 November 1854. Pfeiffer (1854b), in *Malakozoologische Blätter*, cited Newcomb's descriptions, sometimes in his own words, but often repeating Newcomb virtually word for word, and dated the publication of Newcomb's paper as 13 December 1853. However, this was the date of the meeting at which Newcomb's paper was presented, not the publication date. In fact, it appears that Pfeiffer had based his descriptions on an earlier version, published separately, of Newcomb's paper (Newcomb, 1854a) that can be dated as "before June 1854, and probably in that year" (Clarke, 1958). All the names should continue to be ascribed to Newcomb and date from the earlier separate publication.

In 2 additional papers, Newcomb (1855b,c) published descriptions of new species. Pfeiffer (1856c) subsequently also described many of these species. Even though Pfeiffer attributed the names to Newcomb, he did not explicitly attribute the descriptions to him. In addition, the title of Pfeiffer's publication says "Descriptions of . . . new species . . .". Pfeiffer's names therefore are junior homonyms of Newcomb's and not just subsequent descriptions. Newcomb (1866: 211) mentioned this double publishing, indicating that Pfeiffer was unaware of his earlier publication (Newcomb, 1855c).

Subfamily ACHATINELLINAE

Genus ACHATINELLA Swainson

ACHATINELLA Swainson, 1828: 83. Type species: *Monodonta seminigra* Lamarck, 1822 [= *Helix apexfulva* Dixon, 1789], by original designation. [Gulick (1873a: 90) listed "*Achatinella vulpina*, Fér." as the type].

APEX Martens, 1860: 248. Type species: *Helix lugubris* Férussac, 1825 (as "*Achatinella lugubris* Chemn.") [= *Helix apexfulva* Dixon, 1789], by original designation.

Pilsbry & Cooke (1913b: 118) recognized 3 "subgenera or sections": *Bulimella* Pfeiffer, *Achatinellastrum* Pfeiffer and *Achatinella s. str.* Cooke & Kondo (1961) treated them as subgenera.

Pilsbry & Cooke (1913b, 1914a) listed 43 species in the genus *Achatinella* (including *solitaria* Newcomb and *dolium* Pfeiffer (both of doubtful status) but excluding *vestita* Mighels as "impossible to identify positively" and *faba* Pfeiffer as "Unrecognized or undescribed Achatinellidae, etc."). Caum (1928) listed 42 species, omitting *dolium*, *vestita* and *faba*. Welch (1942) reduced *cestus* Newcomb, *vittata* Reeve, *turgida* Newcomb and *leucorrhaphe* Gulick to subspecies of *apexfulva* Dixon, and *swiftii* Newcomb to synonymy with *turgida*. Welch (1942) transferred the subspecies *kahukuensis* Pilsbry & Cooke and *cinerosa* Pfeiffer from *A. valida* Pfeiffer to *A. apexfulva*, but left the other *valida* subspecies of Pilsbry & Cooke (1914a) in *A. valida*. *Achatinella valida s. str.* was not synonymized with *A. apexfulva* as the United States Fish and Wildlife Service (1993: 3) stated. Welch (1954: 79; 1958: 134) reduced *rosea* Swainson and *elegans* Newcomb, respectively, to subspecies of *bulimoides* Swainson. Cooke & Kondo (1961: 292) retained *turgida* as a species but made no other changes. The reductions due to Welch (1942, 1954, 1958) are probably legitimate biologically (although his proliferation of subspecies is not), given a modern species concept that, if applied to the whole genus, would reduce the number of species to 12–16 (Christensen, 1985; United States Fish and Wildlife Service, 1993). However, this has not been done formally, so, accepting all Welch's reductions, except the reduction of *turgida* to a subspecies of *apexfulva* (cf. Cooke & Kondo, 1961: 292), and (from a purely nomenclatural not biological standpoint) including *solitaria*, *dolium*, *vestita* and *faba* as species, gives a total of 39 species (9 in *Achatinella s. str.*, 18 in *Achatinellastrum*, 12 in *Bulimella*). Clearly, this figure confounds 2 very different species concepts and includes a number of species of very doubtful status. Readers should be aware that this catalog is based purely on the most recent revisions, however incomplete. In addition, there are 153 infraspecific taxa (104 in *Achatinella s. str.*, 15 in *Achatinellastrum*, 34 in *Bulimella*), many due to Welch.

We also list 77 other available names and 45 unavailable names, giving a staggering total of 314 names that have been applied to what in reality is probably a genus comprised of a dozen or so valid species.

Welch (1938, 1942, 1954, 1958) mapped in detail the distributions of most of the taxa he treated. For his new taxa, he gave not only the name of the type locality, but also more precise details, including his own coded collection site references and an elevation (e.g., for *A. mustelina collaris*, he gave "W330-1, el. 2,000–2,200 ft."). We do not include these coded map references and altitudes; the reader is referred to Welch's publications for those details. In a few instances, although giving a single named type locality, he gave more than 1 site code, e.g., for *Achatinella mustelina griseipicta* Welch, 1938, he gave "Kapuna Gulch" as the type locality but then indicated 2 sites: "W160A-2a, el. 1,450 ft., W160A-2b, el. 1,450 ft." In such cases, the precise type locality depends on a future lectotype designation.

Subgenus ACHATINELLA Swainson

ACHATINELLA Swainson, 1828: 83 (as genus). Type species: *Monodonta seminigra* Lamarck, 1822 [= *Helix apexfulva* Dixon, 1789], by original designation. [Gulick (1873a: 90) listed "*Achatinella vulpina*, Fér." as the type].

APEX Martens, 1860: 248 (as genus). Type species: *Helix lugubris* Férussac, 1825 (as "*Achatinella lugubris* Chemn.") [= *Helix apexfulva* Dixon, 1789], by original designation.

alba.
> *Achatinella alba* Jay, 1839: 58. *Nom. nud.*
>> Remarks. Synonym of *lorata* Férussac, *teste* Pilsbry & Cooke (1914a: 279) and Welch (1942: 169).

+alba. (O)
> *Achatinella apicata* var. *alba* Sykes, 1900: 299. Oahu: "near head of Kawailoa Gulch".
>> Remarks. Subspecies of *apexfulva* Dixon. Senior synonym of *beata* Pilsbry & Cooke, which was treated by both Pilsbry & Cooke (1914a: 329) and Welch (1942: 169) as a subspecies of *apexfulva* Dixon.

+albipraetexta. (O)
> *Achatinella apexfulva albipraetexta* Welch, 1942: 89, pl. 2, fig. 3, pl. 8, figs. 18–21. Oahu: "North-South Waiawa Ridge".

+albofasciata. (O)
> *Apex albofasciatus* Smith *in* Gulick & Smith, 1873: 78, pl. 9, fig. 21. Hawaiian Islands: [no additional details].
>> Remarks. Subspecies of *apexfulva* Dixon, *teste* Welch (1942: 47). Previously considered a synonym of *simulans* Reeve by Pilsbry & Cooke (1914a: 292).

albospira.
> *Apex albospira* Smith *in* Gulick & Smith, 1873: 77, pl. 10, fig. 8. Oahu: "Ewa".
>> Remarks. Synonym of *swiftii* Newcomb, *teste* Pilsbry & Cooke (1914a: 306); *swiftii* subsequently synonymized with *turgida* Newcomb by Welch (1942: 70). Synonym of *turgida* Newcomb. **N. syn.**

+aloha. (O)
> *Achatinella (A.) apexfulva aloha* Pilsbry & Cooke, 1914a: 330, pl. 60, figs. 15, 15a, 16. Oahu: "Crest of the division ridge between the two branches of the Kaukinehua [= Kaukonahua (Welch, 1942: 160)] stream, above the Wahiawa head-gates cabin, the colony extending to within 3/4 mile of main ridge".

+altiformis. (O)
> *Achatinella mustelina altiformis* Welch, 1938: 63, pl. 4, figs. 14–16. Oahu: "West-Central Makaleha Ridge".

apexfulva. (O)
> *Helix apexfulva* Dixon, 1789: 354. Hawaiian Islands: [no additional details].
>> Remarks. Originally proposed as "*Apex Fulva*". Because it was intended as a single entity (i.e., the yellow-tipped *Helix*), it is available with the orthography "*apexfulva*" (*Code* Art. 11(h)(v)). Dixon published it in combination with *Helix* (p. 354) and *Turbo* (unnumbered plate, fig. 1). Pilsbry & Cooke (1914a: 318) acted as first revisers and selected *Helix apexfulva* as the correct original combination.

+apicata. (O)
> *Achatinella apicata* Pfeiffer, 1856c: 210. Hawaiian Islands ("Sandwich Islands" in publication title): [no additional details].
>> Remarks. Subspecies of *apexfulva* Dixon, *teste* Pilsbry & Cooke (1914a: 324) and Welch (1942: 152).

ACHATINELLIDAE [39]

+aureola. (O)
    Achatinella apexfulva aureola Welch, 1942: 64, pl. 1, figs. 34, 35, pl. 5, figs. 19–21a. Oahu: "Kalauao-Hanaiki Ridge".

+bakeri. (O)
    Achatinella apexfulva bakeri Welch, 1942: 180, pl. 3, fig. 30, pl. 11, figs. 20–20b. Oahu: "Waimea . . . 'Found on a Plateau north of Waihalona [Kawaihalona?] Gulch below North Branch on pua [*Osmanthus*] tree on a clump in open plain'".

beata.
    Achatinella (*A.*) *apexfulva beata* Pilsbry & Cooke, 1914a: 329, pl. 55, fig. 5, pl. 60, figs. 17–17c. Oahu: "Crest of the Poamoho-Helemano division ridge".
        Remarks. Synonym of *alba* Sykes. Welch (1942: 169) treated *alba* Sykes as a junior synonym of *beata*, possibly believing *alba* was preoccupied by *alba* Jay, 1839.

+bicolor. (O)
    Achatinella bicolor Pfeiffer, 1859d: 529. Oahu: "Lehui".
        Remarks. Subspecies of *mustelina* Mighels, *teste* Welch (1938: 96). Originally a Gulick name in the Cuming collection. Previously considered a "race" or "color-pattern" of *mustelina* by Pilsbry & Cooke (1914a: 342–43, 348).

+bruneola. (O)
    Achatinella apexfulva bruneola Welch, 1942: 61, pl. 1, fig. 32, pl. 5, figs. 17–17b. Oahu: "Kalauao-Hanaiki Ridge".

+brunibasis. (O)
    Achatinella mustelina brunibasis Welch, 1938: 82, pl. 6, figs. 1–4c. Oahu: "South Maili-East Pulee West Branch Ridge".

+brunicolor. (O)
    Achatinella mustelina brunicolor Welch, 1938: 67, pl. 5, figs. 14–16. Oahu: "Makua Valley".

+brunosa. (O)
    Achatinella apexfulva brunosa Welch, 1942: 146, pl. 3, fig. 11, pl. 10, figs. 26–28. Oahu: "Central Poamoho-Central Poamoho North Branch Ridge".

+buena. (O)
    Achatinella apexfulva buena Welch, 1942: 35, pl. 1, fig. 8, pl. 4, figs. 14, 14a. Oahu: "Distribution, area?: Nuuanu, west side".

+cervixnivea. (O)
    Achatinella apexfulva "pattern" *cervixnivea* Pilsbry & Cooke, 1914a: 329, pl. 60, figs. 8, 8a. Oahu: "Helemano-Poamoho division ridge".
        Remarks. Subspecies of *apexfulva* Dixon, *teste* Welch (1942: 167).

+cestus. (O)
    Achatinella cestus Newcomb, 1854a: 7, pl. 22, fig. 8 [1854b: 132, pl. 22, fig. 8]. Oahu: "Palolo".
        Remarks. Subspecies of *apexfulva* Dixon, *teste* Welch (1942: 31). Previously treated as a species by Pilsbry & Cooke (1914a: 286).

+christopherseni. (O)
    Achatinella mustelina christopherseni Welch, 1938: 111, pl. 10, figs. 1–3. Oahu: "North-Central Huliwai Ridge".

+chromatacme. (O)
    Achatinella (*A.*) *swiftii chromatacme* Pilsbry & Cooke, 1914a: 316, pl. 59, figs. 5–5b. Oahu: "Waiawa".
        Remarks. Subspecies of *apexfulva* Dixon, *teste* Welch (1942: 84).

+cinerea. (O)
    Achatinella vittata var. *cinerea* Sykes, 1900: 305. Oahu: "Nuuanu".
        Remarks. Subspecies of *apexfulva* Dixon, *teste* Welch (1942: 46). Pilsbry & Cooke (1914a: 291) previously considered it a "race" of *vittata* Reeve.

cinerosa.
  Achatinella cinerosa Pfeiffer, 1855b: 5 [1855c: 2]. Hawaiian Islands: [no additional details].
    Remarks. Synonym of *leucozona* Gulick, *teste* Welch (1942: 181). Previously considered a subspecies of *valida* Pfeiffer by Pilsbry & Cooke (1914a: 336).
+**collaris**. (O)
  Achatinella mustelina collaris Welch, 1938: 91, pl. 7, figs. 10–11c. Oahu: "Mohiakea Gulch".
concavospira. (O)
  Achatinella (Bulimella) concavospira Pfeiffer, 1859c: 30. Hawaiian Islands: [no additional details].
+**coniformis**. (O)
  Apex coniformis Gulick *in* Gulick & Smith, 1873: 80, pl. 9, fig. 23. Oahu: "Kalaikoa and Ahonui . . . Wahiawa and Helemano".
    Remarks. Subspecies of *apexfulva* Dixon, *teste* Welch (1942: 126). Pilsbry & Cooke (1914a: 306) previously treated it as a synonym of *swiftii* Newcomb.
+**cookei**. (O)
  Achatinella (Apex) cookei Baldwin, 1895: 220, pl. 10, fig. 15. Oahu: "Waiau, Ewa".
    Remarks. Subspecies of *apexfulva* Dixon, *teste* Welch (1942: 75). Previously treated as a "race" of *turgida* Newcomb by Pilsbry & Cooke (1914a: 294, 300).
cookii.
  Achatinella (Apex) cookii Baldwin, 1893: 3. *Nom. nud.*
+**dautzenbergi**. (O)
  Achatinella mustelina dautzenbergi Welch, 1938: 140, pl. 13, figs. 12–14. Oahu: "Waianae Mountains" (in publication title).
+**decolor**. (O)
  Achatinella mustelina decolor Welch, 1938: 29, pl. 1, figs. 20–23. Oahu: "Makaleha Valley".
decora. (O)
  Helix (Cochlogena) decora Férussac, 1821c: 60. Hawaiian Islands: [no additional details].
    Remarks. Proposed as an alternative name for *sinistrorsus* Chemnitz.
+**diffusa**. (O)
  Achatinella mustelina diffusa Welch, 1938: 86, pl. 7, figs. 7–9b. Oahu: "West Pulee-North Haleauau Ridge".
dolium. (O)
  Achatinella dolium Pfeiffer, 1855c: 5, pl. 30, fig. 15. Hawaiian Islands: [no additional details].
    Remarks. Pilsbry & Cooke (1914a: 317) stated that it is only known from "the type" and considered it "probably only a color-form" of *swiftii* Newcomb. They indicated that Sykes considered it correctly placed in *Achatinella s. str.* and quoted Sykes who said that Baldwin's Molokai locality was wrong and that it "belongs to Oahu". Welch (1942) did not discuss it in his revision of *apexfulva* Dixon. It has never formally been reduced to infraspecific status or synonymy and from a purely nomenclatural standpoint it is retained as a valid species.
+**duplocincta**. (O)
  Achatinella apexfulva color-form *duplocincta* Pilsbry & Cooke, 1914a: 323, pl. 55, figs. 6–8. Oahu: "Wahiawa . . . Kawailoa, east side".
    Remarks. Subspecies of *apexfulva* Dixon, *teste* Welch (1942: 179). The type locality was discussed further by Welch (1942: 179) who gave "Kawailoa, Waialua", and considered the Wahiawa locality erroneous, although an annotation by C.M. Cooke, Jr. in a copy of Welch's paper in the Bishop Museum indicates that this was the locality given him (Cooke) by Emerson, the collector of the type material.

elongata.
: *Achatinella elongata* Pfeiffer, 1858: 172. *Nom. nud.*
: Remarks. Pilsbry & Cooke (1914a: 350) explained the origin of this name, which Pfeiffer originally published as a synonym of *napus* Pfeiffer.

+**ewaensis**. (O)
: *Achatinella apexfulva ewaensis* Welch, 1942: 115, pl. 2, fig. 15, pl. 7, figs. 23–24a. Oahu: "North-Central Kipapa Ridge".

+**flavida**. (O)
: *Apex flavidus* Gulick *in* Gulick & Smith, 1873: 80, pl. 10, fig. 1. Oahu: "Kalaikoa".
: Remarks. Subspecies of *apexfulva* Dixon, *teste* Welch (1942: 136). Previously treated as a synonym of *swiftii* Newcomb by Pilsbry & Cooke (1914a: 306). Welch (1942: 136) designated a lectotype by inference (*Code* Art. 74), by designating a single specimen (his pl. 10, fig. 17) as the "holotype". The type locality has been verified by reference to the data associated with the lectotype and paralectotype specimens (MCZ 39904).

+**flavitincta**. (O)
: *Achatinella apexfulva flavitincta* Welch, 1942: 94, pl. 2, fig. 7, pl. 8, figs. 30–30b. Oahu: "Waiawa-Panihakea Ridge".

+**forbesiana**. (O)
: *Achatinella forbesiana* Pfeiffer, 1855c: 5, pl. 30, fig. 16. Hawaiian Islands: [no additional details].
: Remarks. Subspecies of *apexfulva* Dixon, *teste* Welch (1942: 28). Previously considered a geographical "race" of *cestus* Newcomb by Pilsbry & Cooke (1914a: 288).

+**fumositincta**. (O)
: *Achatinella apexfulva fumositincta* Welch, 1942: 100, pl. 2, fig. 9, pl. 8, figs. 31, 31a, pl. 9, fig. 1. Oahu: "Panihakea-Kipapa Ridge".

+**fuscostriata**. (O)
: *Achatinella apexfulva fuscostriata* Welch, 1942: 32, pl. 1, fig. 4, pl. 4, figs. 7, 7a. Oahu: "Palolo".

+**glaucopicta**. (O)
: *Achatinella apexfulva glaucopicta* Welch, 1942: 140, pl. 3, fig. 10, pl. 10, figs. 22, 22a. Oahu: "Central Poamoho Stream".

+**globosa**. (O)
: *Achatinella globosa* Pfeiffer, 1855c: 7, pl. 30, fig. 25. Hawaiian Islands: [no additional details].
: Remarks. Subspecies of *apexfulva* Dixon, *teste* Welch (1942: 36). Previously treated as a synonym of *vittata* Reeve by Pilsbry & Cooke (1914a: 289, 291).

+**griseibasis**. (O)
: *Achatinella apexfulva griseibasis* Welch, 1942: 90, pl. 2, fig. 6, pl. 8, figs. 17, 17a. Oahu: "East Waiawa, 'Ridge W. of Engineer's camp, 1/4 mi. mauka [toward the mountains] of the ditch trail from large koa tree down toward camp'".

+**griseipicta**. (O)
: *Achatinella mustelina griseipicta* Welch, 1938: 20, pl. 1, figs. 11–13. Oahu: "Kapuna Gulch".

+**griseitincta**. (O)
: *Achatinella mustelina griseitincta* Welch, 1938: 78, pl. 8, figs. 14–17. Oahu: "East Makaleha Ridge".

+**griseizona**. (O)
: *Achatinella* (*A.*) *concavospira griseizona* Pilsbry & Cooke, 1914a: 353, pl. 61, figs. 10–10b. Oahu: "Palihua iki . . . head of a small valley running north".
: Remarks. Considered a "local form or subspecies" by Pilsbry & Cooke (1914a: 353).

+**gulickii**. (O)
> *Apex gulickii* Smith *in* Gulick & Smith, 1873: 78, pl. 9, fig. 17 [?error, should be fig. 19 (Welch, 1942: 133)]. Oahu: "Kalaikoa and Ahonui but it is sometimes found in valleys to the west as far as Waialei".
> *Remarks.* Subspecies of *apexfulva* Dixon, *teste* Welch (1942: 133). Previously treated as a "color-pattern" of *apicata* Pfeiffer by Pilsbry & Cooke (1914a: 326). Welch (1942: 134) considered Kalaikoa the type locality,.

+**hanleyana**. (O)
> *Achatinella (Bulimella) hanleyana* Pfeiffer, 1856b: 202. Hawaiian Islands (in publication title): [no additional details].
> *Remarks.* Subspecies of *apexfulva* Dixon, *teste* Welch (1942: 37). Previously considered a synonym of *lorata* Férussac by Pilsbry & Cooke (1914a: 278, 281).

+**ihiihiensis**. (O)
> *Achatinella apexfulva ihiihiensis* Welch, 1942: 187, pl. 3, fig. 35, pl. 11, fig. 34. Oahu: "Ihiihi-Kahawainui Ridge".

+**innotabilis**. (O)
> *Apex innotabilis* Smith *in* Gulick & Smith, 1873: 78, pl. 9, fig. 19 [?error, should be fig. 23 (Welch, 1942: 33)]. Hawaiian Islands: [no additional details].
> *Remarks.* Subspecies of *apexfulva* Dixon, *teste* Welch (1942: 33). Previously considered a synonym of *swiftii* Newcomb by Pilsbry & Cooke (1914a: 306).

+**irwini**. (O)
> *Achatinella (A.) leucorrhaphe irwini* Pilsbry & Cooke, 1914a: 302, pl. 59, figs. 9–15a (as "*leucorraphe*"). Oahu: "Division ridges between gulches of Kipapa and Waikakalaua, Waikakalaua and Kalaikoa, and Kalaikoa and Kuakine-hua, above the 1,500 ft. contour, extending up each ridge to within a mile of the main ridge".
> *Remarks.* Subspecies of *apexfulva* Dixon, *teste* Welch (1942: 124).

+**kaalaensis**. (O)
> *Achatinella mustelina kaalaensis* Welch, 1938: 87, pl. 6, figs. 11–12e. Oahu: "South Haleauau Gulch".

+**kahukuensis**. (O)
> *Achatinella (A.) valida kahukuensis* Pilsbry & Cooke, 1914a: 338, pl. 52, figs. 17, 17a. Oahu: "Kahuku, at an elevation 1,500 to 1,750 ft."
> *Remarks.* Subspecies of *apexfulva* Dixon, *teste* Welch (1942: 175).

+**kapuensis**. (O)
> *Achatinella mustelina kapuensis* Welch, 1938: 132, pl. 13, figs. 7–10a. Oahu: "South Kaaikukai Central Branch Manuwaikaalae Ridge below Mauna Kapu".

+**kawaiiki**. (O)
> *Achatinella apexfulva kawaiiki* Welch, 1942: 159, pl. 3, fig. 29, pl. 12, figs. 26, 26a. Oahu: "Kawaiiki-Kawainui Ridge".

+**lathropae**. (O)
> *Achatinella mustelina lathropae* Welch, 1938: 103, pl. 9, figs. 15–17. Oahu: "South Huliwai Gulch".

+**laurani**. (O)
> *Achatinella apexfulva laurani* Welch, 1942: 55, pl. 1, figs. 37, 38, pl. 5, figs. 28–31, pl. 6, fig. 1. Oahu: "Hanaiki-Waimalu Ridge".

+**lemkei**. (O)
> *Achatinella apexfulva lemkei* Welch, 1942: 95, pl. 2, fig. 5, pl. 8, figs. 22–22c. Oahu: "North Waiawa Stream".

+**leucophaea**. (O)
> *Apex leucophaeus* Gulick *in* Gulick & Smith, 1873: 82, pl. 9, fig. 16. Oahu: "Waialei".
> *Remarks.* Subspecies of *valida* Pfeiffer, *teste* Pilsbry & Cooke (1914a: 336).

+**leucorrhaphe**. (O)
>*Apex leucorrhaphe* Gulick *in* Gulick & Smith, 1873: 79, pl. 10, fig. 2. Oahu: "Kalaikoa . . . Waimea".
>>*Remarks.* Subspecies of *apexfulva* Dixon, *teste* Welch (1942: 101). Previously considered a species by Pilsbry & Cooke (1914a: 301).

+**leucozona**. (O)
>*Apex leucozonus* Gulick *in* Gulick & Smith, 1873: 83, pl. 10, fig. 6. Oahu: "Waialei [Waialee]".
>>*Remarks.* Subspecies of *apexfulva* Dixon, *teste* Welch (1942: 181). Previously considered a synonym of *cinerosa* Pfeiffer by Pilsbry & Cooke (1914a: 337, 338). Welch (1942: 181) designated a lectotype by inference (*Code* Art. 74(b)) (MCZ 39906). The type locality has been verified by reference to the data associated with the lectotype specimen.

+**lilacea**. (O)
>*Apex lilaceus* Gulick *in* Gulick & Smith, 1873: 79, pl. 10, fig. 4. Oahu: "Kalaikoa".
>>*Remarks.* Subspecies of *apexfulva* Dixon, *teste* Welch (1942: 137). Previously treated as a "pattern" of *apicata* Pfeiffer by Pilsbry & Cooke (1914a: 327). Welch (1942: 137) designated a lectotype (his pl. 10, fig. 8) by inference (*Code* Art. 74(b)) (MCZ 39905). The type locality has been verified by reference to the data associated with the lectotype specimen.

+**lineipicta**. (O)
>*Achatinella apexfulva lineipicta* Welch, 1942: 98, pl. 2, fig. 8, pl. 8, figs. 1–3a. Oahu: "Waiawa-Panihakea Ridge".

lorata.
>*Helix (Cochlogena) lorata* Férussac, 1821c: 60. *Nom. nud.*

lorata. (O)
>*Helix lorata* Férussac *in* Quoy & Gaimard, 1825d: 479, pl. 68, figs. 8–12. Hawaiian Islands: [no additional details].

lugubris.
>*Turbo lugubris* Chemnitz, 1795: 278, pl. 209, figs. 2059, 2060. Unavailable name; proposed in a non-binominal work.
>>*Remarks.* Considered a synonym of *apexfulva* Dixon by Pilsbry & Cooke (1914a: 318) and Welch (1942: 176).

lugubris.
>*Helix lugubris* Férussac, 1821c: 60. *Nom. nud.*
>>*Remarks.* This name was not made available by bibliographic reference to Chemnitz (1795), which is a non-binominal work (*Code* Art. 11(d)(ii)).

lugubris.
>*Helix lugubris* Férussac *in* Quoy & Gaimard, 1825d: 479. Hawaiian Islands: [no additional details].
>>*Remarks.* Synonym of *apexfulva* Dixon, *teste* Welch (1942: 176).

lymaniana.
>*Achatinella (Bulimella) lymaniana* Baldwin, 1893: 5. *Nom. nud.*

+**lymaniana**. (O)
>*Achatinella (Bulimella) lymaniana* Baldwin, 1895: 219, pl. 10, fig. 14. Oahu: "Waianae Mts."
>>*Remarks.* Subspecies of *mustelina* Mighels, *teste* Pilsbry & Cooke (1914a: 350).

+**mailiensis**. (O)
>*Achatinella mustelina mailiensis* Welch, 1938: 42, pl. 7, figs. 4–6a. Oahu: "Puulu Gulch".

+**makahaensis**. (O)
>*Achatinella mustelina makahaensis* Pilsbry & Cooke, 1914a: 345, pl. 62, figs. 3, 4. Oahu: "in Makaha, back of the home clearing of the Manager of Makaha plantation, and on the edge of the coffee clearing".

+**maxima**. (O)
> *Achatinella mustelina maxima* Welch, 1938: 54, pl. 5, figs. 1–4. Oahu: "South Keaau" [Welch, 1938: 161, in explanation of pl. 5, fig. 1].

+**meadowsi**. (O)
> *Achatinella apexfulva meadowsi* Welch, 1942: 70, pl. 1, fig. 42, pl. 6, figs. 11–12. Oahu: "South Waimano-South Central Waimano Ridge".

+**melanogama**. (O)
> *Achatinella (A.) lorata* form *melanogama* Pilsbry & Cooke, 1914a: 283, pl. 51, figs. 1–3, pl. 52, figs. 8, 8a. Oahu: "Maunawili, on the north side of the main range opposite Manoa valley—the Kailua flank of Mt. Olympus".

+**mixta**. (O)
> *Achatinella mustelina mixta* Welch, 1938: 74, pl. 8, figs. 18–20. Oahu: "North Kalalua Gulch".

multilineata.
> *Achatinella multilineata* Newcomb, 1854a: 12, pl. 22, fig. 23 [1854b: 138, pl. 22, fig. 23]. Oahu: "Kolau poco".
>> *Remarks.* Synonym of *mustelina* Mighels, *teste* Pilsbry & Cooke (1914a: 343). Pilsbry & Cooke (1914a: 343) considered Newcomb's locality "certainly erroneous", the shell being "from the Waianae mountains, and probably the type came from Mokuleia district".

+**muricolor**. (O)
> *Achatinella apexfulva muricolor* Welch, 1942: 30, pl. 1, fig. 1, pl. 4, figs. 1–1b. Oahu: "Niu-Wailupe Ridge".

mustelina. (O)
> *Achatinella mustelina* Mighels, 1845: 21. Oahu: "Waianai".

+**napus**. (O)
> *Achatinella napus* Pfeiffer, 1855c: 5, pl. 30, fig. 19 [1855g: 67]. Hawaiian Islands: [no additional details].
>> *Remarks.* Subspecies of *apexfulva* Dixon, *teste* Welch (1942: 182). Previously considered a synonym of *sordida* Newcomb by Pilsbry & Cooke (1914a: 349–50).

neglectus.
> *Apex neglectus* Smith *in* Gulick & Smith, 1873: 78, pl. 9, fig. 22. Hawaiian Islands: [no additional details].
>> *Remarks.* Synonym of *polymorpha* Gulick, *teste* Welch (1942: 91). Previously considered a synonym of *swiftii* Newcomb by Pilsbry & Cooke (1914a: 306).

+**nigripicta**. (O)
> *Achatinella apexfulva nigripicta* Welch, 1942: 78, pl. 1, fig. 55, pl. 7, figs. 3–3d, 4. Oahu: "Manana-Waiawa Ridge".

+**nobilis**. (O)
> *Achatinella (Bulimella) nobilis* Pfeiffer, 1856b: 202. Oahu: [no additional details].
>> *Remarks.* Subspecies of *lorata* Férussac, *teste* Pilsbry & Cooke (1914a: 283).

+**nocturna**. (O)
> *Achatinella mustelina nocturna* Welch, 1938: 71, pl. 8, figs. 8–11. Oahu: "South Kalalua Gulch".

+**obesiformis**. (O)
> *Achatinella mustelina obesiformis* Welch, 1938: 77, pl. 8, figs 4–5a. Oahu: "Kanewai Gulch".

+**oioensis**. (O)
> *Achatinella apexfulva oioensis* Welch, 1942: 184, pl. 11, figs. 23–23b. Oahu: "Oio-Oio East Branch Ridge".

+**oliveri**. (O)
> *Achatinella apexfulva oliveri* Welch, 1942: 48, pl. 1, fig. 18, pl. 4 figs. 30, 30a, 31. Oahu: "Kamanaiki Stream".

ACHATINELLIDAE [45]

+**ovum.** (O)
 Achatinella (Achatinellastrum) ovum Pfeiffer, 1857a: 334. Oahu: [no additional details].
  Remarks. Subspecies of *apexfulva* Dixon, *teste* Welch (1942: 57). Previously considered a "race" of *turgida* Newcomb by Pilsbry & Cooke (1914a: 294, 297).

+**paalaensis.** (O)
 Achatinella apexfulva paalaensis Welch, 1942: 156, pl. 3, fig. 21, pl. 12, fig. 20. Oahu: "North-South Helemano Ridge".

pallida.
 Achatinella pallida Jay, 1839: 58. *Nom. nud.*

pallida.
 Achatinella pallida Reeve, 1850e: pl. 1, figs. 2a, b. Oahu (as "Wahoo, Sandwich Islands"): [no additional details].
  Remarks. Synonym of *lorata* Férussac, *teste* Pilsbry & Cooke (1914a: 279).

+**parvicolor.** (O)
 Achatinella apexfulva parvicolor Welch, 1942: 56, pl. 1, fig. 40, pl. 6, figs. 15–15c. Oahu: "Waimalu?, . . . somewhere in Waiau and Punanani Gulch" [Welch expressed doubt as to the exact location of the type locality].

+**paumaluensis.** (O)
 Achatinella apexfulva paumaluensis Welch, 1942: 183, pl. 3, fig. 32, pl. 11, figs. 21, 21a. Oahu: "Paumalu-Kaunala Ridge".

+**perplexa.** (O)
 Achatinella (A.) turgida perplexa Pilsbry & Cooke, 1914a: 296, pl. 56, figs. 5–5f. Oahu: "Lateral spurs, and northern ridge of Waimano valley".
  Remarks. Subspecies of *apexfulva* Dixon, *teste* Welch (1942: 72).

perversa.
 Achatinella perversa Swainson, 1828: 84. "Islands of the Pacific Ocean" (in introduction to the genus): [no additional details].
  Remarks. Synonym of *decora* Férussac, *teste* Pilsbry & Cooke (1914a: 331).

pica.
 Achatinella pica Swainson, 1828: 84. "Islands of the Pacific Ocean" (in introduction to the genus): [no additional details].
  Remarks. Synonym of *apexfulva* Dixon, *teste* Pilsbry & Cooke (1914a: 318–22) and Welch (1942: 176).

+**pilsbryi.** (O)
 Achatinella apexfulva pilsbryi Welch, 1942: 52, pl. 1, fig. 29, pl. 5, fig. 27. Oahu: "Aiea-Kalauao Ridge".

+**poamohoensis.** (O)
 Achatinella apexfulva poamohoensis Welch, 1942: 166, pl. 3, figs, 5, 6, pl. 12, figs. 7, 7a, 8. Oahu: "North Kaukonahua-Poamoho Ridge".

+**polymorpha.** (O)
 Apex polymorpha Gulick *in* Gulick & Smith, 1873: 81, pl. 10, fig. 5. Oahu: "Waipio and Wahiawa . . . Kalaikoa and Ahonui".
  Remarks. Subspecies of *apexfulva* Dixon, *teste* Welch (1942: 91). Previously considered a synonym of *swiftii* Newcomb by Pilsbry & Cooke (1914a: 306).

+**popouwelensis.** (O)
 Achatinella mustelina popouwelensis Welch, 1938: 99, pl. 9, figs. 5–7. Oahu: "South Waieli Gulch North Branch".

+**pulchella.** (O)
 Achatinella pulchella Pfeiffer, 1855c: 6, pl. 30, fig. 2 [1855g: 68]. Hawaiian Islands: [no additional details].
  Remarks. Subspecies of *lorata* Férussac, *teste* Pilsbry & Cooke (1914a: 284).

+punicea. (O)
> *Achatinella apexfulva punicea* Welch, 1942: 141, pl. 2, fig. 26, pl. 10, figs. 6–6c. Oahu: "Kaukonahua".

quernea.
> *Achatinella quernea* Pilsbry & Cooke, 1914a: 332. *Nom. nud.*
> > Remarks. Originally a Frick collection name. Synonym of *decora* Férussac, *teste* Pilsbry & Cooke (1914a: 332).

+roseata. (O)
> *Achatinella apexfulva roseata* Welch, 1942: 54, pl. 1, fig. 39, pl. 6, figs. 25–25b. Oahu: "Waimano Stream. 'Just below the bridge on the plantation R. R.,' . . . in the bottom of the gulch".

+roseipicta. (O)
> *Achatinella apexfulva roseipicta* Welch, 1942: 162, pl. 3, fig. 7, pl. 12, figs. 4–4b. Oahu: "North Kaukonahua-Poamoho Ridge".

+rubidilinea. (O)
> *Achatinella apexfulva rubidilinea* Welch, 1942: 83, pl. 1, fig. 51, pl. 7, figs. 16–18. Oahu: "Manana-Waiawa Ridge".

+rubidipicta. (O)
> *Achatinella apexfulva rubidipicta* Welch, 1942: 49, pl. 1, fig. 15, pl. 5, figs. 2–2d. Oahu: "Nuuanu-Kapalama Ridge".

+russi. (O)
> *Achatinella mustelina russi* Welch, 1938: 123, pl. 12, figs. 3–4c. Oahu: "Napepeiauolelo-Palawai Ridge".

seminigra.
> *Monodonta seminigra* Lamarck, 1822: 37. Tahiti: "la mer Pacifique, sur les rivages de l'île d'Othaïti [the Pacific Ocean, on the shores of the island of Tahiti]" [?error = Oahu].
> > Remarks. Synonym of *apexfulva* Dixon, *teste* Pilsbry & Cooke (1914a: 318, 320) and Welch (1942: 176).

+simulacrum. (O)
> *Achatinella* (A.) *turgida simulacrum* Pilsbry & Cooke, 1914a: 299, pl. 56, figs. 13–14d. Oahu: "Waimano-Manana ridge, along the summit trail, above the locality of *A. t. cookei*".
> > Remarks. Subspecies of *apexfulva* Dixon, *teste* Welch (1942: 81).

+simulans. (O)
> *Achatinella simulans* Reeve, 1850c: pl. 2, fig. 15. Type locality not given.
> > Remarks. Subspecies of *apexfulva* Dixon, *teste* Welch (1942: 39). Previously considered a "race" of *vittata* Reeve by Pilsbry & Cooke (1914a: 292).

+simulator. (O)
> *Achatinella cestus* form *simulator* Pilsbry & Cooke, 1914a: 288, pl. 55, figs. 2–4. Oahu: "Palolo".
> > Remarks. Subspecies of *apexfulva* Dixon, *teste* Welch (1942: 34).

sinistrorsus.
> *Turbo lugubris sinistrorsus* Chemnitz, 1795: 307, figs. 3014, 3015. Unavailable name; proposed in a non-binominal work.

+sordida. (O)
> *Achatinella sordida* Newcomb, 1854a: 13, pl. 23, fig. 27 [1854b: 139, pl. 23, fig. 27]. Oahu: "Lettui [= Lihue (Pilsbry & Cooke, 1914a: 349)]".
> > Remarks. Subspecies of *mustelina* Mighels, *teste* Pilsbry & Cooke (1914a: 349).

+steeli. (O)
> *Achatinella apexfulva steeli* Welch, 1942: 143, pl. 2, fig. 25, pl. 10, figs. 3–5c. Oahu: "North-South Kaukonahua Ridge".

+**suturafusca**. (O)
>*Achatinella apexfulva suturafusca* Welch, 1942: 148, pl. 3, fig. 9, pl. 10, figs. 24, 24a. Oahu: "Central Poamoho Stream".

+**suturalba**. (O)
>*Achatinella apexfulva suturalba* Welch, 1942: 151, pl. 3, fig. 12, pl. 10, figs. 29, 29a. Oahu: "North Poamoho Stream".

swiftii.
>*Achatinella swiftii* Newcomb, 1854a: 7, pl. 22, figs. 9, 9a [1854b: 133, pl. 22, figs. 9, 9a]. Oahu: "District of Ewa".
>>*Remarks.* Synonym of *turgida* Newcomb, *teste* Welch (1942: 70). Previously treated as a species by Pilsbry & Cooke (1914a: 306).

+**tuberans**. (O)
>*Apex tuberans* Gulick *in* Gulick & Smith, 1873: 81, pl. 10, fig. 3. Oahu: "Kalaikoa . . . Ahonui . . . Wahiawa and Helemano".
>>*Remarks.* Subspecies of *apexfulva* Dixon, *teste* Welch (1942: 107). Previously considered a synonym of *swiftii* Newcomb by Pilsbry & Cooke (1914a: 306).

tumefactus.
>*Apex tumefactus* Gulick *in* Gulick & Smith, 1873: 82, pl. 9, fig. 20. Oahu: "Wahiawa; . . . Helemano".
>>*Remarks.* Synonym of *decora* Férussac, *teste* Pilsbry & Cooke (1914a: 331).

+**turbiniformis**. (O)
>*Apex turbiniformis* Gulick *in* Gulick & Smith, 1873: 81, pl. 10, fig. 7. Oahu: "Kalaikoa and Lehui".
>>*Remarks.* Subspecies of *concavospira* Pfeiffer, *teste* Pilsbry & Cooke (1914a: 353).

turgida. (O)
>*Achatinella turgida* Newcomb, 1854a: 8, pl. 22, figs 10, 10a [1854b: 134, pl. 22, figs. 10, 10a]. Oahu: "Ewa".
>>*Remarks.* Subspecies of *apexfulva* Dixon, *teste* Welch (1942: 68), but subsequently treated as a species by Cooke & Kondo (1961: 292), following Pilsbry & Cooke (1914a: 294).

valida. (O)
>*Achatinella valida* Pfeiffer, 1855c: 6, pl. 30, fig. 24. Hawaiian Islands: [no additional details].
>>*Remarks.* Pilsbry & Cooke (1914a: 335), by referring to "Pfeiffer's type figure", designated a lectotype. However, although they considered that "Pupukea . . . may be accepted as type locality", this locality was associated with specimens collected by Gulick and they gave no evidence linking it with Pfeiffer's specimen; restriction of the type locality requires further study.

ventrosa.
>*Achatinella ventrosa* Pfeiffer, 1855c: 6, pl. 30, fig. 20 [1855g: 67]. Hawaiian Islands: [no additional details].
>>*Remarks.* Synonym of *lorata* Férussac, *teste* Pilsbry & Cooke (1914a: 279).

+**versicolor**. (O)
>*Apex versicolor* Gulick *in* Gulick & Smith, 1873: 80, pl. 9, fig. 18. Oahu: "Ahonui and Kalaikoa".
>>*Remarks.* Subspecies of *apexfulva* Dixon, *teste* Welch (1942: 130). Previously considered a synonym of *swiftii* Newcomb by Pilsbry & Cooke (1914a: 306).

vespertina.
>*Achatinella (Apex) vespertina* Baldwin, 1893: 4. *Nom. nud.*

+**vespertina**. (O)
>*Achatinella (Apex) vespertina* Baldwin, 1895: 219, pl. 10, fig. 14. Oahu: "Kawailoa".
>>*Remarks.* Subspecies of *apexfulva* Dixon, *teste* Pilsbry & Cooke (1914a: 322), Welch (1942: 174).

vestita. (O)
> *Achatinella vestita* Mighels, 1845: 20. Oahu: "Waianai"; Hawaii: [no additional details].
>> *Remarks.* Questionably considered a synonym of *mustelina* Mighels by Welch (1938: 16) and Pilsbry & Cooke (1914a: 342). Welch (1938: 20), following Pilsbry & Cooke (1914a: 344), "dropped" this species as unidentifiable since the holotype was destroyed by fire. It has never been formally synonymized and from a purely nomenclatural standpoint is retained here as a valid species.

+virgatifulva. (O)
> *Achatinella apexfulva virgatifulva* Welch, 1942: 103, pl. 2, fig. 16, pl. 9, figs. 5–5e. Oahu: "Waikakalaua Stream".

+vittata. (O)
> *Achatinella vittata* Reeve, 1850c: pl. 2, fig. 9. Hawaiian Islands: [no additional details]
>> *Remarks.* Subspecies of *apexfulva* Dixon, *teste* Welch (1942: 42). Previously considered a species by Pilsbry & Cooke (1914a: 289).

+wahiawa. (O)
> *Achatinella apexfulva wahiawa* Welch, 1942: 132, pl. 2, fig. 36, pl. 11, figs. 1–1b. Oahu: "Wahiawa".

+waialaeensis. (O)
> *Achatinella apexfulva waialaeensis* Welch, 1942: 29, pl. 1, fig. 3, pl. 4, figs. 2–4. Oahu: "Waialae Iki-Waialae Nui Ridge".

+waianaeensis. (O)
> *Achatinella mustelina waianaeensis* Welch, 1938: 43, pl. 3, figs. 4–6. Oahu: "Pahole-Kapuna Ridge".

+wailelensis. (O)
> *Achatinella apexfulva wailelensis* Welch, 1942: 188, pl. 3, fig. 36, pl. 11, fig. 35. Oahu: " Wailele Gulch".

+waimaluensis. (O)
> *Achatinella apexfulva waimaluensis* Welch, 1942: 66, pl. 1, fig. 45, pl. 5, figs. 25, 25a. Oahu: "Waimalu-South Central Waimano Ridge".

Subgenus ACHATINELLASTRUM Pfeiffer

HELICTERES Férussac, 1821c: 60 (as an infrageneric grouping of *Helix* (*Cochlogena*)). Unavailable name; genus-group name proposed in the nominative plural.

HELICTERES Beck, 1837: 51 (as *Bulimus* subg.). Type species: *Helix vulpina* Férussac, 1825, by subsequent designation of Herrmansen (1847b: 515).

ACHATINELLASTRUM Pfeiffer, 1854b: 133. Type species: *Achatinella producta* Reeve, 1850, by subsequent designation of Pilsbry & Cooke (1914a: 181).

HELICTER: Incorrect subsequent spelling of *Helicteres* Férussac (Pease, 1862: 6).

HELICTERELLA Gulick, 1873b: 497. *Nom. nud.*

The genus-group name *Helicteres* was first proposed by Férussac (1821c: 60) as an infrageneric grouping of his subgenus *Cochlogena* of *Helix*. However, as is evident from his formation of names of similar levels of infrageneric groupings, it was proposed in the nominative plural, making it unavailable for nomenclatural purposes. Menke (1830: 25) also used the name *Helicteres* in the nominative plural, essentially copying Férussac's (1821c) classification and names. The first to validate the name *Helicteres* was Beck (1837: 51) who listed it as a subgenus of *Bulimus* and included 12 validly named species. The type species of *Helicteres* Beck is currently placed in subg. *Achatinellastrum* Pfeiffer. *Helicteres* Beck, 1837 and *Achatinellastrum* Pfeiffer, 1854 are therefore subjective synonyms. In accordance with common usage, we continue to use *Achatinellastrum*. Action by the ICZN to suppress *Helicteres* Beck is necessary to formalize this usage.

adusta.
> *Achatinella adusta* Reeve, 1850c: pl. 4, fig. 30. Type locality not given.
> *Remarks.* Synonym of *vulpina* Férussac, *teste* Pilsbry & Cooke (1914a: 213).

albescens.
> *Achatinella albescens* Gulick, 1858: 237, pl. 8, fig. 57. Oahu: "Waimea, Pupukea, Waialei, Kahuku, and Hauula".
> *Remarks.* Synonym of *dimorpha* Gulick, *teste* Pilsbry & Cooke (1914a: 258).

+**ampla**. (O)
> *Achatinella ampla* Newcomb, 1854a: 11, pl. 22, fig. 19 [1854b: 137, pl. 22, fig. 19]. Oahu: "Kolau [= Koolau]".
> *Remarks.* Subspecies of *fulgens* Newcomb, *teste* Pilsbry & Cooke (1914a: 198).

analoga.
> *Achatinella analoga* Gulick, 1856: 227, pl. 7, fig. 47. Oahu: "Halawa".
> *Remarks.* Synonym of *vulpina* Férussac, *teste* Pilsbry & Cooke (1914a: 213).

angusta.
> *Achatinella angusta* Paetel, 1873: 105. *Nom. nud.*
> *Remarks.* Synonym of *fulgens* Newcomb, *teste* Pilsbry & Cooke (1914a: 191). Possibly a misspelling of *augusta*.

aplustre.
> *Achatinella aplustre* Newcomb, 1854a: 21, pl. 23, fig. 51 [1854b: 147, pl. 23, fig. 51]. Oahu: "Kolau [= Koolau]".
> *Remarks.* Synonym of *stewartii* Green, *teste* Pilsbry & Cooke (1914a: 205).

augusta.
> *Achatinella augusta* Smith *in* Gulick & Smith, 1873: 74, pl. 9, fig. 7. Oahu: "Waialae, near the east end of Oahu . . . Wailupe and Palolo".
> *Remarks.* Synonym of *fulgens* Newcomb, *teste* Pilsbry & Cooke (1914a: 191).

bellula. (O)
> *Achatinella bellula* Smith *in* Gulick & Smith, 1873: 77, pl. 9, fig. 8. Hawaiian Islands: [no additional details].

bilineata.
> *Achatinella bilineata* Reeve, 1850c: pl. 3, fig. 22. Type locality not given.
> *Remarks.* Synonym of *producta* Reeve, *teste* Pilsbry & Cooke (1914a: 208).

buddii. (O)
> *Achatinella buddii* Newcomb, 1854a: 29, pl. 24, fig. 73 [1854b: 155, pl. 24, fig. 73]. Oahu: "Palolo".

caesia. (O)
> *Achatinella caesia* Gulick, 1858: 234, pl. 8, fig. 53. Oahu: "Waimea".

casta. (O)
> *Achatinella casta* Newcomb, 1854a: 8, pl. 22, fig. 12 [1854b: 134, pl. 22, fig. 12]. Oahu: "Ewa".

castanea.
> *Achatinella castanea* Reeve, 1850c: pl. 3, fig. 24. Hawaiian Islands: [no additional details].
> *Remarks.* Synonym of *vulpina* Férussac, *teste* Pilsbry & Cooke (1914a: 212).

+**cervina**. (O)
> *Achatinella cervina* Gulick, 1858: 241, pl. 8, fig. 62. Oahu: "Kahana".
> *Remarks.* "Race" of *caesia* Gulick, *teste* Pilsbry & Cooke (1914a: 264).

+**cognata**. (O)
> *Achatinella cognata* Gulick, 1858: 240, pl. 8, fig. 60. Oahu: "Hakipu . . . Waikane".
> *Remarks.* "Race" of *caesia* Gulick, *teste* Pilsbry & Cooke (1914a: 264).

+**colorata.** (O)
>*Achatinella colorata* Reeve, 1850c: pl. 3, fig. 18. Hawaiian Islands: [no additional details].
>>Remarks. Subspecies of *vulpina* Férussac, *teste* Pilsbry & Cooke (1914a: 224). Pilsbry & Cooke (1914a: 224–26) indicated Oahu localities.

+**concidens.** (O)
>*Achatinella concidens* Gulick, 1858: 234, pl. 8, fig. 53. Oahu: "Waimea".
>>Remarks. "Color-form" of *caesia* Gulick, *teste* Pilsbry & Cooke (1914a: 265).

concolor.
>*Achatinella concolor* Smith *in* Gulick & Smith, 1873: 75, pl. 9, fig. 1. Oahu: "Ewa".
>>Remarks. Synonym of *casta* Newcomb, *teste* Pilsbry & Cooke (1914a: 236).

consanguinea.
>*Achatinella consanguinea* Smith *in* Gulick & Smith, 1873: 73, pl. 9, fig. 3. Oahu: "Ahuimanu".
>>Remarks. Synonym of *colorata* Reeve, *teste* Pilsbry & Cooke (1914a: 224).

contracta.
>*Achatinella contracta* Gulick, 1858: 239, pl. 8, fig. 59. Oahu: "Kaawa [= Kaaawa] . . . Hauula".
>>Remarks. Synonym of *dimorpha* Gulick, *teste* Pilsbry & Cooke (1914a: 258).

crassidentata.
>*Achatinella crassidentata* Pfeiffer, 1855c: 6, pl. 30, fig. 23. Hawaiian Islands: [no additional details].
>>Remarks. Synonym of *fulgens* Newcomb, *teste* Pilsbry & Cooke (1914a: 191).

cucumis.
>*Achatinella cucumis* Gulick, 1856: 225, pl. 7, fig. 45. Oahu: "Kalihi".
>>Remarks. Synonym of *vulpina* Férussac, *teste* Pilsbry & Cooke (1914a: 213).

cuneus.
>*Achatinella (Achatinellastrum) cuneus* Pfeiffer, 1856b: 205. Hawaiian Islands ("Sandwich Islands" in publication title): [no additional details].
>>Remarks. Synonym of *casta* Newcomb, *teste* Cooke *in* Pilsbry & Cooke (1914a: 236).

curta. (O)
>*Achatinella curta* Newcomb, 1854a: 18, pl. 23, fig. 43 [1854b: 144, pl. 23, fig. 43]. Oahu: "Waialua".

delta.
>*Achatinella delta* Gulick, 1858: 231, pl. 8, fig. 50. Oahu: "Kalaikoa, Ahonui, Wahiawa, and Helemanu".
>>Remarks. Synonym of *curta* Newcomb, *teste* Pilsbry & Cooke (1914a: 252). Pilsbry & Cooke (1914a: 257) stated that "Mr. Gulick's type of *delta* is a Wahiawa shell . . . No. 92,619 A.N.S." Unfortunately, this lot (ANSP 92619) contains numerous specimens, none of which is clearly marked as type. The statement of Pilsbry & Cooke thus does not qualify as a lectotype designation.

diluta.
>*Achatinella diluta* Smith *in* Gulick & Smith, 1873: 74, pl. 9, fig. 14. Hawaiian Islands: [no additional details, but see below].
>>Remarks. Pilsbry & Cooke (1914a: 229) considered *diluta* as a "rare dextral variant" of *olivacea* Reeve, clearly not intending it to have subspecific status. Pilsbry & Cooke (1914a: 213) included *olivacea* Reeve in the synonymy of *vulpina* Férussac. Synonym of *vulpina* Férussac. **N. syn.** Smith (*in* Gulick & Smith, 1873: 75) stated that "Judging from its affinities, we may believe that it comes from the island of Oahu". Pilsbry (*in* Pilsbry & Cooke, 1914a: 229) wrote that "Mr. Gulick has marked his specimen as from "Halawa?", and I imagine that locality is not far wrong."

dimorpha. (O)
>*Achatinella dimorpha* Gulick, 1858: 236, pl. 8, fig. 56. Oahu: "Pupukea".
>>Remarks. Gulick gave 5 localities. Pilsbry & Cooke (1914a: 260) designated a "type specimen" (= lectotype; "no. 56 Boston Soc. coll.") from 1 of these 5 localities, Pupukea.

diversa.
    Achatinella diversa Gulick, 1856: 220, pl. 7, figs. 42a, b. Oahu: "Palolo, Waialae, Wailupe, and Niu".
        Remarks. Synonym of *fulgens* Newcomb, *teste* Pilsbry & Cooke (1914a: 191).

dunkeri.
    Achatinella dunkeri Pfeiffer, 1856c: 208. Hawaiian Islands ("Sandwich Islands" in publication title): [no additional details].
        Remarks. Synonym of *producta* Reeve, *teste* Pilsbry & Cooke (1914a: 208). Originally a Cuming collection name. Pilsbry & Cooke (1914a: 208) indicated Oahu localities for *producta*.

+emmersonii. (O)
    Achatinella emmersonii Newcomb, 1854a: 30, pl. 24, fig. 74 [1854b: 156, pl. 24, fig. 74]. Oahu: "District of Waialua".
        Remarks. Subspecies of *livida* Swainson, *teste* Pilsbry & Cooke (1914a: 247).

ernestina.
    Achatinella (Achatinellastrum) ernestina Baldwin, 1893: 4. *Nom. nud.*

ernestina.
    Achatinella (Achatinellastrum) ernestina Baldwin, 1895: 217, pl. 10, figs. 5, 6. Oahu: "Nuuanu Valley".
        Remarks. Synonym of *vulpina* Férussac, *teste* Pilsbry & Cooke (1914a: 213).

formosa.
    Achatinella formosa Gulick, 1858: 235, pl. 8, fig. 55. Oahu: "Waimea".
        Remarks. Synonym of *caesia* Gulick, *teste* Pilsbry & Cooke (1914a: 264).

fulgens. (O)
    Achatinella fulgens Newcomb, 1854a: 5 [1854b: 131]. Oahu: "Niu".

fuscolineata.
    Achatinella fuscolineata Smith *in* Gulick & Smith, 1873: 75, pl. 9, fig. 2. Oahu: "Kailua . . . nearly all the valleys between Palolo and Halawa".
        Remarks. Synonym of *versipellis* Gulick, *teste* Pilsbry & Cooke (1914a: 196).

fuscozona.
    Achatinella fuscozona Smith *in* Gulick & Smith, 1873: 76, pl. 9, fig. 9. Oahu: "Makiki . . . Palolo".
        Remarks. Synonym of *buddii* Newcomb, *teste* Pilsbry & Cooke (1914a: 190).

glauca.
    Achatinella glauca Gulick, 1858: 232, pl. 8, fig. 51. Oahu: "Kawailoa".
        Remarks. Synonym of *recta* Newcomb, *teste* Pilsbry & Cooke (1914a: 249), whose citation of Gulick, "Ann. Lyc. N.H. vi, p. 60", is incorrect.

+gulickiana. (O)
    Achatinella (Achatinellastrum) lehuiensis gulickiana Pilsbry & Cooke, 1914a: 273, pl. 42, fig. 4. Oahu: "Waianae Range: Mokuleia".

+herbacea. (O)
    Achatinella herbacea Gulick, 1858: 233, pl. 8, fig. 52. Oahu: "In the forests between the streams of Waimea and Kawailoa".
        Remarks. Subspecies of *livida* Swainson, *teste* Pilsbry & Cooke (1914a: 251).

hybrida.
    Achatinella hybrida Newcomb, 1854a: 21, pl. 23, fig. 52 [1854b: 147, pl. 23, fig. 52]. Oahu: "Kolau [= Koolau]".
        Remarks. Synonym of *producta* Reeve, *teste* Pilsbry & Cooke (1914a: 208).

johnsoni.
    Achatinella johnsoni Newcomb, 1854a: 21, pl. 23, fig. 50 [1854b: 147, pl. 23, fig. 50]. Oahu: "Kolau [= Koolau]".
        Remarks. Synonym of *stewartii* Green, *teste* Pilsbry & Cooke (1914a: 205).

juddii.
    Achatinella (Achatinellastrum) juddii Baldwin, 1893: 4. *Nom. nud.*

juddii. (O)
>Achatinella (Achatinellastrum) juddii Baldwin, 1895: 216, pl. 10, figs. 3, 4. Oahu: "Halawa".

juncea. (O)
>Achatinella juncea Gulick, 1856: 230, pl. 7, fig. 49. Oahu: "Kalaikoa, Wahiawa, and Helemanu".
>>Remarks. Although Gulick (1856: 230) gave 3 localities, Pilsbry & Cooke (1914a: 242) stated that Gulick's figure was of a specimen marked "type" by him and that it was from Wahiawa, thereby designating the figured specimen as the lectotype.

lehuiensis. (O)
>Achatinella lehuiensis Smith in Gulick & Smith, 1873: 76, pl. 9, fig. 4. Oahu: "Lehui".

ligata.
>Achatinella ligata Smith in Gulick & Smith, 1873: 76, pl. 9, fig. 13. Oahu: "Waimalu".
>>Remarks. Synonym of casta Newcomb, teste Pilsbry & Cooke (1914a: 236).

liliacea.
>Achatinella (Achatinellastrum) liliacea Pfeiffer, 1859c: 31. Hawaiian Islands: [no additional details].
>>Remarks. Synonym of fulgens Newcomb, teste Pilsbry & Cooke (1914a: 191).

+littoralis. (O)
>Achatinella (Achatinellastrum) caesia littoralis Pilsbry & Cooke, 1914a: 266, pl. 44, figs. 17–20. Oahu: "about 1 1/2 miles east of Kahuku".

livida. (O)
>Achatinella livida Swainson, 1828: 85. "Islands of the Pacific Ocean" (in introduction to the genus): [no additional details].

longispira.
>Achatinella longispira Smith in Gulick & Smith, 1873: 73, pl. 9, fig. 5. Oahu: "Halawa . . . Ahuimanu".
>>Remarks. Synonym of vulpina Férussac, teste Pilsbry & Cooke (1914a: 213).

+margaretae. (O)
>Achatinella (Achatinellastrum) casta margaretae Pilsbry & Cooke, 1914a: 240, pl. 42, figs. 9, 10. Oahu: "Kolokukahau peak, at the head of Waiau valley on the Waimalu division ridge, elevation 1450 ft."

+meineckei. (O)
>Achatinella lehuiensis meineckei Pilsbry & Cooke, 1921: 110, pl. 4, figs. 6–9. Oahu: "Waianae mountains in Haleauau valley, where the trail ascending Kaala leaves the stream".
>>Remarks. Pilsbry & Cooke (1921: 109–10) used both the spellings *meinickei* and *meineckei*. Although it is virtually certain that the former was an inadvertent error, Pilsbry & Cooke did not explicitly state that the taxon was named for W.H. Meinecke. We therefore consider *meineckei* the correct original spelling (*Code* Art. 24(c)). Considered as "var." *meineckei* by Neal (1928: 34–49) but as a subspecies by Cooke & Kondo (1961: 291).

meinickei.
>Achatinella lehuiensis meinickei Pilsbry & Cooke, 1921: 109. Incorrect original spelling of *meineckei* Pilsbry & Cooke.

multizonata.
>Achatinella (Achatinellastrum) multizonata Baldwin, 1893: 5. Nom. nud.

multizonata.
>Achatinella (Achatinellastrum) multizonata Baldwin, 1895: 215, pl. 10, figs. 1, 2. Oahu: "Nuuanu Valley".
>>Remarks. Synonym of bellula Smith, teste Pilsbry & Cooke (1914a: 230).

olesonii.
>   *Achatinella (Achatinellastrum) olesonii* Baldwin, 1893: 5. *Nom. nud.*
>   Remarks. Included among "Unrecognized species" by Pilsbry & Cooke (1914a: 369) but included here because of its original publication in subgenus *Achatinellastrum*.

olivacea.
>   *Achatinella olivacea* Reeve, 1850c: pl. 3, fig. 20. Type locality not given.
>   Remarks. Synonym of *vulpina* Férussac, *teste* Pilsbry & Cooke (1914a: 213).

papyracea. (O)
>   *Achatinella papyracea* Gulick, 1856: 229, pl. 7, fig. 48. Oahu: "Kalaikoa".
>   Remarks. Gulick gave 3 localities. Pilsbry & Cooke (1914a: 418, in the explanation of pl. 54, fig. 8) designated a lectotype ("no. 48 Boston Soc.") by inference (*Code* Art. 74), and (p. 243) stated that this specimen was from 1 of Gulick's localities, Kalaikoa.

phaeozona. (O)
>   *Achatinella phaeozona* Gulick, 1856: 214, pl. 7, fig. 40. Oahu: "Keowaawa".

plumata.
>   *Achatinella plumata* Gulick, 1856: 217, pl. 7, fig. 41. Oahu: "Niu".
>   Remarks. Synonym of *fulgens* Newcomb, *teste* Pilsbry & Cooke (1914a: 191).

prasinus.
>   *Achatinella prasinus* Reeve, 1850c: pl. 4, fig. 27. Type locality not given.
>   Remarks. Synonym of *vulpina* Férussac, *teste* Pilsbry & Cooke (1914a: 213).

+producta. (O)
>   *Achatinella producta* Reeve, 1850c: pl. 2, fig. 13. Hawaiian Islands: [no additional details].
>   Remarks. Subspecies of *stewartii* Green, *teste* Pilsbry & Cooke (1914a: 207).

pygmaea.
>   *Achatinella pygmaea* Smith *in* Gulick & Smith, 1873: 75, pl. 9, fig. 11. Oahu: "Waipio".
>   Remarks. Synonym of *casta* Newcomb, *teste* Pilsbry & Cooke (1914a: 236).

+recta. (O)
>   *Achatinella recta* Newcomb, 1854a: 19, pl. 23, fig. 45 [1854b: 145, pl. 23, fig. 45]. Oahu: "Waialua".
>   Remarks. Subspecies of *livida* Swainson, *teste* Pilsbry & Cooke (1914a: 248).

reevei.
>   *Achatinella reevei* Adams, 1851a: 128 [1851b: 44] (by bibliographic reference to Reeve's monograph of *Achatinella* [= 1850c, pl. 5, fig. 25]). Type locality not given.
>   Remarks. Synonym of *livida* Swainson, *teste* Pilsbry & Cooke (1914a: 246).

rhodorhaphe.
>   *Achatinella rhodorhaphe* Smith *in* Gulick & Smith, 1873: 74, pl. 9, fig. 10. Oahu: ". . . Helemano, on Oahu. . . . Ahonui, Wahiawa, Opaiula, and Kawailoa".
>   Remarks. Synonym of *curta* Newcomb, *teste* Pilsbry & Cooke (1914a: 252).

scitula.
>   *Achatinella scitula* Gulick, 1858: 241, pl. 8, fig. 61. Oahu: "Hakipu".
>   Remarks. Synonym of *cognata* Gulick, *teste* Pilsbry & Cooke (1914a: 264).

senistra.
>   *Achatinella colorata* var. *senistra* Schmeltz, 1865: 25. *Nom. nud.*
>   Remarks. A collection name not formally described. Not treated in the *Manual of Conchology*.

solitaria. (O)
>   *Achatinella solitaria* Newcomb, 1854a: 24, pl. 24, fig. 60 [1854b: 150, pl. 24, fig. 60]. Oahu: "Palolo".
>   Remarks. According to Pilsbry & Cooke (1914a: 204), who indicated that only 2 shells were known, "it is not likely that *A. solitaria* is a valid species or subspecies". They considered it close to or the same as *fulgens* Newcomb but retained it as a valid species. We retain it as a valid species, pending further research.

spaldingi. (O)
    Achatinella (Achatinellastrum) spaldingi Pilsbry & Cooke, 1914a: 271, pl. 42, figs. 1–3. Oahu: "Waianae Range: Pukuloa, one-half mile above the Mountain House, back of Leilehua".

stewartii. (O)
    Achatina stewartii Green, 1827: 47, pl. 4, figs. 1–4. Oahu: [no additional details].

suturalis.
    Achatinella (Achatinellastrum) vulpina pattern *suturalis* Pilsbry & Cooke, 1914a: 214, 221, pl. 40, figs. 9–10b. Oahu: "central ridge of Kahauiki".
        Remarks. Pilsbry & Cooke (1914a: 214, 221) established *suturalis* as a "pattern", clearly not intending it to have subspecific status. Synonym of *vulpina* Férussac. **N. syn.**

thaamuni.
    Achatinella (Achatinellastrum) thaamuni Pilsbry & Cooke, 1914a: 273, pl. 42, figs. 5, 6. Incorrect original spelling of *thaanumi* Pilsbry & Cooke.

thaanumi. (O)
    Achatinella (Achatinellastrum) thaanumi Pilsbry & Cooke, 1914a: 274, pl. 42, figs. 5, 6. Oahu: "Waianae range: a gulch of Mt. Kaala running into Haleauau gulch".
        Remarks. Pilsbry & Cooke (1914a) spelled it *"thaamuni"* on p. 273 but *"thaanumi"* on p. 274, without explicitly naming it for Ditlev Thaanum, the presumable namesake, cited as the collector. We here select *thaanumi* as the correct original spelling.

+tricolor. (O)
    Achatinella tricolor Smith *in* Gulick & Smith, 1873: 76, pl. 9, fig. 6. Oahu: "Ioleka, in Heia".
        Remarks. Subspecies of *vulpina* Férussac, *teste* Pilsbry & Cooke (1914a: 226).

trilineata.
    Achatinella trilineata Gulick, 1856: 226, pl. 7, fig. 46. Oahu: "Palolo, Waialae, Wailupe, and Niu".
        Remarks. Synonym of *fulgens* Newcomb, *teste* Pilsbry & Cooke (1914a: 192).

undulata.
    Achatinella undulata Newcomb, 1855b: 218. Oahu: "Waialua".
        Remarks. Synonym of *curta* Newcomb, *teste* Pilsbry & Cooke (1914a: 252).

undulata.
    Achatinella undulata Pfeiffer, 1856c: 208. Oahu: [no additional details].
        Remarks. Primary junior homonym of *undulata* Newcomb (see comments regarding Pfeiffer and Newcomb in the introduction to the family Achatinellidae). Synonym of *curta* Newcomb. **N. syn.**

ustulata.
    Achatinella ustulata Pfeiffer, 1859d: 534. Unavailable name; proposed in synonymy with *colorata* Reeve, not validated before 1961.
        Remarks. Achatinella ustulata Pfeiffer, 1859 preoccupied by *Achatinella ustulata* Gulick, 1856 (now in *Partulina*).

varia.
    Achatinella varia Gulick, 1856: 222, pl. 7, fig. 43. Oahu: "Palolo, Waialae, and Wailupe".
        Remarks. Synonym of *fulgens* Newcomb, *teste* Pilsbry & Cooke (1914a: 191).

venulata.
    Achatinella venulata Newcomb, 1854a: 20, pl. 23, figs. 48, 48a [1854b: 146, pl. 23, figs. 48, 48a]. Oahu: "Kolau [= Koolau]".
        Remarks. Synonym of *producta* Reeve, *teste* Pilsbry & Cooke (1914a: 211), who stated that the shells were "from Mt. Tantalus".

+versipellis. (O)
    Achatinella versipellis Gulick, 1856: 224, pl. 7, figs. 44a, b. Oahu: "Pohakunui, in Kailua".
        Remarks. Subspecies of *fulgens* Newcomb, *teste* Pilsbry & Cooke (1914a: 196).

ACHATINELLIDAE [55]

virens.
 Achatinella virens Gulick, 1858: 254, pl. 8, fig. 73. Oahu: "Halawa and Nuuanu".
  Remarks. Synonym of *vulpina* Férussac, *teste* Pilsbry & Cooke (1914a: 213).
vulpina.
 Helix (Cochlogena) vulpina Férussac, 1821c: 60. *Nom. nud.*
vulpina. (O)
 Helix vulpina Férussac *in* Quoy & Gaimard, 1825d: 477, pl. 68, figs. 13, 14. Hawaiian Islands: [no additional details].
zonata.
 Achatinella zonata Gulick, 1858: 238, pl. 8, fig. 58. Oahu: "Waimea, Pupukea, Waialei, Kahuku, and Hauula . . . Kaawa [= Kaaawa]".
  Remarks. Synonym of *dimorpha* Gulick, *teste* Pilsbry & Cooke (1914a: 258).

Subgenus BULIMELLA Pfeiffer

BULIMELLA Pfeiffer, 1854b: 119. Type species: *Achatinella rugosa* Newcomb, 1854, by subsequent designation of Martens (1860: 245).

abbreviata. (O)
 Achatinella abbreviata Reeve, 1850c: pl. 3, fig. 19. Type locality not given.
  Remarks. Pilsbry & Cooke (1913b: 123) considered Palolo the type locality,.
+albalabia. (O)
 Achatinella bulimoides albalabia Welch, 1954: 73, pl. 1, fig. 7. Oahu: "Poamoho?"
+arnemani. (O)
 Achatinella bulimoides arnemani Welch, 1958: 174, pl. 13, figs. 4–6. Oahu: "Punaluu-Kahana Ridge . . . Summit of Puu Piei".
bacca.
 Achatinella bacca Reeve, 1850e: pl. 6, fig. 45. Type locality not given.
  Remarks. Synonym of *abbreviata* Reeve, *teste* Pilsbry & Cooke (1913b: 123).
bulimoides. (O)
 Achatinella bulimoides Swainson, 1828: 85. Oahu: "Kawailoa".
  Remarks. Neotype designated by Welch (1954: 92).
byronensis.
 Helicteres byronensis Beck, 1837: 51. Unjustified emendation of *byronii* Wood, 1828.
  Remarks. Objective synonym of *byronii* Wood, *teste* Pilsbry & Cooke (1913b: 134).
byronii. (O)
 Helix byronii Wood, 1828: 22, pl. 7, fig. 30. Tahiti (as "Otaheite"): [?error (Pilsbry & Cooke, 1913b: 134)].
  Remarks. Pilsbry & Cooke (1913b: 134, 135) selected "Ahonui as the type locality", but no lectotype designation was made. Wood published *byronii* in combination with both *Helix* and *Achatina*. We here select *Helix byronii* as the correct original combination. Perhaps a manuscript name of Gray, since the name was attributed to Gray by Beck (1837: 51 [as "*byronensis*"]) and Gulick (1858: 244).
+caesiapicta. (O)
 Achatinella bulimoides caesiapicta Welch, 1954: 95, pl. 2, figs. 14–19. Oahu: "Kawailoa".
candida.
 Achatinella candida Pfeiffer, 1855b: 4 [1855c: 2]. Hawaiian Islands: [no additional details].
  Remarks. Synonym of *elegans* Newcomb, *teste* Welch (1958: 134). Previously considered a "form" of *ovata* Newcomb by Pilsbry & Cooke (1913b: 162).
capax.
 Achatinella (Bulimella) byronii rugosa subvariety *capax* Pilsbry & Cooke, 1913b: 137, pl. 31, figs. 7–7b. Unavailable name; infrasubspecific.

+circulospadix. (O)
>*Achatinella bulimoides circulospadix* Welch, 1954: 85, pl. 1, figs. 4–5a. Oahu: "Kaukonahua".

clementina.
>*Achatinella (Achatinellastrum) clementina* Pfeiffer, 1856b: 205. Hawaiian Islands ("Sandwich Islands" in publication title): [no additional details].
>>Remarks. Synonym of *abbreviata* Reeve, *teste* Pilsbry & Cooke (1913b: 123).

corrugata.
>*Achatinella corrugata* Gulick, 1858: 248, pl. 8, fig. 66. Oahu: "Hakipu".
>>Remarks. Synonym of *decipiens* Newcomb, *teste* Pilsbry & Cooke (1913b: 145).

decipiens. (O)
>*Achatinella decipiens* Newcomb, 1854a: 27, pl. 24, fig. 68 [1854b: 153, pl. 24, fig. 68]. Oahu: "Kahana".

+dextroversa. (O)
>*Achatinella (Bulimella) sowerbyana dextroversa* Pilsbry & Cooke, 1914a: 179, pl. 35, figs. 8–13. Oahu: "Pupukea".

+elegans. (O)
>*Achatinella elegans* Newcomb, 1854a: 23, pl. 24, fig. 57 [1854b: 149, pl. 24, fig. 57]. Oahu: "Hauula".
>>Remarks. Subspecies of *bulimoides* Swainson, *teste* Welch (1958: 134). Previously treated as a species by Pilsbry & Cooke (1913b: 166).

faba. (O)
>*Achatinella (Bulimella) faba* Pfeiffer, 1859c: 30. Hawaiian Islands: [no additional details].
>>Remarks. Subgeneric placement doubtful (Pilsbry & Cooke, 1914a: 368). We include it here on the basis of its original publication in subgenus *Bulimella*. The true status of this species may never be determined since it has never been figured and no type material has been found (see Pilsbry & Cooke, 1914a: 368). However, it has never been reduced to infraspecific status or synonymy so we retain it as a valid species.

+fricki. (O)
>*Achatinella fricki* Pfeiffer, 1855b: 5 [1855c: 3]. Hawaiian Islands: [no additional details].
>>Remarks. Subspecies of *bulimoides* Swainson, *teste* Welch (1958: 142). Previously considered a "variety" of *glabra* Newcomb by Pilsbry & Cooke (1913b: 165, 166).

+fulvula. (O)
>*Achatinella bulimoides fulvula* Welch, 1954: 82, pl. 1, fig. 19. Oahu: "North-South Helemano Ridge".

fuscobasis. (O)
>*Bulimella fuscobasis* Smith *in* Gulick & Smith, 1873: 77, pl. 9, fig. 15. Oahu: "High up on Mount Kaala on the Mokuleia side" [?error; "probably . . . Palolo valley" (Pilsbry & Cooke, 1913b: 171)].

+glabra. (O)
>*Achatinella glabra* Newcomb, 1854a: 13, pl. 22, fig. 25 [1854b: 139, pl. 22, fig. 25]. Oahu: "Kolau [= Koolau] poko" [Oahu not explicitly mentioned; ?error = "Waialee" (Welch, 1954: 98)].
>>Remarks. Subspecies of *bulimoides* Swainson, *teste* Pilsbry & Cooke (1913b: 164) and Welch (1954: 97).

inelegans.
>*Achatinella (Bulimella) elegans* var. *inelegans* Pilsbry & Cooke, 1913b: 168, pl. 33, fig. 12. Oahu: "Kaliuwaa".
>>Remarks. Synonym of *vidua* Pfeiffer, *teste* Welch (1958: 151).

+kaipapauensis. (O)
>*Achatinella bulimoides kaipapauensis* Welch, 1958: 162, pl. 11, figs. 14–16. Oahu: "Kaipapau Gulch".

+kalekukei. (O)
>*Achatinella bulimoides kalekukei* Welch, 1954: 99, pl. 2, fig. 29. Oahu: "Waimea . . . 'Crest of the Kamananui division Ridge . . .'".

+kaliuwaaensis. (O)
>*Achatinella (Bulimella) decipiens kaliuwaaensis* Pilsbry & Cooke, 1913b: 150, pl. 32, figs. 1–1b. Oahu: "Eastern ravines of Kaliuwaa".

+laiensis. (O)
>*Achatinella (Bulimella) sowerbyana laiensis* Pilsbry & Cooke, 1914a: 178, pl. 34, figs. 15, 15a. Oahu: "Laie, division ridge above the Castle cut trail".

lila. (O)
>*Achatinella (Bulimella) lila* Pilsbry *in* Pilsbry & Cooke, 1913b: 139, pl. 31, figs. 15–15d. Oahu: "crest of the Waimano-Manana ridge at junction with the main range, running from the summit down about three-fourths of a mile along the ridge trail".

limbata.
>*Achatinella limbata* Gulick, 1858: 252, pl. 8, figs. 70a, b. Oahu: "Ahonui and Kalaikoa".
>>*Remarks.* Synonym of *byronii* Wood, *teste* Pilsbry & Cooke (1913b: 134).

luteostoma.
>*Achatinella (Bulimella) luteostoma* Baldwin, 1893: 5. *Nom. nud.*

luteostoma.
>*Achatinella (Bulimella) luteostoma* Baldwin, 1895: 217, pl. 10, figs. 7, 8. Oahu: "Palolo to Niu".
>>*Remarks.* Synonym of *fuscobasis* Smith, *teste* Pilsbry & Cooke (1913b: 170).

lyonsiana.
>*Achatinella (Bulimella) lyonsiana* Baldwin, 1893: 5. *Nom. nud.*

+lyonsiana. (O)
>*Achatinella (Bulimella) lyonsiana* Baldwin, 1895: 218, pl. 10, figs. 9–11. Oahu: "Konahuanui Mt. . . . at an altitude of about 3,000 feet above sea level".
>>*Remarks.* Subspecies of *fuscobasis* Smith, *teste* Pilsbry & Cooke (1913b: 172) and Cooke & Kondo (1961: 289).

macrostoma.
>*Achatinella macrostoma* Pfeiffer, 1855b: 5 [1855c: 2]. Hawaiian Islands: [no additional details].
>>*Remarks.* Synonym of *viridans* Mighels, *teste* Pilsbry & Cooke (1913b: 126).

mahogani.
>*Achatinella mahogani* Gulick, 1858: 254, pl. 8, fig. 72. Oahu: "Ahonui and Kalaikoa".
>>*Remarks.* Synonym of *pulcherrima* Swainson, *teste* Pilsbry & Cooke (1913b: 141).

melanostoma.
>*Achatinella melanostoma* Newcomb, 1854a: 6, pl. 22, fig. 7 [1854b: 132, pl. 22, fig. 7]. Oahu: "Ewa".
>>*Remarks.* Pilsbry & Cooke (1913b: 134) included *melanostoma* Newcomb in the synonymy of *byronii* Wood and (p. 141–42) in that of *pulcherrima* Swainson, but discussed it only under the latter.

+mistura. (O)
>*Achatinella (Bulimella) bulimoides mistura* Pilsbry & Cooke, 1913b: 156, pl. 33, figs. 5, 5c, 6, 7. Oahu: "In Kaliuwaa valley and the ridge eastward along the Castle trail . . . On the edge of Punaluu valley, where the Castle trail passes over the crest into Kaliuwaa . . .. In Kaliuwaa valley, about three-fourths of a mile east from the following colony . . . At the bottom of Kaliuwaa, along the stream back of a cabin".

monacha.
> *Achatinella monacha* Pfeiffer, 1855b: 6 [1855c: 3]. Hawaiian Islands: [no additional details]
>> *Remarks.* Synonym of *fricki* Pfeiffer, *teste* Welch (1958: 142, as "*monarcha*"). Previously considered a "form" of *mustelina* Mighels by Pilsbry & Cooke (1914a: 344).

multicolor.
> *Achatinella multicolor* Pfeiffer, 1855c: 4, pl. 30, fig. 11 [1855g: 64]. Hawaiian Islands: [no additional details].
>> *Remarks.* Synonym of *pulcherrima* Swainson, *teste* Pilsbry & Cooke (1913b: 143).
>> *Achatinella multicolor* figured in pl. 30, figs. 11 and 11a; Pilsbry & Cooke (1913b: 143) restricted *multicolor* to just fig. 11.

+nigricans. (O)
> *Achatinella (Bulimella) byronii nigricans* Pilsbry & Cooke, 1913b: 138, pl. 31, figs. 10–12. Oahu: "Waimano-Manana ridge at about 1400 ft. elevation, in a very small area along the summit trail . . . alongside the trail, about 1000 yards down the ridge from the locality of *A. turgida cookei*".

nivosa.
> *Achatinella nivosa* Newcomb, 1854a: 6, pl. 22, fig. 6 [1854b: 132, pl. 22, fig. 6]. Oahu: "Niu".
>> *Remarks.* Synonym of *abbreviata* Reeve, *teste* Pilsbry & Cooke (1913b: 123).

+nympha. (O)
> *Achatinella nympha* Gulick, 1858: 251, pl. 8, fig. 69. Oahu: "Ahonui, Wahiawa, Helemanu, Kawailoa, and Waimea".
>> *Remarks.* Subspecies of *pulcherrima* Swainson, *teste* Pilsbry & Cooke (1913b: 144).

+obliqua. (O)
> *Achatinella obliqua* Gulick, 1858: 245, pl. 8, fig. 63. Oahu: "Kahana".
>> *Remarks.* Subspecies of *bulimoides* Swainson, *teste* Pilsbry & Cooke (1913b: 158) and Welch (1958: 178).

oomorpha.
> *Achatinella oomorpha* Gulick, 1858: 246, pl. 8, fig. 64. Oahu: "Kahana".
>> *Remarks.* Synonym of *obliqua* Gulick, *teste* Pilsbry & Cooke (1913b: 158–59) and Welch (1958: 179).

+oswaldi. (O)
> *Achatinella bulimoides oswaldi* Welch, 1958: 170, pl. 13, figs. 1–1b, 2, 17–17b. Oahu: "Kahana-Punaluu Ridge . . . in a small glen facing the sea below a precipice".

+ovata. (O)
> *Achatinella ovata* Newcomb, 1853: 22 [1854a: 4, pl. 22, figs. 2, 2a; 1854b: 130, pl. 22, figs. 2, 2a]. Oahu: "Waiauai [= Waianae; error = Kahana (Newcomb, 1854a: 4; 1854b: 130; Pilsbry & Cooke, 1913b: 161)]".
>> *Remarks.* Subspecies of *bulimoides* Swainson, *teste* Pilsbry & Cooke (1913b: 160) and Welch (1958: 185).

oviformis.
> *Achatinella oviformis* Newcomb, 1855c: 147. Nom. nud.
>> *Remarks.* Synonym of *sowerbyana* Pfeiffer, *teste* Pilsbry & Cooke (1913b: 176).

oviformis.
> *Achatinella oviformis* Pfeiffer, 1856c: 208. Oahu: [no additional details].
>> *Remarks.* Synonym of *sowerbyana* Pfeiffer, *teste* Pilsbry & Cooke (1913b: 176). See comments regarding Pfeiffer and Newcomb in the introduction to the family Achatinellidae.

+papakokoensis. (O)
> *Achatinella bulimoides papakokoensis* Welch, 1958: 146, pl. 12, figs. 13–14a. Oahu: "Papakoko . . . Punaluu, mauka [towards the mountains] of the road [Kamehameha Highway] opposite A. Young's, in a plowed field".

+**planospira**. (O)
    Achatinella planospira Pfeiffer, 1855b: 6 [1855c: 3]. Hawaiian Islands: [no additional details].
        Remarks. "Variety (?)" of *decipiens* Newcomb, *teste* Pilsbry & Cooke (1913b: 147). Thwing (1907: 189) indicated an Oahu locality.

pulcherrima. (O)
    Achatinella pulcherrima Swainson, 1828: 86. "Islands of the Pacific Ocean" (in introduction to the genus): [no additional details].
        Remarks. Pilsbry & Cooke (1913b: 141) took "Kawaihalona" (Oahu) as the type locality, but no lectotype designation was made.

pupukanioe. (O)
    Achatinella (*Bulimella*) *pupukanioe* Pilsbry & Cooke, 1913b: 174, pl. 35, figs. 14–17. Oahu: "Crest of the Waimano-Manana ridge, 1 to 1 1/2 miles from its junction with the main range".

radiata.
    Achatinella radiata Pfeiffer, 1846a: 89. Hawaiian Islands: [no additional details].
        Remarks. Primary junior homonym of *Achatinella radiata* Gould, 1845 (now in *Partulina*). Synonym of *viridans* Mighels, *teste* Pilsbry & Cooke (1913b: 126).

+**rosea**. (O)
    Achatinella bulimoides var. *rosea* Swainson, 1828: 85. "Islands of the Pacific Ocean" (in introduction to the genus): [no additional details].
        Remarks. Subspecies of *bulimoides* Swainson, *teste* Welch (1954: 79). Previously treated as a species by Pilsbry & Cooke (1913b: 151).

+**rosealimbata**. (O)
    Achatinella bulimoides rosealimbata Welch, 1954: 73, pl. 1, figs. 9–11, 18. Oahu: "North Poamoho Stream".

+**roseoplica**. (O)
    Achatinella (*Bulimella*) *sowerbyana roseoplica* Pilsbry & Cooke, 1914a: 180, pl. 34, fig. 12. Oahu: "Opaeula, above forest-fence line".

+**rotunda**. (O)
    Achatinella rotunda Gulick, 1858: 249, pl. 8, fig. 67. Oahu: "Kaawa [= Kaaawa] and Kahana".
        Remarks. Subspecies of *bulimoides* Swainson, *teste* Pilsbry & Cooke (1913b: 163), Welch (1958: 189).

rubiginosa.
    Achatinella rubiginosa Newcomb, 1854a: 28, pl. 24, fig. 69 [1854b: 154, pl. 24, fig. 69]. Oahu: "Palolo".
        Remarks. Synonym of *taeniolata* Pfeiffer, *teste* Pilsbry & Cooke (1913b: 130).

+**rufapicta**. (O)
    Achatinella bulimoides rufapicta Welch, 1958: 175, pl. 13, figs. 7–10, 13–16. Oahu: "Punaluu-Kahana Ridge . . . up third glen mauka [towards the mountains] of Piei. At 5th spur, about 200 ft. mauka of black rock scarp . . . Crest of ridge [Punaluu-Kahana] at 5th spur from Piei".

+**rugosa**. (O)
    Achatinella rugosa Newcomb, 1854a: 12, pl. 22, figs. 22, 22a [1854b: 138, pl. 22, figs. 22, 22a]. Oahu: "Ewa".
        Remarks. Subspecies of *byronii* Wood, *teste* Pilsbry & Cooke (1913b: 135).

rutila.
    Achatinella rutila Newcomb, 1854a: 12, pl. 22, fig. 21 [1854b: 138, pl. 22, fig. 21]. Oahu: "Niu".
        Remarks. Synonym of *viridans* Mighels, *teste* Pilsbry & Cooke (1913b: 126).

sowerbyana. (O)
>*Achatinella sowerbyana* Pfeiffer, 1855c: 4, pl. 30, fig. 14. Hawaiian islands: [no additional details].
>>*Remarks.* Pilsbry & Cooke (1913b: 176) felt that "Pfeiffer's type" likely came from "some valley of the Kaipapau-Kaliuwaa region".

+spadicea. (O)
>*Achatinella spadicea* Gulick, 1858: 247, pl. 8, fig. 65. Oahu: "Kahana".
>>*Remarks.* Subspecies of *bulimoides* Swainson, *teste* Pilsbry & Cooke (1913b: 157) and Welch (1954: 87).

subvirens.
>*Achatinella subvirens* Newcomb, 1854a: 10, pl. 22, fig. 18 [1854b: 136, pl. 22, fig. 18]. Oahu: "Niu".
>>*Remarks.* Synonym of *viridans* Mighels, *teste* Pilsbry & Cooke (1913b: 129).

+swainsoni. (O)
>*Achatinella swainsoni* Pfeiffer, 1855c: 4, pl. 30, fig. 13. Hawaiian Islands: [no additional details].
>>*Remarks.* Subspecies of *decipiens* Newcomb, *teste* Pilsbry & Cooke (1913b: 150).

taeniolata. (O)
>*Achatinella taeniolata* Pfeiffer, 1846c: 38. Hawaiian Islands: [no additional details].
>>*Remarks.* Pilsbry & Cooke (1913b: 131) stated that "Palolo may be taken as type locality", but no lectotype designation was made.

+thurstoni. (O)
>*Achatinella (Bulimella) sowerbyana thurstoni* Pilsbry & Cooke, 1914a: 177, pl. 34, figs. 13–14b. Oahu: "Kahuku, 1,500–1,700 ft. elevation".

torrida.
>*Achatinella torrida* Gulick, 1858: 250, pl. 8, fig. 68. Oahu: "Kahana, Kaawa [= Kaaawa], and Waikane".
>>*Remarks.* Synonym of *decipiens* Newcomb, *teste* Pilsbry & Cooke (1913b: 145).

+vidua. (O)
>*Achatinella vidua* Pfeiffer, 1855b: 6 [1855c: 3]. Hawaiian Islands: [no additional details].
>>*Remarks.* Subspecies of *bulimoides* Swainson, *teste* Welch (1958: 151). Pilsbry & Cooke (1913b: 163) considered it "merely a very small or stunted specimen of *ovata*".

viridans. (O)
>*Achatinella viridans* Mighels, 1845: 20. Oahu: [no additional details].

waimanoensis.
>*Achatinella (Bulimella) byronii rugosa* subvariety *waimanoensis* Pilsbry & Cooke, 1913b: 137, pl. 31, figs 9–9d. Unavailable name; infrasubspecific.

+wheatleyana. (O)
>*Achatinella (Bulimella) elegans* var. *wheatleyana* Pilsbry & Cooke, 1913b: 168, pl. 28, figs. 11–11b. Oahu: "Punaluu".
>>*Remarks.* Subspecies of *bulimoides* Swainson, *teste* Welch (1958: 147).

wheatleyi.
>*Achatinella wheatleyi* Newcomb, 1855c: 147. *Nom. nud.*
>>*Remarks.* See also Newcomb (1858: 324) and Pilsbry & Cooke (1913b: 169).

+wilderi. (O)
>*Achatinella (Bulimella) fuscobasis wilderi* Pilsbry *in* Pilsbry & Cooke, 1913b: 173, pl. 41, figs. 1–3a. Oahu: "Summit of Lanihuli, at head of the Nuuanu-Kalihi ridge . . . Also "Mauna Kope" a peak at head of the Kalihi-Moanalua ridge".

ACHATINELLIDAE

Incertae Sedis in Genus ACHATINELLA *s.l.*

Pilsbry & Cooke (1914a: 366–69) indicated (as "unrecognized species") a number of *nomina nuda* in *Achatinella s.l.* from Paetel's, Jay's and Baldwin's catalogs (Baldwin, 1893; Jay, 1839; Paetel, 1873, 1883, 1887–1890; Schaufuss, 1869). Among these, we place "*octavula*" in *Leptachatina* (Amastridae), following the general consensus indicated in the Remarks section under that name; "*olesonii*" in subgenus *Achatinellastrum* since this is how it was first published; "*cinnamomea*" appears to be *cinnamomea* Pfeiffer, 1857, now in *Newcombia*.

agatha.
    Achatinella agatha Schaufuss, 1869: 83. *Nom. nud.*
anacardiensis.
    Achatinella anacardiensis Paetel, 1883: 153. *Nom. nud.*
bensonia.
    Achatinella bensonia Schaufuss, 1869: 83. *Nom. nud.*
cingulata.
    Achatinella cingulata Paetel, 1873: 105. *Nom. nud.*
circulata.
    Achatinella circulata Paetel, 1883: 154. *Nom. nud.*
compressa.
    Achatinella compressa Paetel, 1873: 105. *Nom. nud.*
gravis.
    Achatinella gravis Paetel, 1873: 105. *Nom. nud.*
havaiana.
    Achatinella havaiana Schaufuss, 1869: 83. *Nom. nud.*
hawaiana.
    Achatinella hawaiana Paetel, 1883: 154. *Nom. nud.*
ignominiosus.
    Achatinella ignominiosus Paetel, 1873: 105. *Nom. nud.*
impressa.
    Achatinella impressa Paetel, 1873: 105. *Nom. nud.*
magnifica.
    Achatinella magnifica Schaufuss, 1869: 83. *Nom. nud.*
scamnata.
    Achatinella scamnata Schaufuss, 1869: 84. *Nom. nud.*
semitecta.
    Achatinella semitecta Paetel, 1890: 275. *Nom. nud.*
sinistra.
    Achatinella vulpina var. *sinistra* Schaufuss, 1869: 84. *Nom. nud.*
subovata.
    Achatinella subovata Schaufuss, 1869: 84. *Nom. nud.*
torquata.
    Achatinella torquata Schaufuss, 1869: 84. *Nom. nud.*
turbinata.
    Achatinella turbinata Jay, 1839: 58. *Nom. nud.*

Genus NEWCOMBIA Pfeiffer

NEWCOMBIA Pfeiffer, 1854b: 117 (as *Achatinella* subg.). Type species: *Achatinella plicata* Pfeiffer, 1848 (as "Migh.") [= *Bulimus liratus* Pfeiffer, 1853], by subsequent designation of Martens (1860: 249).
NEWCOMBIANA: Incorrect subsequent spelling of *Newcombia* Pfeiffer (Thwing, 1907: 138).

Gulick (1873a: 91) incorrectly gave the type species as *Newcombia cumingi* Newcomb. Pilsbry & Cooke (1912: 1–13) listed 9 species. Caum (1928) listed the same 9 species, although Pilsbry & Cooke (1914a: 358) had previously removed *carinella* Baldwin to *Partulina* (section *Perdicella*) and (p. 355) synonymized *cinnamomea* Pfeiffer and *pfeifferi* Newcomb. Caum (1928: 15) listed *carinella* under both *Newcombia* and *Partulina*. Cooke & Kondo (1961) made no change. The Hawaiian fauna currently consists of 7 species and 5 infraspecific taxa.

canaliculata.
> *Achatinella* (*Newcombia*) *canaliculata* Baldwin, 1893: 8. *Nom. nud.*

canaliculata. (Mo)
> *Achatinella* (*Newcombia*) *canaliculata* Baldwin, 1895: 226. Molokai: "Halawa".

cinnamomea.
> *Achatinella cinnamomea* Pfeiffer, 1857b: 230. Hawaiian Islands: [no additional details].
>> *Remarks.* Synonym of *pfeifferi* Newcomb, *teste* Pilsbry & Cooke (1914a: 355).

costata.
> *Newcombia costata* Borcherding, 1901: 57. Molokai: "Halawa".
>> *Remarks.* Synonym of *gemma* Pfeiffer, *teste* Pilsbry & Cooke (1912: 4).

cumingi. (M)
> *Achatinella cumingi* Newcomb, 1853: 25 [1854a: 24, pl. 24, fig. 59; 1854b: 150, pl. 24, fig. 59 (as "*cumingii*")]. Maui: "Hale-a-ka-la".

+decorata. (Mo)
> *Newcombia cinnanomea* var. *decorata* Pilsbry & Cooke, 1912: 12, pl. 14, figs. 2–4. Molokai: "Ahaino and Kupeke".

+gemma. (Mo)
> *Achatinella gemma* Pfeiffer, 1857b: 230. Hawaiian Islands: [no additional details].
>> *Remarks.* Subspecies of *plicata* Pfeiffer [= *lirata* Pfeiffer], *teste* Pilsbry & Cooke (1912: 3).

+honomuniensis. (Mo)
> *Newcombia pfeifferi* var. *honomuniensis* Pilsbry & Cooke, 1912: 12, pl. 14, figs. 6, 7. Molokai: "Honomuni".

lirata. (Mo)
> *Bulimus liratus* Pfeiffer, 1853a: 261. Molokai: [no additional details; Pfeiffer (1848b: 235)].
>> *Remarks.* Proposed as a new name for *Achatinella plicata* Pfeiffer, 1848, due to secondary homonymy in *Bulimus*; valid by *Code* Art. 59(b).

newcombianus.
> *Bulimus newcombianus* Pfeiffer, 1853a: 261. Hawaiian Islands: [no additional details].
>> *Remarks.* Synonym of *pfeifferi* Newcomb, *teste* Pilsbry & Cooke (1912: 13). Also published by Pfeiffer (1853c: 414), citing his original description.

perkinsi. (Mo)
> *Newcombia perkinsi* Sykes, 1896: 130. Molokai: "Molokai Mountains".

pfeifferi. (Mo)
>   *Achatinella pfeifferi* Newcomb, 1853: 25 [1854a: 24, pl. 24, fig. 58; 1854b: 150, pl. 24, fig. 58]. Molokai: [no additional details].

philippiana. (Mo)
>   *Achatinella philippiana* Pfeiffer in Dohrn & Pfeiffer, 1857: 89. Hawaiian Islands: [no additional details].

plicata.
>   *Achatinella plicata* Pfeiffer, 1848c: 235. Molokai: [no additional details].
>   >   *Remarks*. Secondary junior homonym, replaced by *lirata* Pfeiffer; *lirata* valid by Code Art. 59(b). Originally a Mighels name in the Cuming collection.

sulcata. (Mo)
>   *Achatinella sulcata* Pfeiffer, 1857b: 231. Hawaiian Islands: [no additional details].

+ualapuensis. (Mo)
>   *Newcombia cinnamomea ualapuensis* Pilsbry & Cooke, 1912: 12, pl. 2, figs. 11, 12. Molokai: "Ualapue; Kahaanui".
>   >   *Remarks*. Pilsbry & Cooke (1914a: 355) considered *ualapuensis* a subspecies of *pfeifferi* Newcomb not *cinnamomea* Pfeiffer.

+wailauensis. (Mo)
>   *Newcombia canaliculata wailauensis* Pilsbry & Cooke, 1912: 7, pl. 14, fig. 5. Molokai: "Wailau".

Genus PARTULINA Pfeiffer

PARTULINA Pfeiffer, 1854b: 114 (as *Achatinella* subg.). Type species: *Partula virgulata* Mighels, 1845 (as "*Achatinella*"), by subsequent designation of Martens (1860: 243).

PARTULINELLA Hyatt in Pilsbry & Cooke, 1914a: 392 (as a new "subdivision of *Partulina*"; Pilsbry in footnote on p. 392). Type species: *Achatinella marmorata* Gould, 1847 (as "*P. marmorata*"), by original designation (by Pilsbry in footnote on p. 392). **N. syn.**

Cooke & Kondo (1961) treated *Baldwinia*, *Eburnella*, *Partulinella* and *Partulina s. str.* as subgenera on p. 18 but as "sections" on pp. 281–84. The subgenus *Partulinella* Hyatt, 1914 was never defined in a rigorous way. It was not recognized by Pilsbry & Cooke (1912–1914), although Cooke & Kondo (1961) and Vaught (1989: 79) did recognize it. It was first established in "Part II" of Hyatt's (*in* Pilsbry & Cooke, 1914a) appendix in the *Manual of Conchology*, "Part II" being "preliminary statements" derived from the "more or less disconnected notes" of Hyatt, edited by A.G. Mayer after Hyatt's death. Hyatt did not take into account the subgenera ("sections") of the genus *Partulina* of Pilsbry & Cooke (1912–1914) and now generally accepted. He included in *Partulinella* taxa from both the subgenera *Partulina s. str.* and *Baldwinia*. We feel unable to retain *Partulinella* as a valid genus and, for the sake of stability, synonymize *Partulinella* Hyatt with *Partulina* Pfeiffer, leaving all taxa in the subgenera they were referred to by Pilsbry & Cooke (1912–1914). In the remarks section under each of these names we indicate their placement in *Partulinella* by Hyatt.

Pilsbry & Cooke (1912–1914) recognized 52 nominate species (including *aptycha* Pfeiffer (cf. Pilsbry & Cooke, 1912: 54, 1914a: 363; contrary to Cooke *in* Pilsbry & Cooke, 1913b: 145); but excluding *mucida* Baldwin (cf. Pilsbry & Cooke, 1914a: 361; contrary to Pilsbry & Cooke, 1912: 34)). Caum (1928) listed the same 52 species plus *mucida*. Cooke & Kondo (1952: 346) added *arnemanni*, but raised the section *Perdicella* to generic status, thereby removing 9 species from *Partulina*. Cooke & Kondo (1961: p. 18, but not in the main systematic part of the text) treated the other sections of Pilsbry & Cooke (1912: 15, 1914a: 392) as subgenera of *Partulina*; they made no other changes. Retaining *aptycha* but not *mucida* as nomenclaturally valid species gives a total of 44 species (6 in *Baldwinia*, 8 in *Eburnella*, 30 in *Partulina s. str.*). There are also 41 infraspecific taxa (7 in *Baldwinia*, 11 in *Eburnella*, 23 in *Partulina s. str.*).

Subgenus BALDWINIA Ancey

BALDWINIA Ancey, 1899: 270. Type species: *Achatinella physa* Newcomb, 1854 (as "1855"), by subsequent designation of Pilsbry & Cooke (1913a: 90).

RUGOSELLA Coen, 1945: 25 (as *Partulina* sect.). Unavailable name; published after 1930 without fixation of a type species (*Code* Art. 13(b)).

RUGOSELLA Mienis, 1987: 729 (as *Partulina* subg.). Type species: *Achatinella physa* Newcomb, 1854 (as "*Partulina*"), by original designation.

Ancey (1899: 270) indicated 3 "type" species for *Baldwinia*, giving their authors but not dates. Pilsbry & Cooke (1913a: 90) chose the first name in Ancey's list, *Achatinella physa* Newcomb, citing the date as 1855. Following Art. 67(f), this is an incorrect citation of *Achatinella physa* Newcomb, 1854. *Achatinella physa* Newcomb, 1854 is therefore the type species of *Baldwinia*. Newcomb (1854a: 26; 1854b: 152; 1855b: 218; 1866: 214) confused 2 species in his treatments of "*physa*". In the 2 later treatments (1855b, 1866) he considered his original description (1854a,b) to have been of an immature specimen. Sykes (1900: 312) pointed out this confusion, realizing that the specimen first treated as *physa* by Newcomb (1854a,b) in fact belonged to a different species from that subsequently considered by Newcomb (1855b, 1866) as *physa*. Sykes (1900: 312) therefore proposed a "nom. nov.", *Achatinella (Partulina) confusa*, for the taxon treated as *physa* in Newcomb's later 2 publications (1855b, 1866). Although it is this taxon ("*physa* Newcomb, 1855" = *confusa* Sykes, 1900) that Pilsbry & Cooke (1913a: 90) intended as the type species of *Baldwinia* Ancey, their intention is irrelevant nomenclaturally.

+candida. (H)
 Partulina (Baldwinia) horneri var. *candida* Pilsbry & Cooke, 1914a: 365, pl. 54, fig. 12 [by bibliographic reference to Pilsbry & Cooke (1913a: 108, pl. 17, figs. 3, 4)]. Hawaii: "Above Kukuihaele, Hamakua".
  Remarks. Described as "color-race . . . (c)" by Pilsbry & Cooke (1913a: 108).

confusa. (H)
 Achatinella (Partulina) confusa Sykes, 1900: 312 [available by bibliographic reference to Newcomb (1855b: 218)]. Hawaii: [no additional details].

dubia. (O)
 Achatinella dubia Newcomb, 1853: 23 [1854a: 26, pl. 24, fig. 65; 1854b: 152, pl. 24, fig. 65]. Oahu: [no additional details].
  Remarks. Hyatt (*in* Pilsbry & Cooke, 1914a: 393) placed *dubia* in *Partulinella*.

+errans. (H)
 Partulina (Baldwinia) physa var. *errans* Pilsbry & Cooke, 1913a: 111, pl. 17, figs. 14–16. Hawaii: "Near Pahoa, Puna".
  Remarks. Pilsbry & Cooke (1913a: 111) described *errans* as "*P. physa errans*, n. var." in their heading but in their text indicated that it should be "ranked as a subspecies".

fucosa.
    Achatinella fucosa Lyons, 1891: 105. Oahu: [no additional details].
        Remarks. Synonym of *dubia* Newcomb, *teste* Pilsbry & Cooke (1913b: 114).
        Originally a Frick collection name.

+fuscospira. (H)
    Partulina (Baldwinia) horneri var. *fuscospira* Pilsbry & Cooke, 1914a: 365, pl. 54, fig. 13 [available by bibliographic reference to Pilsbry & Cooke (1913a: 108)]. Hawaii: "Above Kukuihaele".
        Remarks. Described as "color-race . . . (d)" by Pilsbry & Cooke (1913a: 108).

+fuscozonata. (H)
    Partulina (Baldwinia) horneri var. *fuscozonata* Pilsbry & Cooke, 1914a: 365 [available by bibliographic reference to Pilsbry & Cooke (1913a: 108, pl. 17, figs. 2, 5)]. Hawaii: [no additional details].
        Remarks. Described as "color-race . . . (b)" by Pilsbry & Cooke (1913a: 108).

grisea. (M)
    Achatinella grisea Newcomb, 1854a: 26, pl. 24, fig. 66 [1854b: 153, pl. 24, fig. 66]. Maui (as "Mani"): "Makawao".
        Remarks. Hyatt (*in* Pilsbry & Cooke, 1914a: 392) placed *grisea* in *Partulinella.*

hawaiiensis.
    Achatinella (Partulina) hawaiiensis Baldwin, 1893: 6. *Nom. nud.*

hawaiiensis.
    Achatinella (Partulina) hawaiiensis Baldwin, 1895: 225, pl. 10, figs. 24–26. Hawaii: "Hamakua".
        Remarks. Synonym of *physa* Newcomb, *teste* Pilsbry & Cooke (1913a: 109). Hyatt (*in* Pilsbry & Cooke, 1914a: 393) placed *hawaiiensis* in *Partulinella.*

horneri.
    Achatinella (Partulina) horneri Baldwin, 1893: 6. *Nom. nud.*

horneri. (H)
    Achatinella (Partulina) horneri Baldwin, 1895: 224, pl. 10, figs. 20–22. Hawaii: "Hamakua".
        Remarks. Hyatt (*in* Pilsbry & Cooke, 1914a: 392) placed *horneri* in *Partulinella.*

+kapuana. (H)
    Partulina horneri var. *kapuana* Gouveia & Gouveia, 1920: 53, pl. 15. Hawaii: "Waialohe, Kapua, South Kona, Hawaii: About one-half mile above Government Road".

+konana. (H)
    Partulina (Baldwinia) physa konana Pilsbry & Cooke, 1914a: 365, pl. 54, figs. 5, 5a. Hawaii: "North Kona at Honoula".

morbida.
    Achatinella (Bulimella) morbida Pfeiffer, 1859c: 30. Hawaiian Islands: [no additional details]
        Remarks. Synonym of *dubia* Newcomb, *teste* Pilsbry & Cooke (1913b: 114).

+perantiqua. (O)
    Partulina dubia perantiqua Cooke & Kondo, 1952: 344, fig. 6c. Oahu: "Kahuku Point".

pexa.
    Achatinella pexa Gulick, 1856: 197, pl. 6, fig. 26. Hawaiian Islands: [no additional details].
        Remarks. Synonym of *dubia* Newcomb, *teste* Pilsbry & Cooke (1913b: 116).

phaeostoma.
    Partulina physa var. *phaeostoma* Ancey, 1904a: 121. Hawaii: [no additional details].
        Remarks. Synonym of *confusa* Sykes, *teste* Pilsbry & Cooke (1913a: 107).

physa. (H)
>Achatinella physa Newcomb, 1854a: 26, pl. 24, fig. 64 [1854b: 152, pl. 24, fig. 64]. Hawaii: "Mouna [= Mauna] Kea".
>Remarks. Hyatt (*in* Pilsbry & Cooke, 1914a: 393) placed *physa* in *Partulinella*.

platystyla.
>Achatinella platystyla Gulick, 1856: 196, pl. 6, fig. 25. Oahu: "Kawailoa".
>Remarks. Synonym of *dubia* Newcomb, *teste* Pilsbry & Cooke (1913b: 114).

procera.
>Achatinella physa var. procera Ancey, 1904b: 69. *Nom. nud.*
>Remarks. "Form" of *confusa* Sykes, *teste* Pilsbry & Cooke (1913a: 107).

thaanumiana. (M)
>Partulina (Baldwinia) thaanumiana Pilsbry *in* Pilsbry & Cooke, 1913a: 112, pl. 18, figs. 6, 7. Maui: "West Maui: Waiehu Gulch".

Subgenus EBURNELLA Pease

EBURNELLA Pease, 1870b: 647 (as genus). Type species: *Achatinella variabilis* Newcomb, 1854 (as "*P. variabilis* (Newc.)"), by subsequent designation of Pilsbry & Cooke (1913a: 67).

Gulick (1873a: 91) selected "*Partulina variabilis*, Newc." as an example of *Eburnella* without explicitly designating it as the type.

anceyana.
>Achatinella (Partulina) anceyana Baldwin, 1893: 6. *Nom. nud.*

anceyana. (M)
>Achatinella (Partulina) anceyana Baldwin, 1895: 223, pl. 10, fig. 16. Maui: "Makawao".

+bella. (Mo)
>Achatinella bella Reeve, 1850c: pl. 3, fig. 17. Molokai: [no additional details].
>Remarks. Subspecies of *mighelsiana* Pfeiffer, *teste* Pilsbry & Cooke (1913a: 79).

cooperi.
>Partulina cooperi Baldwin, 1906b: 135. Maui: "Hana, East Maui".
>Remarks. Synonym of *wailuaensis* Sykes, *teste* Pilsbry & Cooke (1913a: 72).

+dixoni. (Mo)
>Achatinellastrum dixoni Borcherding, 1906: 83, pl. 8, figs. 11–14. Molokai: "Kawela et Kaamola [= Kawela and Kaamola]".
>Remarks. "Color-form" of *polita* Newcomb, *teste* Pilsbry & Cooke (1913a: 82).

+flemingi. (M)
>Partulina flemingi Baldwin, 1906a: 111. Maui: "Nahiku, East Maui".
>Remarks. Subspecies of *porcellana* Newcomb, *teste* Pilsbry & Cooke (1913a: 71).

fulva.
>Achatinella fulva Pfeiffer, 1856c: 208. Hawaiian Islands (as "Sandwich Islands" in publication title): [no additional details].
>Remarks. Synonym of *variabilis* Newcomb, *teste* Pilsbry & Cooke (1913a: 84). Originally a Newcomb name in the Cuming collection.

+fulvicans. (M)
>Partulina fulvicans Baldwin, 1906b: 135. Maui: "Kipahulu valley, Hana, East Maui".
>Remarks. Subspecies of *porcellana* Newcomb, *teste* Pilsbry & Cooke (1913a: 73).

germana. (M)
>Achatinella germana Newcomb, 1854a: 25, pl. 24, fig. 61 [1854b: 151, pl. 24, fig. 61]. Maui (as "Mani"): "Makawao".

+hayseldeni. (L)
>Partulina hayseldeni Baldwin, 1896: 31. Lanai: [no additional details].
>Remarks. "Variety" of *semicarinata* Newcomb, *teste* Pilsbry & Cooke (1913a: 88).

+**hepatica**. (Mo)
>   *Achatinellastrum hepaticum* Borcherding, 1906: 83, pl. 8, figs. 15, 16. Molokai: "Waileia et Kawela [= Waileia and Kawela]".
>   *Remarks.* "Color-form" of *polita* Newcomb, *teste* Pilsbry & Cooke (1913a: 83).

+**lactea**. (L)
>   *Achatinella lactea* Gulick, 1856: 198, pl. 6, fig. 27. Lanai: [no additional details].
>   *Remarks.* Variety of *variabilis* Newcomb, *teste* Pilsbry & Cooke (1913a: 86).

+**latizona**. (Mo)
>   *Achatinellastrum latizona* Borcherding, 1906: 82, pl. 8, fig. 10. Molokai: "Kaamola".
>   *Remarks.* "Color-form" of *polita* Newcomb, *teste* Pilsbry & Cooke (1913a: 82).

+**martensi**. (Mo)
>   *Achatinellastrum mighelsiana* var. *martensi* Borcherding, 1906: 80, pl. 7, figs. 14, 16, 18, 20. Molokai: "Kawela".

mighelsiana. (Mo)
>   *Achatinella mighelsiana* Pfeiffer, 1848a: 231 [1848c: 238]. Molokai: [no additional details].
>   *Remarks.* We have not been able to establish an exact date of publication of *mighelsiana* in the *Monographia Heliceorum Viventium. Volumen secundum* (Pfeiffer, 1848c: 238). The volume was published in parts, with 1848 on the title page. We therefore provisionally consider the *Proc. Zool. Soc. Lond.* paper, published on 29 March 1848, to constitute the original publication of *mighelsiana*.

mutabilis. (M)
>   *Partulina mutabilis* Baldwin, 1908: 68. Maui: "Waichu [= Waiehu] Valley, West Maui".

nattii. (M)
>   *Achatinella (Achatinellastrum) nattii* Hartman, 1888b: 34, pl. 1, fig. 3. Maui: "Makawao, E. Maui".
>   *Remarks.* The description of *nattii* was published on 10 April 1888 and referred to pl. 1, fig. 3. The explanation of pl. 1 (on p. 56; published on 24 April 1888) has "*nealii*", which we treat as an incorrect subsequent spelling of *nattii*.

+**polita**. (Mo)
>   *Achatinella polita* Newcomb, 1853: 24 [1854a: 16, pl; 23, fig. 37; 1854b: 142, pl. 23, fig. 37]. Molokai: [no additional details].
>   *Remarks.* Subspecies of *mighelsiana* Pfeiffer, *teste* Pilsbry & Cooke (1913a: 80).

porcellana. (M)
>   *Achatinella porcellana* Newcomb, 1854a: 20, pl. 23, fig. 47 [1854b, pl. 23, fig. 47]. Maui: "E. Mani [= Maui]".

saccata.
>   *Achatinella (Achatinellastrum) saccata* Pfeiffer, 1859c: 30. Hawaiian Islands: [no additional details].
>   *Remarks.* Synonym of *lactea* Gulick, *teste* Pilsbry & Cooke (1914a: 364).

semicarinata. (L)
>   *Achatinella semicarinata* Newcomb, 1854a: 30, pl. 24, fig. 76 [1854b: 156, pl. 24, fig. 76]. Lanai (as "Ranai"): [no additional details].

variabilis. (L)
>   *Achatinella variabilis* Newcomb, 1854a: 28, pl. 24, fig. 70 [1854b: 154, pl. 24, fig. 70]. Lanai (as "Ranai"): [no additional details].

+**wailuaensis**. (M)
>   *Achatinella (Achatinellastrum) wailuaensis* Sykes, 1900: 328, pl. 11, figs. 19, 20. Maui: "Wailua".
>   *Remarks.* Subspecies of *porcellana* Newcomb, *teste* Pilsbry & Cooke (1913a: 72).

Subgenus PARTULINA Pfeiffer

PARTULINA Pfeiffer, 1854b: 114 (as *Achatinella* subg.). Type species: *Partula virgulata* Mighels, 1845 (as "*Achatinella*"), by subsequent designation of Martens (1860: 243).

PARTULINELLA Hyatt *in* Pilsbry & Cooke, 1914a: 392 (as "subdivision" of *Partulina*; Pilsbry in footnote on p. 392). Type species: *Achatinella marmorata* Gould, 1847 (as "*P. marmorata*"), by original designation (by Pilsbry in footnote on p. 392). **N. syn.**

See comments under genus *Partulina* (above) regarding the status of subgenus *Partulinella* Hyatt.

adamsi.
> *Achatinella adamsi* Newcomb, 1853: 19 [1854a: 11, pl. 22, fig. 20; 1854b: 137, pl. 22, fig. 20 (as "*adamsii*")]. Maui: "Makawao".
> *Remarks.* Synonym of *marmorata* Gould, *teste* Pilsbry & Cooke (1912: 43).

+ampulla. (M)
> *Achatinella ampulla* Gulick, 1856: 200, pl. 7, fig. 29. Maui: "Honukawai".
> *Remarks.* Subspecies of *tappaniana* Adams, *teste* Pilsbry & Cooke (1912: 57).

aptycha. (?M)
> *Achatinella aptycha* Pfeiffer 1855c: 1, pl. 30, fig. 1. Hawaiian Islands: [no additional details].
> *Remarks.* Pilsbry & Cooke (1912: 54) dealt with this as a species of *Partulina s. str.*, considering it close to *gouldi* Newcomb or *baileyana* Gulick. They quoted Sykes as indicating that it is "probably from Maui", but nevertheless stated that it was a "lost" species. Cooke (*in* Pilsbry & Cooke, 1913b: 145) considered it a "form" of *nympha* Gulick (*Achatinella* subg. *Bulimella*). Pilsbry & Cooke (1914a: 363) again dealt with it as a species of *Partulina s. str.*

arnemanni. (O)
> *Partulina arnemanni* Cooke & Kondo, 1952: 346, fig. 6a. Oahu: "Kahuku, in front of airfield".

+attenuata. (M)
> *Achatinella attenuata* Pfeiffer, 1855c: 4, pl. 30, fig. 12. Maui (as "Mani"): [no additional details].
> *Remarks.* Subspecies of *terebra* Newcomb, *teste* Pilsbry & Cooke (1912: 63).

+baileyana. (M)
> *Achatinella baileyana* Gulick, 1856: 202, pl. 7, figs. 31a, b. Maui: "Wailuku mountain".
> *Remarks.* Pilsbry & Cooke (1912: 52) treated *baileyana* as a distinct variety of *splendida* Newcomb, even though including it in their synonymy of *splendida* (p. 51).

carnicolor. (M)
> *Partulina carnicolor* Baldwin, 1906a: 112. Maui: "Nahiku, East Maui".
> *Remarks.* "Might better be classed as a subspecies of *eburnea*" (Pilsbry & Cooke, 1914a: 364). This is not a definitive statement so we retain *carnicolor* as a valid species.

+compta. (Mo)
> *Partulina compta* Pease, 1869a: 175. Molokai: [no additional details].
> *Remarks.* Subspecies of *dwightii* Newcomb, *teste* Pilsbry & Cooke (1912: 36).

+concomitans. (Mo)
> *Partulina dwightii* var. *concomitans* Hyatt *in* Pilsbry & Cooke, 1912: 37, pl. 8, figs. 9–13. Molokai: "Makakupaia; Kawela; Makolelau".

corusca.
> *Achatinella corusca* Pfeiffer, 1859d: 525. *Nom. nud.*
> *Remarks.* A Gulick collection name.

ACHATINELLIDAE [69]

+corusca. (M)
> *Achatinella corusca* Pilsbry & Cooke, 1912: 62, pl. 15, fig. 9. Maui: "Wailuku . . . Waihee".
>> Remarks. A Gulick collection name. Pilsbry & Cooke (1912: 62) treated *corusca* as a distinct variety of *terebra* Newcomb, even though including it in their synonymy of *terebra*.

crassa. (L)
> *Achatinella crassa* Newcomb, 1854a: 29, pl. 24, fig. 71 [1854b: 155, pl. 24, fig. 71]. Lanai (as "Ranai"): [no additional details].
>> Remarks. Hyatt (*in* Pilsbry & Cooke, 1914a: 392) placed *crassa* in *Partulinella*.

crocea. (M)
> *Achatinella crocea* Gulick, 1856: 210, pl. 7, fig. 36. Maui: "Waihee".

densilineata.
> *Partula densilineata* Reeve, 1850b: pl. 2, fig. 9. Type locality not given.
>> Remarks. Synonym of *radiata* Gould, *teste* Pilsbry & Cooke (1912: 49).

dextra.
> *Achatinella splendida* var. *dextra* Schmeltz, 1865: 26. *Nom. nud.*
>> Remarks. Not treated in the Manual of Conchology.

dolei. (M)
> *Achatinella (Partulina) dolei* Baldwin, 1895: 221, pl. 10, figs. 17, 18. Maui: "Honomanu".
>> Remarks. Hyatt (*in* Pilsbry & Cooke, 1914a: 394) placed *dolei* in *Partulinella*.

+dubiosa. (M)
> *Achatinella tappaniana* var. *dubiosa* Adams, 1851a: 126 [1851b: 42]. Hawaiian Islands: [no additional details].

dwighti.
> *Achatinella dwighti* Pfeiffer, 1856c: 207. Molokai: [no additional details].
>> Remarks. Primary junior homonym of *dwightii* Newcomb (see comments regarding Pfeiffer and Newcomb in the introduction to the family Achatinellidae). Synonym of *dwightii* Newcomb. **N. syn.**

dwightii. (Mo)
> *Achatinella dwightii* Newcomb, 1855c: 145. Molokai: [no additional details].
>> Remarks. Hyatt (*in* Pilsbry & Cooke, 1914a: 394) placed *dwightii* in *Partulinella*.

+eburnea. (M)
> *Achatinella eburnea* Gulick, 1856: 199, pl. 6, figs. 28a, b. Maui: "Honuaula, E. Maui".
>> Remarks. Subspecies of *tappaniana* Adams, *teste* Pilsbry & Cooke (1912: 57). Hyatt (*in* Pilsbry & Cooke, 1914a: 394) placed *tappaniana* and *eburnea* in *Partulinella*.

+fasciata. (M)
> *Achatinella fasciata* Gulick, 1856: 201, pl. 7, fig. 30. Maui: "Honukawai".
>> Remarks. Subspecies of *tappaniana* Adams, *teste* Pilsbry & Cooke (1912: 56).

fusoidea. (M)
> *Achatinella fusoidea* Newcomb, 1855c: 144. Maui: "E. Maui".

gouldi.
> *Bulimus gouldi* Pfeiffer, 1846d: 116. *Nom. nud.*

gouldi.
> *Bulimus gouldi* Pfeiffer, 1848b: 74. Hawaiian Islands: [no additional details].
>> Remarks. Synonym of *radiata* Gould, *teste* Pilsbry & Cooke (1912: 49).

gouldi.
> *Achatinella gouldi* Newcomb, 1853: 21 [1854a: 4, pl. 22, fig. 1 (as "*gouldii*"); 1854b: 129, pl. 22, fig. 1 (as "*gouldii*")]. Maui: "Wailuku valley".
>> Remarks. Secondary junior homonym of *gouldi* Pfeiffer. Replaced by *talpina* Gulick from synonymy. Synonym of *talpina* Gulick.

+halawaensis. (Mo)
> *Achatinella virgulata* var. *halawaensis* Borcherding, 1906: 52, pl. 1, figs. 13–16. Molokai: "Halawa".

+idae. (Mo)
> *Partulina idae* Borcherding, 1901: 52. Molokai: "Kalae, Kealia".
> *Remarks.* Pilsbry & Cooke (1912: 31) treated *idae* both as a "var." and a "local race" of *rufa* Newcomb, even though including it in their synonymy of *rufa* (p. 30).

induta. (M)
> *Achatinella induta* Gulick, 1856: 207, pl. 7, figs. 34a, e. Maui: "Mountain ridges of Wailuku, Maui".

insignis.
> *Bulimus insignis* Reeve, 1850e: pl. 1, species 3. *Nom. nud.*
> *Remarks.* Originally a Mighels collection name. Synonym of *virgulata* Mighels, *teste* Pilsbry & Cooke (1912: 26).

kaaeana. (M)
> *Partulina kaaeana* Baldwin, 1906a: 113. Maui: "Mt. Helu, 4000 ft. alt., West Maui".

+kaluaahacola. (Mo)
> *Partulina (P.) virgulata* var. *kaluaahacola* Pilsbry & Cooke, 1914a: 359, pl. 26, figs. 3, 3a. Molokai: "Kaluaaha, at about 1,700 ft."
> *Remarks.* Described as a new variety in the description heading but as a "local race" in the body of the description, and subsequently (Pilsbry & Cooke, 1914a: 363) as "probably a mere color-form". We retain *kaluaahacola* as an infraspecific taxon, pending further revision.

+kamaloensis. (Mo)
> *Partulina (P.) redfieldi* (as "*redfieldii*") *kamaloensis* Pilsbry & Cooke, 1914a: 362, pl. 26, figs. 4, 4a. Molokai: "Between the branch ravines above the Kamalo amphitheatre and below the old irrigation ditch".

+kaupakaluana. (M)
> *Partulina (P.) nivea* var. *kaupakaluana* Pilsbry & Cooke, 1912: 60, pl. 13, fig. 16. Maui: "Kaupakalua".

lemmoni. (M)
> *Partulina lemmoni* Baldwin, 1906a: 112. Maui: "Nahiku, East Maui".

+lignaria. (M)
> *Achatinella lignaria* Gulick, 1856: 209, pl. 7. fig. 35. Maui: "Wailuku".
> *Remarks.* Subspecies of *terebra* Newcomb, *teste* Pilsbry & Cooke (1912: 63). Hyatt (*in* Pilsbry & Cooke, 1914a: 394) placed *lignaria* in *Partulinella*.

+longior. (M)
> *Partulina (P.) terebra* var. *longior* Pilsbry & Cooke, 1912: 63, pl. 15, fig. 12. Maui: "Wailuku; also Waiehu".

lutea.
> *Achatinella lutea* Pfeiffer, 1876b: 217. *Nom. nud.*
> *Remarks.* Originally a Gulick manuscript name, printed in the synonymy of *tappaniana* Adams.

+macrodon. (Mo)
> *Partulina macrodon* Borcherding, 1901: 56. Molokai: "Makakupeia".
> *Remarks.* "Color-var." of *mucida* Baldwin, *teste* Pilsbry & Cooke (1912: 35).

marmorata. (M)
> *Achatinella marmorata* Gould, 1847b: 200. Maui: "Haleakala Mountains".

+meyeri. (Mo)
> *Partulina meyeri* Borcherding, 1901: 55. Molokai: "Pelekunu".
> *Remarks.* Subspecies of *tessellata* Newcomb, *teste* Pilsbry & Cooke (1912: 29).

montagui. (O)
>*Partulina montagui* Pilsbry, 1913: 40. Oahu: "in a superficial road cutting at the junction of Manoa road with the upper road, back of Rocky Hill, which terminates the western ridge of Manoa valley".

mucida.
>*Achatinella (Partulina) mucida* Baldwin, 1893: 7. *Nom. nud.*

+mucida. (Mo)
>*Achatinella (Partulina) mucida* Baldwin, 1895: 222, pl. 10, fig. 23. Molokai: "Makakupaia".
>>*Remarks.* "Form or race" of *dwightii* Newcomb, *teste* Pilsbry & Cooke (1914a: 361).

+multistrigata. (Mo)
>*Partulina (P.) theodorei* var. *multistrigata* Pilsbry & Cooke, 1912: 34, pl. 9, fig. 17. Molokai: [no additional details].
>>*Remarks.* Subspecies of *proxima* Pease, *teste* Pilsbry & Cooke (1914a: 360).

myrrhea.
>*Achatinella myrrhea* Pfeiffer, 1859d: 517. *Nom. nud.*
>>*Remarks.* Originally a Gulick manuscript name. Synonym of *gouldi* Newcomb [= *talpina* Gulick], *teste* Pilsbry & Cooke (1912: 52).

nivea.
>*Achatinella (Partulina) nivea* Baldwin, 1893: 7. *Nom. nud.*

nivea. (M)
>*Achatinella (Partulina) nivea* Baldwin, 1895: 222, pl. 10, fig. 19. Maui: "Makawao to Huelo".

+occidentalis. (Mo)
>*Partulina (P.) dwightii* var. *occidentalis* Pilsbry & Cooke, 1914a: 361, pl. 26, fig. 6. Molokai: "Sand dunes of Moomomi (on the north coast almost due north of Mauna Loa); also summit of Mauna Loa".

perdix. (M)
>*Achatinella perdix* Reeve, 1850e: pl. 6, figs. 43a, b. Type locality not given.
>>*Remarks.* Pilsbry & Cooke (1912: 45) indicated "W. Maui: Lahaina" as the type locality, but no lectotype was selected. Hyatt (*in* Pilsbry & Cooke, 1914a: 392) placed *perdix* in *Partulinella*.

perfecta.
>*Partulina gouldi* var. *perfecta* Pilsbry *in* Pilsbry & Cooke, 1912: 54, pl. 11, figs. 18–21. Maui: "Wailuku".
>>*Remarks.* Considered a synonym of *gouldi* Newcomb by Pilsbry & Cooke (1914a: 363); *gouldi* Newcomb is a junior homonym of *gouldi* Pfeiffer and a synonym of *talpina* Gulick. Synonym of *talpina*. **N. syn.**

perforata.
>*Achatinella perforata* Pfeiffer, 1859d: 525. *Nom. nud.*
>>*Remarks.* Listed by Pfeiffer as a synonym of *terebra* Newcomb. Originally a Gulick collection name, treated as a synonym of *lignaria* Gulick by Pilsbry & Cooke (1912: 64).

plumbea. (M)
>*Achatinella plumbea* Gulick, 1856: 213, pl. 7, fig. 39. Maui: "Kula, E. Maui".
>>*Remarks.* Hyatt (*in* Pilsbry & Cooke, 1914a: 392) placed *plumbea* in *Partulinella*.

proxima. (Mo)
>*Helicter proximus* Pease, 1862: 6. Molokai: [no additional details].
>>*Remarks.* Hyatt (*in* Pilsbry & Cooke, 1914a: 392) placed *proxima* in *Partulinella*.

pyramidalis.
>*Achatinella pyramidalis* Gulick, 1856: 204, pl. 7, fig. 32. Maui: "Lahaina".
>>*Remarks.* Synonym of *perdix* Reeve, *teste* Pilsbry & Cooke (1914a: 363).

radiata. (?M)
>   *Achatinella radiata* Gould, 1845: 27. Hawaiian Islands ("Sandwich Islands" in publication title): [no additional details].
>   *Remarks.* Senior homonym of *radiata* Pfeiffer (in *Achatinella* subg. *Bulimella*).

redfieldi. (?M, Mo)
>   *Achatinella redfieldi* Newcomb, 1853: 22 [1854a: 6, pl. 22, fig. 5; 1854b: 131, pl. 22, fig. 5]. Maui: "Wailuku".
>   *Remarks.* Localities given as "Molokai and E. Mani [= Maui]" by Newcomb (1854a: 6; 1854b: 131). Hyatt (*in* Pilsbry & Cooke, 1914a: 392) placed *redfieldi* in *Partulinella*.

rohri.
>   *Achatinella rohri* Pfeiffer, 1846c: 38. Hawaiian Islands: [no additional details].
>   *Remarks.* Synonym of *virgulata* Mighels, *teste* Pilsbry & Cooke (1912: 26). The description was published in the *Proc. Zool. Soc. Lond.* on 31 July 1846 (Waterhouse, 1937). The name *rohri* was listed and the above description referred to by Pfeiffer (1846e: 58) in his *Symbolae ad historiam heliceorum, Sectio tertia*, also published in 1846 (precise date not known). Pfeiffer (1846d: 115) synonymized *rohri* with *virgulata* Mighels and moved it to *Bulimus* in the *Zeitschrift für Malakozoologie* issue for August 1846 (taken here as 31 August 1846). The *Proc. Zool. Soc. Lond.* description is therefore treated here as the first publication of the name *rohri*.

rufa. (Mo)
>   *Achatinella rufa* Newcomb, 1853: 21 [1854a: 4, pl. 22, fig. 1; 1854b: 130, pl. 22, fig. 3]. Molokai: [no additional details].
>   *Remarks.* Hyatt (*in* Pilsbry & Cooke, 1914a: 392) placed *rufa* in *Partulinella*.

+schauinslandi. (Mo)
>   *Partulina schauinslandi* Borcherding, 1901: 54. Molokai: "Kaluahauoni, Waileia".
>   *Remarks.* "Color-var." of *proxima* Pease, *teste* Pilsbry & Cooke (1912: 33). Pilsbry & Cooke (1912: 32) did not include *schauinslandi* Borcherding in the synonymy of *proxima*. We retain it as a valid infraspecific taxon, pending revisionary study.

solida.
>   *Achatinella solida* Pfeiffer, 1859d: 516. *Nom. nud.*
>   *Remarks.* Originally a Gulick manuscript name. Listed by Pfeiffer in synonymy with *splendida* Newcomb. Also listed but never validated by Bland & Binney (1873: 332), Pfeiffer & Clessin (1879: 304), Hartman (1888a: 15), Ancey (1889a: 236) and Pilsbry & Cooke (1912: 51). It is listed here in *Partulina* as it was first published in this genus; Pilsbry & Cooke (1915a: 84) included "*Auriculella solida* Gulick, BLAND" in the synonymy of *Auriculella pulchra* Pease.

splendida. (M)
>   *Achatinella splendida* Newcomb, 1853: 20 [1854a: 5, pl. 22, fig. 4; 1854b: 131, pl. 22, fig. 4]. Maui: "Wailuku".
>   *Remarks.* Hyatt (*in* Pilsbry & Cooke, 1914a: 392) placed *splendida* in *Partulinella*.

subpolita. (Mo)
>   *Partulina (P.) subpolita* Pilsbry & Cooke, 1912: 24. Molokai: [no additional details].
>   *Remarks.* First published in a key to the Achatinellinae of Molokai; subsequently described formally by Hyatt & Pilsbry (*in* Pilsbry & Cooke, 1914a: 359, pl. 23, fig. 1).

talpina. (M)
>   *Achatinella talpina* Gulick, 1856: 212, pl. 7, fig. 38. Maui: "Wailuku, Maui".
>   *Remarks.* Generally considered, for example by Pilsbry & Cooke (1912: 52) and Caum (1928: 16), as a junior synonym of *gouldi* Newcomb, 1853, but *gouldi* Newcomb, 1853 is a junior homonym of *gouldi* Pfeiffer, 1848. Since *talpina* Gulick is the oldest available synonym of *gouldi* Newcomb, *talpina* becomes the valid name of the taxon.

tappaniana. (M)
>   *Achatinella tappaniana* Adams, 1851a: 126 [1851b: 42]. Hawaiian Islands: [no additional details].
>   *Remarks.* Hyatt (*in* Pilsbry & Cooke, 1914a: 394) placed *tappaniana* in *Partulinella*.

terebra. (M)
>*Achatinella terebra* Newcomb, 1854a: 18, pl. 23, fig. 40 [1854b: 144, pl. 23, fig. 40]. Maui: "W. Mani [= Maui]".

tessellata. (Mo)
>*Achatinella tessellata* Newcomb, 1853: 19 [1854a: 13, pl. 23, fig. 26; 1854b: 139, pl. 23, fig. 26]. Molokai: [no additional details].
>>*Remarks*. Hyatt (*in* Pilsbry & Cooke, 1914a: 392) placed *tessellata* in *Partulinella*.

theodorei. (Mo)
>*Achatinella theodorei* Baldwin, 1895: 226, pl. 10, fig. 27. Molokai: "Kawela".

theodorii.
>*Achatinella theodorii* Baldwin, 1893: 7. *Nom. nud.*

tuba.
>*Achatinella tuba* Pfeiffer, 1859d: 523. *Nom. nud.*
>>*Remarks*. Originally a Gulick manuscript name, listed in the synonymy of *tappaniana* Adams by Pfeiffer. Considered an "error for *lutea*" by Pilsbry & Cooke (1914a: 367).

+undosa. (M)
>*Achatinella undosa* Gulick, 1856: 205, pl. 7, fig. 33. Maui: "Mountain ridges of Waihee".
>>*Remarks*. Pilsbry & Cooke (1912: 46) treated *undosa* as a distinct variety of *perdix* Reeve, even though they included it in their synonymy of *perdix* (p. 45).

ustulata. (M)
>*Achatinella ustulata* Gulick, 1856: 211, pl. 7, fig. 37. Maui: "Beautiful Valley".

virgulata. (?O, Mo)
>*Partula virgulata* Mighels, 1845: 20. Oahu: "Waianai [= Waianae; ?error]".
>>*Remarks*. In his original description Mighels' only statement of locality is "Waianai" (i.e., Oahu). Baldwin (1893: 7) and Pilsbry & Cooke (1912: 25, 1914a: 358) only mentioned localities on Molokai. They did not discuss Mighels' locality, which is almost certainly wrong. Confirmation of the type locality requires further research.

winniei. (M)
>*Partulina winniei* Baldwin, 1908: 67. Maui: "Kahakukuloa, West Maui".

Genus PERDICELLA Pease

PERDICELLA Pease, 1870b: 648. Type species: *Achatinella helena* Newcomb, 1853, by subsequent designation of Sykes (1900: 329).

Gulick (1873a: 91) gave *Partulina* "*mauiensis* Newc." as an example of *Perdicella* without explicitly designating it as the type. Pilsbry & Cooke (1912) recognized 7 species (in the section *Perdicella* of the genus *Partulina*). Pilsbry & Cooke (1914a: 356–58) added *thwingi* and included *carinella* from *Newcombia* (still using *Perdicella* as a section of *Partulina*), giving a total of 9 species. There is a single infraspecific taxon. Neal (1928) suggested raising *Perdicella* from section to genus but did not explicitly do so. Cooke & Kondo (1961) raised it from section to genus, but made no change in the species.

+balteata. (Mo)
>*Partulina* (*Perdicella*) *helena* var. *balteata* Pilsbry & Cooke, 1912: 17, pl. 4, fig. 7. Type locality not given.

carinella. (M)
>*Newcombia carinella* Baldwin, 1906b: 136. Maui: "Nahiku, East Maui".

fulgurans. (M)
>*Perdicella fulgurans* Sykes, 1900: 329, pl. 11, fig. 5. Maui: "E. Maui, Makawao to Huelo".

helena. (Mo)
>*Achatinella helena* Newcomb, 1853: 27 [1854a: 25, pl. 24, fig. 63; 1854b: 151, pl. 24, fig. 63]. Molokai: [no additional details].

kuhnsi. (M)
>*Partulina (Perdicella) kuhnsi* Pilsbry *in* Pilsbry & Cooke, 1912: 22, pl. 14, figs. 8, 12–15. Maui: "West Maui: Honokohua".

maniensis. (M)
>*Achatinella maniensis* Pfeiffer, 1856c: 207. Maui (as "Mani"): [no additional details].
>>*Remarks*. Originally a Newcomb manuscript name. Pfeiffer (1859d: 563) and Pilsbry & Cooke (1912: 20) explicitly (and other authors implicitly) considered *maniensis* a typographical error for "*mauiensis*".

manoensis.
>*Perdicella manoensis* Pease, 1870b: 648. *Nom. nud.*
>>*Remarks*. Possibly a typographical error for *maniensis* Pfeiffer. Pease attributed the name to Newcomb.

mauiensis.
>*Partulina (Perdicella) mauiensis* Pfeiffer, 1859d: 563. Unjustified emendation of *maniensis* Pfeiffer, 1856.
>>*Remarks*. Objective synonym of *maniensis* Pfeiffer.

minuscula.
>*Achatinella minuscula* Pfeiffer, 1857b: 231. Hawaiian Islands: [no additional details].
>>*Remarks*. Synonym of *helena* Newcomb, *teste* Pilsbry & Cooke (1912: 18).

ornata. (M)
>*Achatinella ornata* Newcomb, 1854a: 23, pl. 24, fig. 55 [1854b: 149, pl. 24, fig. 55]. Maui: "E. Mani [= Maui] . . . in a deep ravine, at the back of Lahaina".
>>*Remarks*. Lahaina is on west Maui. Pilsbry & Cooke (1914a: 356) considered "E. Maui" to be an "oversight". Correct identification of the type locality therefore requires further research.

thwingi. (M)
>*Partulina (Perdicella) thwingi* Pilsbry & Cooke, 1914a: 357, pl. 54, figs. 6–6b. Maui: "East Maui: Auwahi".

zebra. (M)
>*Achatinella zebra* Newcomb, 1855c: 142. Maui: "East Maui".

zebrina. (M)
>*Achatinella (Newcombia) zebrina* Pfeiffer, 1856b: 202. Hawaiian Islands ("Sandwich Islands" in publication title): [no additional details].

Subfamily AURICULELLINAE

Genus AURICULELLA Pfeiffer

AURICULELLA Pfeiffer, 1855c: 1 (as *Achatinella* sect.). Type species: *Partula auricula* Férussac, 1821 (as "*Auriculella*"), by subsequent designation of Gulick (1873: 91).

FRICKELLA Pfeiffer, 1855c: 2 (as "Gruppe" of *Achatinella*). Type species: *Achatinella amoena* Pfeiffer, 1855, by monotypy.

Pilsbry & Cooke (1915a) recognized 31 species, although numbering only 29 (species numbers 20 and 21 occurring twice). Caum (1928) listed 30 species, missing *chamissoi* Pfeiffer. In addition, there are 5 infraspecific taxa. Cooke & Kondo (1961) made no taxonomic/systematic changes but referred *Auriculella* to the tribe Auriculellini.

ambusta. (O)
>*Auriculella ambusta* Pease, 1868: 345. Hawaiian Islands ("îles Hawaï" in publication title): [no additional details].

amoena. (O)
>*Achatinella (Frickella) amoena* Pfeiffer, 1855b: 3 [1855c: 2]. Hawaiian Islands: [no additional details].

armata. (H)
>*Bulimus armatus* Mighels, 1845: 19. Hawaii: [no additional details].
>>*Remarks.* Senior synonym of *westerlundiana* Ancey, *teste* Johnson (1949: 221), contrary to Pilsbry & Cooke (1915a: 78).

auricula. (O)
>*Partula auricula* Férussac, 1821c: 70. "Sans doute les îles de la mer du Sud? [= without doubt the south sea islands?]".

brunnea. (Mo, ?L)
>*Auriculella brunnea* Smith *in* Gulick & Smith, 1873: 88, pl. 10, fig. 23. Molokai and Lanai: [no additional details].
>>*Remarks.* Pilsbry & Cooke (1915a: 105) did not record it from Lanai.

+cacuminis. (O)
>*Auriculella diaphana cacuminis* Pilsbry & Cooke, 1915a: 77, pl. 24, figs. 11, 12. Oahu: "'Mauna Kope,' at the head of the Kalihi-Moana-lua ridge".

canalifera. (Mo)
>*Auriculella canalifera* Ancey, 1904a: 121, pl. 7, fig. 11. Molokai: "Halawa".

castanea. (O, ?Mo, ?M)
>*Tornatellina castanea* Pfeiffer, 1853c: 524. Hawaiian Islands: [no additional details].

cerea. (Mo)
>*Achatinella cerea* Pfeiffer, 1855b: 3 [1855c: 2]. Hawaiian Islands: [no additional details].

chamissoi. (O, H)
>*Achatinella (Auriculella) chamissoi* Pfeiffer, 1855f: 98. Hawaiian Islands: [no additional details].

crassula. (?Mo, M)
>*Auriculella crassula* Smith *in* Gulick & Smith, 1873: 88, pl. 10, fig. 22. Maui: "Makawao".

diaphana. (O)
>*Auriculella diaphana* Smith *in* Gulick & Smith, 1873: 87, pl. 10, fig. 25. Oahu: "Olomana and Kailua . . . Palolo, Makiki, and Kalihi".

dumartroyii.
>*Partula dumartroyii* Souleyet, 1842: 102. Hawaiian Islands: [no additional details].
>>*Remarks.* Synonym of *auricula* Férussac, *teste* Pilsbry & Cooke (1915a: 78). Pilsbry & Cooke (1915a: 79) had "no doubt that this shell came from Nuuanu".

expansa. (M)
>*Auriculella expansa* Pease, 1868: 343, pl. 14, fig. 8. Hawaiian Islands ("îles Hawaï" in publication title): [no additional details].

flavida. (Mo)
>*Auriculella flavida* Cooke *in* Pilsbry & Cooke, 1915a: 103, pl. 26, figs. 8, 9. Molokai: "Kamalo".

jucunda.
 Auriculella jucunda Bland & Binney, 1873: 331. *Nom. nud.*
  Remarks. Originally attributed to Smith by Bland & Binney (1873: 331) but apparently only a manuscript name (Ancey, 1889a: 236; Sykes, 1900: 379).
+jucunda. (M)
 Auriculella jucunda Pilsbry & Cooke, 1915a: 108, pl. 18, figs. 6, 7. Maui (in section heading): [no additional details].
  Remarks. Although not formally described by Pilsbry & Cooke (1915a: 108), they validated the name as a "form" of *uniplicata* Pease.
kuesteri. (?O)
 Tornatellina kuesteri Pfeiffer, 1855e: 295. Type locality not given.
  Remarks. Pilsbry & Cooke (1915a: 101) speculated that it might be from Oahu.
lanaiensis. (L)
 Auriculella lanaiensis Cooke *in* Pilsbry & Cooke, 1915a: 107, pl. 19, figs. 12–16. Lanai: [no additional details].
lurida.
 Achatinella (Auriculella) lurida Pfeiffer, 1856a: 166. *Nom. nud.*
  Remarks. Synonym of *castanea* Pfeiffer, *teste* Pilsbry & Cooke (1915a: 94).
malleata. (O)
 Auriculella malleata Ancey, 1904a: 120, pl. 7, fig. 12. Oahu: "In cacumine montis Kaala (4000' s.m.), insulae Oahu [= at the summit of Mount Kaala (4000 ft. above sea level), on the island of Oahu]".
minuta. (O)
 Auriculella minuta Cooke & Pilsbry *in* Pilsbry & Cooke, 1915a: 90, pl. 25, figs. 5–9. Oahu: "Nuuanu, Palolo".
montana. (O)
 Auriculella montana Cooke *in* Pilsbry & Cooke, 1915a: 82, pl. 27, fig. 9. Oahu: "Mt. Konahuanui".
newcombi. (Mo)
 Balea newcombi Pfeiffer, 1854a: 67. Hawaiian Islands: [no additional details].
obeliscus.
 Achatinella (Auriculella) obeliscus Pfeiffer, 1856b: 206. *Nom. nud.*
  Remarks. Listed by various authors and included in the synonymy of *newcombi* Pfeiffer (e.g., Pilsbry & Cooke, 1915a: 102), but never validated.
+obliqua. (O)
 Auriculella obliqua Ancey, 1892: 721. Oahu: [no additional details].
  Remarks. Subspecies of *ambusta* Pease, *teste* Pilsbry & Cooke (1915a: 88).
olivacea. (O)
 Auriculella olivacea Cooke *in* Pilsbry & Cooke, 1915a: 81, pl. 27, figs. 10, 11. Oahu: "Mt. Olympus, at an elevation of about 2,500 feet; Konahuanui, at an elevation of 3,300 feet".
owaihiensis.
 Auricula owaihiensis Chamisso, 1829: 639, pl. 36, fig. 1. Oahu: "in dumetis interioris insulae O-Wahu regni O-Waihiensis [= in thickets in the interior of the island of Oahu of the Kingdom of Hawaii]".
  Remarks. Synonym of *auricula* Férussac, *teste* Pilsbry & Cooke (1915a: 78).
patula.
 Auriculella patula Smith *in* Gulick & Smith, 1873: 88, pl. 10, fig. 24. Hawaiian Islands: [no additional details].
  Remarks. Synonym of *auricula* Férussac, *teste* Pilsbry & Cooke (1915a: 79).
pellucida.
 Auriculella pellucida Gulick, 1905: 220. *Nom. nud.*

+pellucida. (O)
    Auriculella auricula form *pellucida* Pilsbry & Cooke, 1915a: 80, pl. 24, fig. 10. Oahu: "Kaliuwaa, Punaluu and Hauula".

perkinsi.
    Auriculella perkinsi Sykes, 1900: 377, pl. 11, figs. 17, 18. Oahu: "ridges round Nuuanu, and Mount Tantalus".
        Remarks. Synonym of *auricula* Férussac, *teste* Pilsbry & Cooke (1915a: 79).

perpusilla. (O)
    Auriculella perpusilla Smith *in* Gulick & Smith, 1873: 87, pl. 10, fig. 26. Oahu: "Kahalu".

perversa. (O)
    Auriculella perversa Cooke *in* Pilsbry & Cooke, 1915a: 90, pl. 25, figs. 3, 4. Oahu: "Nuuanu; Kuliouou".

petitiana. (?O)
    Tornatellina petitiana Pfeiffer, 1847b: 149. Type locality not given.
        Remarks. Pilsbry & Cooke (1915a: 95) listed it among the Oahu species.

ponderosa.
    Auriculella ponderosa Ancey, 1889a: 225. Maui: [no additional details].
        Remarks. Synonym of *crassula* Smith, *teste* Pilsbry & Cooke (1915a: 109).

+porcellana. (M)
    Auriculella expansa var. *porcellana* Ancey, 1889a: 226. Maui: "la partie orientale de l'île de Maui [= east Maui]".
        Remarks. Pilsbry & Cooke (1915a: 110) considered *porcellana* as "probably a synonym of *A. crassula*". This statement is not definitive and we retain *porcellana* as an infraspecific taxon of *expansa* Pease, pending further revision.

pulchra. (O)
    Auriculella pulchra Pease, 1868: 346, pl. 14, fig. 6. Hawaiian Islands ("îles Hawaï" in publication title): [= ?"upper Waolani valley including the valley flank of Waolani Peak" (Pilsbry & Cooke, 1915a: 85)].

serrula. (O)
    Auriculella serrula Cooke *in* Pilsbry & Cooke, 1915a: 93, pl. 25, figs. 13, 14. Oahu: "Mt Konahuanui, at about 3,000 feet; Kuliouou".

sinistrorsa.
    Auricula sinistrorsa Chamisso, 1829: 640, pl. 36, fig. 2. Oahu: "cum praecedente [= with the preceding [species]", i.e., *Auriculella owaihiensis*; thus the type locality is "the interior of the island of Oahu"].
        Remarks. Synonym of *auricula* Férussac, *teste* Pilsbry & Cooke (1915a: 78).

solida.
    Auriculella tenuis var. *solida* Ancey, 1889a: 230. Oahu: "Tantalus".
        Remarks. Synonym of *castanea* Pfeiffer, *teste* Pilsbry & Cooke (1915a: 94).

solidissima.
    Auriculella solidissima Bland & Binney, 1873: 331. *Nom. nud.*
        Remarks. Synonym of *crassula* Smith, *teste* Pilsbry & Cooke (1915a: 109). See also Ancey (1889a: 236).

straminea. (O)
    Auriculella straminea Cooke *in* Pilsbry & Cooke, 1915a: 77, pl. 24, fig. 13. Oahu: "Mt. Tantalus".

tantalus. (O)
    Auriculella tantalus Pilsbry & Cooke, 1915a: 97, pl. 24, figs. 15, 16. Oahu: "Pauoa side of Tantalus along the Castle trail".

tenella. (O)
    Auriculella tenella Ancey, 1889a: 232. Oahu: "Waianae".

tenuis. (O)
> *Auriculella tenuis* Smith *in* Gulick & Smith, 1873: 87, pl. 10, fig. 27. Oahu: "Kalaikoa, on Wahiawa, Helemano and Kawailoa on Oahu".
>> *Remarks.* The apparent grammatical or printing errors in Smith's statement of localities introduce some confusion. Pilsbry & Cooke (1915a: 98) simply stated "Oahu: Wahiawa, Helemano and Kawailoa".

triplicata.
> *Auriculella triplicata* Pease, 1868: 346. Hawaiian Islands ("îles Hawaï" in publication title): [no additional details].
>> *Remarks.* Synonym of *auricula* Férussac, *teste* Pilsbry & Cooke (1915a: 79), who also stated that "Pease's specimens undoubtedly came from Palolo Valley" on Oahu.

turritella. (O)
> *Auriculella turritella* Cooke *in* Pilsbry & Cooke, 1915a: 92, pl. 25, fig. 15. Oahu: "Konahuanui, at an elevation of about 3,000 feet".

uniplicata. (M, ?Mo)
> *Auriculella uniplicata* Pease, 1868: 344, pl. 14, figs. 7, 7a. Maui: [no additional details].
>> *Remarks.* Smith (*in* Gulick & Smith, 1873: 88) gave the locality as "Lahaina, on West Maui". Sykes (1900: 378) gave Molokai localities as well as West Maui, but Pilsbry & Cooke (1915a: 108), while adding more detailed Maui localities, expressed doubts about the identification of the Molokai material.

westerlundiana.
> *Auriculella westerlundiana* Ancey, 1889a: 218. Hawaii: "District de Kona, dans la partie méridionale et occidentale de l'île d'Hawaii [= Kona district, in the southern and western part of the island of Hawaii]".
>> *Remarks.* Synonym of *armata* Mighels, *teste* Johnson (1949: 221).

Genus GULICKIA Cooke

GULICKIA Cooke *in* Pilsbry & Cooke, 1915a: 112. Type species: *Gulickia alexandri* Cooke, 1915, by original designation.

Cooke (*in* Pilsbry & Cooke, 1915a: 112) described both the genus and its single species, *G. alexandri*. Cooke & Kondo (1961) made no taxonomic change but referred *Gulickia* to the tribe Auriculellini.

alexandri. (M)
> *Gulickia alexandri* Cooke *in* Pilsbry & Cooke, 1915a: 112, pl. 28, fig. 7. Maui: "W. Maui: Maunahooma . . . Honokowai".
>> *Remarks.* Although Cooke (*in* Pilsbry & Cooke, 1915a: 113) gave 2 localities without explicitly stating which was the type locality, he did designate a type specimen (BPBM 14148). The label associated with this specimen says "Hahakea West Maui" but the catalog entry says "valley E. of Maunahooma, above, Japanese camp, 800–1000 ft." Identification of the type locality therefore requires further study.

Subfamily PACIFICELLINAE, n. stat.

The family-group name Pacificellidae was established by Steenberg (1925: 195) for Odhner's genus *Pacificella*. The Hawaiian species formerly referred to the genus *Tornatellinops* are here referred to *Pacificella* (see remarks under genus *Pacificella*). Steenberg's family-group name predates Lamellideinae Cooke & Kondo, 1961 and Pacificellinae is therefore the appropriate name for this subfamily.

Genus LAMELLIDEA Pilsbry

LAMELLINA Pease, 1861: 439. Type species: *Lamellina serrata* Pease, 1861 [not Hawaiian], by monotypy. [Preoccupied, Bory de Saint-Vincent, 1826].

LAMELLARIA Liardet, 1876: 101. Type species: *Lamellaria perforata* Liardet, 1876 [= *Partula pusilla* Gould, 1847] [not Hawaiian], by monotypy. [Preoccupied, Montagu, 1816].

LAMELLIDEA Pilsbry, 1910: 123 (as *Tornatellina* sect.). Type species: *Pupa peponum* Gould, 1847 (as "*Tornatellina*"), by original designation.

Lamellidea was erected by Pilsbry (1910: 123) as a section of *Tornatellina* but subsequently synonymized with *Lamellina* Pease, 1861 by Pilsbry & Cooke (1915b: 150). Pilsbry & Cooke (1916: 273), having realized that *Lamellina* was preoccupied by *Lamellina* Bory de Saint-Vincent, 1826, re-established *Lamellidea*. Liardet (1876: 101) had earlier used *Lamellaria* for his new species *perforata*, which was included by Cooke & Kondo (1961: 185) in the synonymy of *Lamellidea pusilla* (Gould). Although it is possible that *Lamellaria* was a simple error for *Lamellina*, as suggested by Pilsbry & Cooke (1915b: 199), it may have been a proposal of a new replacement name for the preoccupied *Lamellina*, Liardet not realizing that *Lamellaria* was also preoccupied by *Lamellaria* Montagu, 1816; Oken, 1817; and d'Orbigny, 1842.

Cooke & Kondo (1961) recognized 8 species, all referred to *Tornatellina* by Pilsbry & Cooke (1915b) and Caum (1928), with 1 (*oblonga* Pease) being introduced. The native Hawaiian fauna thus consists of 7 species. In addition, there are 2 infraspecific taxa. Cooke & Kondo (1961: 18) included *Lamellidea* in the tribe Lamellideini.

Subgenus ELAMELLIDEA Cooke & Kondo

ELAMELLIDEA Cooke & Kondo, 1961: 211. Type species: *Tornatellina tantalus* Pilsbry & Cooke, 1915 (as "*Lamellidea*"), by original designation.

tantalus. (O)
Tornatellina tantalus Pilsbry & Cooke, 1915b: 172, pl. 40, figs. 8–10. Oahu: "southwestern rim of Tantalus bowl, outside".

Subgenus LAMELLIDEA Pilsbry

LAMELLINA Pease, 1861: 439. Type species: *Lamellina serrata* Pease, 1861 [not Hawaiian], by monotypy. [Preoccupied, Bory de Saint-Vincent, 1826].

LAMELLARIA Liardet, 1876: 101. Type species: *Lamellaria perforata* Liardet, 1876 [= *Partula pusilla* Gould, 1847] [not Hawaiian], by monotypy. [Preoccupied, Montagu, 1816].

LAMELLIDEA Pilsbry, 1910: 123 (as *Tornatellina* sect.). Type species: *Pupa peponum* Gould, 1847 (as "*Tornatellina*"), by original designation.

cylindrica. (K, O)
Tornatellina cylindrica Sykes, 1900: 381, pl. 11, fig. 28. Oahu: "Waianae Mts."; Kauai: "Makaweli".
  Remarks. Pilsbry & Cooke (1915b: 153) considered "Waianae Mts." the type locality, but no lectotype designation was made. We consider "*cylindrata*" in Cooke (1907: 14) to be an incorrect spelling.

dentata.
Tornatellina dentata Pease, 1871b: 460. Hawaii: [no additional details].
  Remarks. Synonym of *oblonga* Pease [not Hawaiian], *teste* Pilsbry & Cooke (1915b: 162).

extincta.
> *Tornatellina extincta* Ancey, 1890: 341. Maui: "Dans l'isthme central de l'île de Maui, qui réunit la partie orientale à la portion occidentale de l'île [= In the central isthmus of the island of Maui that unites the eastern part to the western portion of the island]".
> *Remarks.* Synonym of *gracilis* Pease, *teste* Pilsbry & Cooke (1915b: 159).

gayi. (K)
> *Tornatellina gayi* Cooke & Pilsbry *in* Pilsbry & Cooke, 1915b: 172, pl. 42, fig. 3. Kauai: "Makaweli".

gracilis. (Kure, Laysan, Lisianski, Nihoa, N, K, O, Mo, M, L, Ka, H)
> *Tornatellina gracilis* Pease, 1871b: 460. Kauai: [no additional details].

+kamaloensis. (Mo)
> *Tornatellina polygnampta kamaloensis* Pilsbry & Cooke, 1915b: 156, pl. 40, figs. 4, 5. Molokai: "Western ravines of Kamalo near the old irrigation ditch".

+kilohanana. (Mo)
> *Tornatellina cylindrica kilohanana* Pilsbry & Cooke, 1915b: 154, pl. 40, figs. 3, 6. Molokai: "Kilohana, near the leper settlement trail".

lanceolata. (O)
> *Tornatellina lanceolata* Cooke & Pilsbry *in* Pilsbry & Cooke, 1915b: 158, pl. 43, figs. 4–6. Oahu: "Nuuanu, Tantalus".

oblonga. (K, O, Mo, M, H; introduced)
> *Tornatellina oblonga* Pease, 1865: 673. "Islands of the central Pacific" (in publication title): [no additional details].
> *Remarks.* Included here only because *dentata* Pease, described from the Hawaiian Islands, is now considered a junior synonym of *oblonga* Pease. It was probably introduced to the Hawaiian Islands by Polynesian travelers prior to the arrival of Europeans (Cooke & Kondo, 1961: 201–02).

peponum. (O, M, H)
> *Pupa peponum* Gould, 1847a: 197. Hawaii: "Hilo"; Oahu: [no additional details; see remarks].
> *Remarks.* Sykes (1900: 382) restricted *peponum* to specimens in Gould's (1852) pl. 7, figs. 104 and 104d; with figs. 104a–c becoming *Tornatellina confusa* Sykes (here listed under *Tornatellides*); and fig. 104e being another species, perhaps *Tornatellides euryomphala*. Gould said that the specimens came from Hilo and Oahu, but Pilsbry & Cooke (1915b: 156) noted that there is a "Type no. 5506" in the U.S. National Museum; Johnson (1964: 125) called it a "holotype". Examination of this specimen will be necessary to determine the type locality and correspondence with Gould's figures. The outcome of this examination may have an implication for the status of *confusus*.

polygnampta. (M)
> *Tornatellina polygnampta* Pilsbry & Cooke, 1915b: 155, pl. 41, figs. 1–5, 7, 8. Maui: "Makawao and Kaupakalua".

Genus PACIFICELLA Odhner

PACIFICELLA Odhner, 1922: 249. Type species: *Pacificella variabilis* Odhner, 1922 [not Hawaiian], by monotypy.
TORNATELLINOPS: Cooke & Kondo, 1961, not Pilsbry & Cooke, 1915, misidentification.

Pilsbry & Cooke (1915b: 169) established *Tornatellinops* as a section of *Tornatellina*. Caum (1928) did not list it as either genus or section. Cooke & Kondo (1961) treated it as a genus in the tribe Tornatelloptini and included 2 species and a subspecies, previously referred to *Tornatellina s. str.* by Pilsbry & Cooke (1915b) and to *Tornatellina* by Caum (1928). The status of *Tornatellinops* was clarified by

Climo (1974: 578–79; see also Pilsbry & Cooke, 1933: 59–60), who stated that the Hawaiian taxa do not belong in *Tornatellinops* but with the "group of species around "*Tornatellinops*" *variabilis* (Odhner)". However, *variabilis* Odhner is the type species of *Pacificella* so the Hawaiian species-group names are therefore all new combinations with *Pacificella*. As a result of this, the family-group name Pacificellinae Steenberg, 1925 replaces Lamellideinae Cooke & Kondo, 1961, as indicated above.

baldwini. (K, O, M, H)
> *Tornatellina baldwini* Ancey, 1889a: 238. Oahu: "Tantalus" [?error = "Manoa"]. **N. comb.**
> *Remarks.* Ancey gave the locality as "Tantalus". Pilsbry & Cooke (1915b: 143) designated a specimen as "Ancey's type", but said that it came from Manoa. Both the label associated with this specimen in the BPBM collection (BPBM 18422) and the catalogue entry say "Manoa".

mcgregori. (H)
> *Tornatellina mcgregori* Pilsbry & Cooke, 1915b: 144, pl. 35, figs. 14, 15. Hawaii: "Hilo". **N. comb.**
> *Remarks.* Cooke & Kondo (1961: 168) cited the catalog number of the "types" incorrectly as 86479 in ANSP; the number given by Pilsbry & Cooke (1915b: 144), 85387, is correct. In addition, Cooke & Kondo (1961: 168) indicated incorrect locality details.

+subrugosa. (M)
> *Tornatellina baldwini subrugosa* Pilsbry & Cooke, 1915b: 143, pl. 35, figs. 7, 10. Maui: [no additional details]. **N. comb.**

Subfamily TORNATELLIDINAE

Genus PHILOPOA Cooke & Kondo

PHILOPOA Cooke & Kondo, 1961: 262. Type species: *Philopoa singularis* Cooke & Kondo, 1961, by original designation.

Cooke & Kondo (1961) erected *Philopoa* in the tribe Tornatellariini to accommodate the single species *P. singularis*.

singularis. (Nihoa)
> *Philopoa singularis* Cooke & Kondo, 1961: 263, figs. 111a–h. Nihoa: "northwest part of island, just below highest peak, dry open ridge . . . alt. about 900 ft."

Genus TORNATELLARIA Pilsbry

TORNATELLARIA Pilsbry, 1910: 123 (as *Tornatellides* sect.). Type species: *Tornatellina newcombi* Pfeiffer, 1857, by original designation.

Cooke & Kondo (1961) included *Tornatellaria* in the tribe Tornatellariini.

abbreviata. (M)
> *Tornatellina abbreviata* Ancey, 1903: 298, pl. 12, figs. 7, 8. Maui: "Kaupakalua".

adelinae. (O)
> *Tornatellaria adelinae* Pilsbry & Cooke, 1915b: 256, pl. 54, figs. 3, 4. Oahu: "Mt. Tantalus, on the Castle Trail, Panoa [= Pauoa] slope".

anceyana. (M)
    Tornatellaria anceyana Cooke & Pilsbry *in* Pilsbry & Cooke, 1916: 263, pl. 55, fig. 4. Maui: "East Maui: Kaupakalua".

baldwiniana. (M)
    Tornatellaria baldwiniana Cooke & Pilsbry *in* Pilsbry & Cooke, 1916: 270, pl. 55, fig. 5. Maui: "W. Maui: Maunahooma".

cincta. (?O, Mo, M, ?L, ?H)
    Tornatellina cincta Ancey, 1903: 297, pl. 12, figs. 5, 6. Oahu: [no additional details]; Molokai: [no additional details]; Maui: "Makawao (partie Est de Maui [= east Maui]) . . . vallée d'Iao . . . Kaupakalua"; Hawaii: [no additional details].
    Remarks. Cooke (*in* Pilsbry & Cooke, 1916: 263) "selected the Makawao lot (no. 18500 Bishop Mus.) as the type". Unfortunately this lot contains more than 1 specimen and so cannot be considered a lectotype designation.

convexior. (H)
    Tornatellaria convexior Pilsbry & Cooke, 1916: 267, pl. 55, fig. 8. Hawaii: "Olaa".

+hawaiiensis. (H)
    Tornatellaria abbreviata var. *hawaiiensis* Pilsbry & Cooke, 1916: 269, pl. 55, fig. 14. Hawaii: "Kukuihaele".

henshawi. (H)
    Tornatellina henshawi Ancey, 1903: 299, pl. 12, figs. 9, 10. Hawaii: "Hamakua".
    Remarks. Ancey gave Hamakua and Olaa as localities without explicitly designating 1 of them as the type locality. Pilsbry & Cooke (1916: 264) designated a lectotype (BPBM 18436), which comes from Hamakua.

+illibata. (Mo)
    Tornatellaria sykesii var. *illibata* Cooke & Pilsbry *in* Pilsbry & Cooke, 1916: 266, pl. 55, fig. 7 (as "*sykesi*"). Molokai: "Kilohana".

lilae. (O)
    Tornatellaria lilae Cooke & Pilsbry *in* Pilsbry & Cooke, 1915b: 256, pl. 54, figs. 1, 2. Oahu: "ridge west of Kolekole Pass on the Leilehua side".

newcombi. (K, O)
    Tornatellina newcombi Pfeiffer, 1857a: 335. Hawaiian Islands: [no additional details].
    Remarks. Pilsbry & Cooke (1916: 259) considered "Waianae Mountains above Waialua" (Oahu) as the type locality.

occidentalis. (O)
    Tornatellaria occidentalis Pilsbry & Cooke, 1916: 257, pl. 54, figs. 5, 6. Oahu: "western ridge of Popouwela, Waianae Mountains".

sharpi. (H)
    Tornatellaria sharpi Pilsbry & Cooke, 1916: 270, pl. 55, fig. 10. Hawaii: "crest of Kilawea [= Kilauea] crater, about half a mile south of the Volcano House".

smithi. (H)
    Tornatellaria smithi Cooke & Pilsbry *in* Pilsbry & Cooke, 1916: 269, pl. 55, fig. 11. Hawaii: "Kaiwiki".
    Remarks. Cooke & Pilsbry gave more than 1 locality but their designated "type" (BPBM 14195) comes from Kaiwiki.

sykesii. (H)
    Tornatellaria sykesii Cooke & Pilsbry *in* Pilsbry & Cooke, 1916: 265, pl. 55, fig. 6. Hawaii: "Olaa".
    Remarks. Cooke & Pilsbry gave more than 1 locality but their designated "type" (BPBM 14194) comes from Olaa.

thaanumi. (Mo)
    Tornatellides thaanumi Cooke & Pilsbry *in* Pilsbry & Cooke, 1915b: 215, pl. 47, figs. 1, 2, 4. Molokai: "Mapulehu".
    Remarks. Transferred to *Tornatellaria* by Cooke & Kondo (1961: 247, 267).

trochoides. (Mo, L, H)
>*Tornatellina trochoides* Sykes, 1900: 383, pl. 11, fig. 31. Lanai: "Lanai Mountains".

umbilicata. (Mo, M)
>*Auriculella umbilicata* Ancey, 1889a: 232. Maui: "Lahaina".

Genus TORNATELLIDES Pilsbry

TORNATELLIDES Pilsbry, 1910: 123. Type species: *Tornatellina simplex* Pease, 1865 [= *Strobilus oblongus* Anton, 1838] [not Hawaiian], by original designation.

Cooke & Kondo (1961) referred the genus to the tribe Tornatellidini. The gender of *Tornatellides* is masculine (*Code* Art. 30(b)).

Subgenus AEDITUANS Cooke & Kondo

AEDITUANS Cooke & Kondo, 1961: 258. Type species: *Tornatellides neckeri* Cooke & Kondo, 1961, by original designation.

neckeri. (Necker)
>*Tornatellides (Aedituans) neckeri* Cooke & Kondo, 1961: 258, figs. 110a–f. Necker: "Summit Hill".

Subgenus TORNATELLIDES Pilsbry

TORNATELLIDES Pilsbry, 1910: 123 (as genus). Type species: *Tornatellides simplex* Pease, 1865 [= *Strobilus oblongus* Anton, 1838] [not Hawaiian], by original designation.

+acicula. (O, Mo, M, L)
>*Tornatellides perkinsi acicula* Pilsbry & Cooke, 1915b: 225, pl. 49, figs. 3, 4. Oahu: "Punaluu".

+ada. (O)
>*Tornatellides macromphala ada* Pilsbry & Cooke, 1915b: 229, pl. 49, figs. 17, 18. Oahu: "Glen Ada, north side of Nuuanu Valley".

+anisoplax. (O)
>*Tornatellides idae* var. *anisoplax* Pilsbry & Cooke, 1915b: 217, pl. 47, fig. 7. Oahu: "western ridge of Popowela".

attenuatus. (O)
>*Tornatellides attenuatus* Cooke & Pilsbry *in* Pilsbry & Cooke, 1915b: 219, pl. 48, figs. 3, 4. Oahu: "Manoa".

bellus. (M)
>*Tornatellides bellus* Cooke & Pilsbry *in* Pilsbry & Cooke, 1915b: 241, pl. 53, figs. 4, 5. Maui: "West Maui: Maunahooma near Lahaina".

brunneus. (O)
>*Tornatellides brunneus* Cooke & Pilsbry *in* Pilsbry & Cooke, 1915b: 238, pl. 51, figs. 5, 6. Oahu: "Nuuanu near the Pali".

bryani. (Laysan)
>*Tornatellides bryani* Cooke & Pilsbry *in* Pilsbry & Cooke, 1915b: 210, pl. 53, figs. 9, 10. Laysan Island: [no additional details].

comes. (Mo)
>*Tornatellides comes* Pilsbry & Cooke, 1915b: 225, pl. 49, figs. 7, 8. Molokai: "Western ravine of Kamalo, near the old ditch trail".

compactus. (H)
> *Tornatellina compacta* Sykes, 1900: 380, pl. 11, fig. 1. Hawaii: "Mauna Loa at 2000 feet".

confusus. (?K, H)
> *Tornatellina confusa* Sykes, 1900: 380 [valid by bibliographic reference to Gould (1852: figs. 104a–c)]. Hawaii: "Hilo"; or Oahu: [no additional details]; Kauai: "Makaweli" (original type localities).
>> Remarks. *Tornatellina confusa* Sykes was originally based on Gould's material of *peponum* (figs. 104a–c), the material of which is either from Hilo (Hawaii) or Oahu. In addition, *confusa* was based on specimens collected by Perkins from Makaweli, Kauai. Pilsbry & Cooke (1915b: 209) considered Gould's types lost and designated a "type" (pl. 46, fig. 1) from Hilo. However, they did not confirm that the Kauai specimens were lost. Hence, their type [= neotype] designation is invalid.

cyphostyla. (H)
> *Tornatellina cyphostyla* Ancey, 1904b: 70, pl. 5, figs. 22, 23. Hawaii: "Palihoukapapa, on the Hamakua slope of Mauna-Kea, Kawii [= Hawaii], at an elevation of 4,000 feet" (in the introduction of the paper).

diptyx. (Mo)
> *Tornatellides diptyx* Pilsbry & Cooke, 1915b: 217, pl. 47, figs. 8, 9. Molokai: "Western ravine of Kamalo".

drepanophorus. (K)
> *Tornatellides drepanophora* Cooke & Pilsbry *in* Pilsbry & Cooke, 1915b: 249, pl. 52, figs. 11–13. Kauai: "Puukapele".

euryomphala. (M)
> *Tornatellina euryomphala* Ancey, 1889a: 239. Maui: "Point culminant de la partie occidentale de Maui [= highest point of west Maui]".

forbesi. (M)
> *Tornatellides forbesi* Cooke & Pilsbry *in* Pilsbry & Cooke, 1915b: 247, pl. 52, figs. 8, 9. Maui: "West Maui: Waikapu".

frit. (Mo)
> *Tornatellides frit* Pilsbry & Cooke, 1915b: 226, pl. 49, fig. 6. Molokai: "Western ravine of Kamalo".

idae. (O)
> *Tornatellides idae* Cooke & Pilsbry *in* Pilsbry & Cooke, 1915b: 216, pl. 47, figs. 3, 5, 6. Oahu: "Palehua, in the Waianae Mts."

inornatus. (Mo)
> *Tornatellides inornatus* Pilsbry & Cooke, 1915b: 214, pl. 46, fig. 13. Molokai: "Western ravine of Kamalo, near the old ditch trail".

insignis. (O)
> *Tornatellides insignis* Pilsbry & Cooke, 1915b: 220, pl. 48, figs. 10–13. Oahu: "in shell-deposits on ledges of the 'coral bluff,' 1 1/2 miles west of Kahuku".

irregularis. (M)
> *Tornatellides irregularis* Cooke & Pilsbry *in* Pilsbry & Cooke, 1915b: 234, pl. 50, figs. 4–6. Maui: "West Maui: Top of Mt. Kukui at about 6,000 ft, elevation".

kahoolavensis. (Kah)
> *Tornatellides kahoolavensis* Cooke & Pilsbry *in* Pilsbry & Cooke, 1915b: 211, pl. 46, figs. 3, 4, 7. Kahoolawe: "Hakioawa".

kahukuensis. (O)
> *Tornatellides kahukuensis* Pilsbry & Cooke, 1915b: 208, pl. 46, figs. 11, 12. Oahu: "ledges near base of the 'coral bluff', 1 1/2 miles west of Kahuku".

+kailuanus. (O)
> *Tornatellides procerulus kailuanus* Pilsbry & Cooke, 1915b: 207, pl. 45, fig. 9. Oahu: "Kaelepulu, Kailua, in ledges and around the base of a low coral bluff".

kamaloensis. (Mo)
>*Tornatellides kamaloensis* Pilsbry & Cooke, 1915b: 207, pl. 45, figs. 10, 11. Molokai: "northwestern ravine of Kamalo, above the old Kamalo ditch trail".

kilauea. (H)
>*Tornatellides kilauea* Pilsbry & Cooke, 1915b: 208, pl. 45, figs. 12, 13. Hawaii: "crest of Kilauea crater, about half a mile south of the Volcano House".

konaensis. (H)
>*Tornatellides konaensis* Cooke & Pilsbry *in* Pilsbry & Cooke, 1915b: 212, pl. 46, figs. 5, 6. Hawaii: "Kona".
>>Remarks. Pilsbry & Cooke (1915b: 212) gave 3 localities but their designated "type" (BPBM 36247) comes from Kona.

leptospira. (O)
>*Tornatellides leptospira* Cooke & Pilsbry *in* Pilsbry & Cooke, 1915b: 243, pl. 51, figs. 11–13. Oahu: "Nuuanu near the Pali".

macromphala. (K, O, Mo, M, L, Kah, H)
>*Tornatellina macromphala* Ancey, 1903: 296, pl. 12, figs. 3, 4. Maui: "Kaupakalua . . . Keanae"; Oahu: "près Honolulu [= near Honolulu]".
>>Remarks. Pilsbry & Cooke (1915b: 228) considered the "type lot" to be from "Kaupakalua, East Maui", but no lectotype designation was made.

macroptychia. (M)
>*Tornatellina macroptychia* Ancey, 1903: 305, pl. 12, figs. 21, 22. Maui: "Kaupakalua".

micromphala. (Mo)
>*Tornatellides micromphala* Pilsbry & Cooke, 1915b: 229, pl. 49, figs. 12, 13. Molokai: "Western ravine of Kamalo".

moomomiensis. (Mo)
>*Tornatellides moomomiensis* Pilsbry & Cooke, 1915b: 222, pl. 48, figs. 14, 15. Molokai: "Moomomi, at base of the bluff and up to about 600 ft.; also back of the dunes, about a quarter of a mile inland".

+nanus. (H)
>*Tornatellides forbesi nanus* Cooke & Pilsbry *in* Pilsbry & Cooke, 1915b: 248, pl. 52, figs. 5, 10. Hawaii: "Reed's Island".

oahuensis. (O)
>*Tornatellides oahuensis* Cooke & Pilsbry *in* Pilsbry & Cooke, 1915b: 222, pl. 48, figs. 8, 9. Oahu: "Kahuku, at a low elevation".

oncospira. (H)
>*Tornatellides oncospira* Cooke & Pilsbry *in* Pilsbry & Cooke, 1915b: 214, pl. 46, fig. 8. Hawaii: "Kaiwiki".

oswaldi. (O)
>*Tornatellides (T.) oswaldi* Cooke & Kondo, 1961: 257, figs. 109a–e. Oahu: "Moanalua Valley, on spur leading to Puu Kahu-auli".

perkinsi. (K)
>*Tornatellina perkinsi* Sykes, 1900: 382, pl. 11, fig. 14. Kauai: "Kaholuamano at 4000 feet".

pilsbryi. (O)
>*Tornatellides pilsbryi* Cooke, 1914: 79. Oahu: "Popouwela, in the Waianae Mts."

plagioptyx. (Mo)
>*Tornatellides plagioptyx* Pilsbry & Cooke, 1915b: 242, pl. 53, figs. 1–3. Molokai: "Western ravine of Kamalo, near the old ditch trail".

popouelensis. (O)
>*Tornatellides popouelensis* Pilsbry & Cooke, 1915b: 235, pl. 50, figs. 9–12. Oahu: "Western ridge of Popouwela, in the Waianae Mountains at the 'Endodonta locality'".

prionoptychia. (O)
    Tornatellides prionoptychia Cooke & Pilsbry *in* Pilsbry & Cooke, 1915b: 246, pl. 52, figs. 3, 4. Oahu: "Nuuanu, near the Pali".

procerulus. (N, K, O, Mo, M, L, H)
    Tornatellina procerula Ancey, 1903: 302, pl. 12, figs. 13, 14. Maui: "Kaupakalua".

productus. (K)
    Tornatellina macromphala var. *producta* Ancey, 1903: 297. Kauai: "Kipu".
        Remarks. Raised to species status by Pilsbry & Cooke (1915b: 226).

+puukolekolensis. (Mo)
    Tornatellides procerulus puukolekolensis Pilsbry & Cooke, 1915b: 207, pl. 45, figs. 7, 8. Molokai: "in the bottom of a small ravine east of Puu Kolekole".

pyramidatus. (H)
    Tornatellina pyramidata Ancey, 1903: 304, pl. 12, figs. 19, 20. Hawaii: "Olaa".

ronaldi. (O)
    Tornatellides ronaldi Cooke & Pilsbry *in* Pilsbry & Cooke, 1915b: 234, pl. 50, figs. 7, 8. Oahu: "Palehua in the Waianae Mts."

serrarius. (O)
    Tornatellides serrarius Pilsbry & Cooke, 1915b: 247, pl. 52, figs. 6, 7. Oahu: "Glen Ada".

spaldingi. (O)
    Tornatellides spaldingi Cooke & Pilsbry *in* Pilsbry & Cooke, 1915b: 230, pl. 53, figs. 6–8. Oahu: "Kaaawa".

stokesi. (O)
    Tornatellaria stokesi Pilsbry & Cooke, 1916: 259, pl. 54, fig. 8. Oahu: "Glen Ada, Nuuanu".

subangulatus. (M)
    Tornatellina subangulata Ancey, 1903: 303, pl. 12, figs. 15, 16. Maui: "Kaupakalua".

terebra. (Mo, M, L, H)
    Tornatellina terebra Ancey, 1903: 303, pl. 12, figs. 17, 18. Maui: "Kaupakalua".

virgula.
    Tornatellina virgula Cooke, 1907: 13. *Nom. nud.*

virgula. (M)
    Tornatellides virgula Cooke & Pilsbry *in* Pilsbry & Cooke, 1915b: 241, pl. 51, figs. 10, 14. Maui: "East Maui: Kaupakalua".

vitreus. (?Hawaiian)
    Tornatellina vitrea Dohrn, 1863: 162. Type locality not given.
        Remarks. Dohrn (1863: 162) implied that this species was from the Hawaiian Islands. Pilsbry & Cooke (1915b: 203–04) considered it a species of *Tornatellides*, but said that it "must be discarded" as it appears to be unidentifiable.

waianaensis. (O)
    Tornatellides waianaensis Pilsbry & Cooke, 1915b: 237, pl. 51, figs. 3, 4. Oahu: "western ridge of Popouwela, Waianae Mountains, in the '*Endodonta* locality'".

Subgenus WAIMEA Cooke & Pilsbry

WAIMEA Cooke & Pilsbry *in* Pilsbry & Cooke, 1915b: 250. Type species: *Tornatellina rudicostata* Ancey, 1904 (as "*Tornatellides*"), by monotypy.

rudicostatus. (H)
    Tornatellina rudicostata Ancey, 1904b: 70, pl. 5, figs. 20, 21. Hawaii: "Palihoukapapa, on the Hamakua slope of Mauna-Kea, Kawaii [= Hawaii], at an elevation of 4,000 feet" (in the introduction of the paper).

Subfamily TORNATELLININAE

Genus ELASMIAS Pilsbry

ELASMIAS Pilsbry, 1910: 122. Type species: *Tornatellina aperta* Pease, 1865 [not Hawaiian], by original designation.

Pilsbry & Cooke (1915a: 115–19) recognized 3 species plus 1 subspecies of *fuscum*. Caum (1928: 54–55) listed the 3 species only. Cooke & Kondo (1961) made no change. Cooke & Kondo (1961: 18) included *Elasmias* in the tribe Tornatellinini but elsewhere (p. 218) referred it to a separate tribe, Elasmiatini, which does not appear in their table on p. 17–18.

anceyanum. (M)
    Elasmias anceyanum Pilsbry & Cooke, 1915a: 118, pl. 31, figs. 7, 8. Maui: "West Maui: Maunahooma".

fuscum. (O, Mo, M, H)
    Tornatellina fusca Ancey, 1903: 306, pl. 12, figs. 23, 24. Hawaii: "Puna".

luakahaense. (K, O)
    Elasmias luakahaense Pilsbry & Cooke, 1915a: 117, pl. 29, figs. 7–10. Oahu: "Nuuanu, Luakaha".

+obtusum. (O)
    Elasmias fuscum obtusum Pilsbry & Cooke, 1915a: 117, pl. 28, fig. 5. Oahu: "western ridge of Popouwela".

Incertae Sedis in ACHATINELLIDAE

Pilsbry & Cooke (1914a: 366–69) dealt with a number of names that they seemed to think were probably Achatinellidae but that they could not place definitively within the family, listing them among their "Unrecognized or undescribed Achatinellidae, etc." One additional *nomen nudum* is included here.

clausinus. (?O, H)
    Bulimus clausinus Mighels, 1845: 20. Hawaii: [no additional details].
    Remarks. Pilsbry & Cooke (1914a: 366) considered it "possibly an Oahuan *Bulimella*" but "certainly undeterminable". Johnson (1949: 224) said "This species has never been recognized".

hawaiiensis.
    Tornatellina hawaiiensis Cooke, 1907: 13. *Nom. nud.*
    Remarks. A manuscript name of Ancey. The genus *Tornatellina* is currently considered restricted to Juan Fernandez taxa (Cooke & Kondo, 1961: 235). We are not sure of the correct placement of this name. There appears to be no evidence from the literature or from the Bishop Museum collections that it is the same as *Tornatellaria abbreviata* var. *hawaiiensis* Pilsbry & Cooke, 1916.

kanaiensis. (?not Hawaiian)
    Bulimus kanaiensis Pfeiffer, 1857a: 332. Kauai (as"Kanai"): [no additional details].
    Remarks. May not be Hawaiian (Sykes, 1900: 399).

kauaiensis.
    Bulimus kauaiensis Pfeiffer, 1859d: 469. Unjustified emendation of *kanaiensis* Pfeiffer, 1857.
    Remarks. Objective synonym of *kanaiensis* Pfeiffer.

leucozonalis.
    Helicteres leucozonalis Beck, 1837: 51. *Nom. nud.*
    Remarks. Locality given as "I. oc. pacif."

sandwichensis. (?not Hawaiian)
>*Spiraxis sandwichensis* Pfeiffer, 1857a: 335. Hawaiian Islands: [no additional details].
>>*Remarks.* Possibly the "*Perdicella sandwichensis* (Pfr.)" listed by Pease (1870b: 648), and the "*Sandwichensis*, Pfr." listed in *Partulina* by Baldwin (1893: 7). May not be Hawaiian (Sykes, 1900: 399). Caum (1928: 61) placed it in the "Stenogyridae", stating that it was not known to occur in the Hawaiian Islands.

sulphuratus.
>*Helicteres sulphuratus* Beck, 1837: 51. *Nom. nud.*
>>*Remarks.* Locality given as "I. oc. pacif."

Family AMASTRIDAE

The family Amastridae is endemic to the Hawaiian Islands. Its phylogenetic relationships are not clear, although the family may be close to the Holarctic Cochlicopidae (Tillier, 1989). Little is known of the basic biology of Amastridae, but in general they appear to have been ground-dwelling, in contrast to the sympatric but generally arboreal Achatinellinae. The family includes the genus *Carelia*, endemic to the islands of Kauai and Niihau, and the largest land snails in the Hawaiian fauna, some species reaching 80 mm or so in shell height. Most amastrids, including *Carelia*, are now considered extinct, although *Leptachatina lepida* Cooke, 1910 was recently found alive by 1 of us (RHC).

The *Manual of Conchology* (Hyatt & Pilsbry, 1910–1911; Pilsbry & Cooke, 1914–1916) recognized 2 subfamilies, Amastrinae and Leptachatininae, with genera allocated as follows (see also Cooke, 1932): Amastrinae: *Amastra, Carelia, Laminella, Planamastra, Pterodiscus* (= *Tropidoptera*); Leptachatininae: *Armsia, Leptachatina, Pauahia*. Vaught (1989) did not recognize subfamilies and included *Pauahia* as a subgenus of *Leptachatina* (contrary to Pilsbry & Cooke, 1914b: 15). In addition, the *Manual of Conchology* used a complex hierarchical arrangement of sections within subgenera of the genus *Amastra*. Vaught's (1989) treatment simplified this arrangement, raising the sections *Amastrella, Heteramastra, Metamastra* and *Paramastra* to subgenera of *Amastra*, and raising *Cyclamastra* and *Kauaia* to generic level, with the section *Armiella* becoming a subgenus of *Kauaia*. We follow a combination of these classifications, recognizing the 2 subfamilies Amastrinae and Leptachatininae with genera allocated to them as in the *Manual of Conchology*. Within the Amastrinae, we follow Zilch (1959a) and Vaught (1989) in treating *Amastrella, Heteramastra, Metamastra* and *Paramastra* as subgenera of *Amastra*. We retain *Cyclamastra* as a subgenus, following the *Manual of Conchology*. We treat *Kauaia* and *Armiella* as subgenera of *Amastra*. Within the Leptachatininae, we follow the *Manual of Conchology* for the treatment of genera and subgenera, with *Pauahia* treated as a genus.

The dates of publication of the papers by Newcomb (1854a,b) and Pfeiffer (1854b), both describing new species of Achatinellidae and Amastridae, are discussed above under the family Achatinellidae. We follow Clarke (1958) in attributing those names proposed by both Newcomb and Pfeiffer to Newcomb.

Our interpretation of the intended specific, subspecific or synonymic status of taxa as treated in the *Manual of Conchology* follows our interpretation of the treatment of the Achatinellidae (see comments under family Achatinellidae). We provide full annotation, as appropriate, in instances where our interpretation of the *Manual* differs from this.

If *semicostata* Pfeiffer, which may not be a Hawaiian taxon, is excluded, the Hawaiian fauna consists of 325 species and 111 infraspecific taxa, as follows: *Amastra* —149 species and 69 infraspecific taxa; *Carelia* — 21 species and 13 infraspecific taxa; *Laminella* — 14 and 14; *Planamastra* — 3 and 1; *Tropidoptera* (= *Pterodiscus*) — 5 and 2; *Armsia* — 1 species only; *Leptachatina* — 129 and 12; *Pauahia* — 3 species only. Solem (1990: 29) indicated 331 species and 213 infraspecific taxa. The reason for the major discrepancy between the latter figure and our figure of 111 is not clear; it may be due to typographical errors, although Solem may have incorporated an estimate of numbers of undescribed taxa in museum collections, as he did for the Endodontidae (cf. numbers of Endodontidae given by Solem (1976: 3; 1990: 29)).

Subfamily AMASTRINAE

Genus AMASTRA Adams & Adams

AMASTRA Adams & Adams, 1855c: 137 (as *Achatinella* subg.). Type species: *Achatinella magna* Adams, 1851 (as "*Amastra*"), by subsequent designation of Gulick (1873a: 91).

Subgenus AMASTRA Adams & Adams

AMASTRA Adams & Adams, 1855c: 137 (as *Achatinella* subg.). Type species: *Achatinella magna* Adams, 1851 (as "*Amastra*"), by subsequent designation of Gulick (1873a: 91).

abberans.
> *Amastra* (*Amastra*) *bigener* var. *abberans* Hyatt & Pilsbry, 1911c: 300, pl. 44, fig. 8. Type locality not given.
> > Remarks. Synonym of *affinis* Newcomb, *teste* Pilsbry & Cooke (1914b: 43).

affinis. (M)
> *Achatinella affinis* Newcomb, 1854a: 16, pl. 23, fig. 35 [1854b: 142, pl. 23, fig. 35]. Maui: "Kula, E. Mani [= Maui]".

albocincta. (Mo)
> *Amastra albocincta* Pilsbry & Cooke, 1914b: 40, pl. 3, figs. 11, 12. Molokai: "northwestern ravine of Kamalo, just above the ditch trail, on a steep, wooded slope".

assimilis. (M)
> *Achatinella assimilis* Newcomb, 1854a: 22, pl. 23, fig. 53 [1854b: 148, pl. 23, fig. 53]. Maui: "W. Mani [= Maui]".

+atroflava. (Mo)
> *Amastra* (*Amastra*) *mucronata atroflava* Hyatt & Pilsbry, 1911c: 272, pl. 40, figs. 13, 14, pl. 41, figs. 9–13. Molokai: "Kawela and Kamalo".

aurostoma. (L)
> *Amastra aurostoma* Baldwin, 1896: 31. Lanai: [no additional details].

baldwiniana. (M)
> *Amastra* (*Amastra*) *baldwiniana* Hyatt & Pilsbry, 1911c: 292, pl. 43, figs. 4, 5. Maui: "West Maui: Wailuku".
> > Remarks. Pilsbry & Cooke (1914b: 42) stated that the types were from "Lahaina" not Wailuku; verification of the type locality requires further study.

baldwinii.
> *Achatinella baldwinii* Newcomb, 1854a: 29, pl. 24, fig. 72 [1854b: 155, pl. 24, fig. 72]. Lanai (as "Ranai"): [no additional details].
> > Remarks. Synonym of *magna* Adams, *teste* Hyatt & Pilsbry (1911b: 237–39).

AMASTRIDAE [91]

+balteata. (L)
> *Amastra (Amastra) magna* var. *balteata* Hyatt & Pilsbry, 1911b: 240, pl. 26, figs. 7, 8. Lanai: [no additional details].

bigener.
> *Amastra (A.) affinis* var. *bigener* Hyatt *in* Hyatt & Pilsbry, 1911c: 300, pl. 44, fig. 7. Maui: [no additional details].
> > *Remarks.* Synonym of *affinis* Newcomb, *teste* Pilsbry & Cooke (1914b: 43).

biplicata. (L)
> *Achatinella biplicata* Newcomb, 1854a: 30, pl. 24, fig. 75 [1854b: 156, pl. 24, fig. 75]. Lanai (as "Ranai"): [no additional details].

borcherdingi. (Mo)
> *Amastra (A.) borcherdingi* Hyatt & Pilsbry, 1911c: 266, pl. 40, figs. 2–6, pl. 41, figs. 1–4. Molokai: [no additional details].

cinderella.
> *Amastra (A.) affinis* var. *cinderella* Hyatt *in* Hyatt & Pilsbry, 1911c: 300, pl. 45, figs. 1, 2. Maui: "Kula".
> > *Remarks.* Synonym of *affinis* Newcomb, *teste* Pilsbry & Cooke (1914: 43).

+citrea. (Mo)
> *Amastra citrea* Sykes, 1896: 129. Molokai: [no additional details].
> > *Remarks.* Subspecies of *mucronata* Newcomb, *teste* Hyatt & Pilsbry (1911c: 271).

conifera. (M)
> *Amastra conifera* Smith *in* Gulick & Smith, 1873: 85, pl. 10, fig. 11. Maui: "Kula, East Maui".

deshaysii.
> *Achatinella deshaysii* Morelet, 1857: 27. Hawaiian Islands: [no additional details].
> > *Remarks.* Synonym of *assimilis* Newcomb, *teste* Hyatt & Pilsbry (1911c: 306).

+dimissa. (Mo)
> *Amastra (A.) modesta dimissa* Hyatt & Pilsbry, 1911c: 276, pl. 40, figs. 11, 12. Molokai: [no additional details].

+dissimiliceps. (Mo)
> *Amastra (A.) nubifera dissimiliceps* Hyatt & Pilsbry, 1911c: 275, pl. 28, figs. 19, 20. Molokai: [no additional details].

durandi. (L)
> *Amastra durandi* Ancey, 1897a: 178. Oahu: "Waianae" [?error = "Lanai" (Hyatt & Pilsbry, 1911c: 245)].

elegantula. (Mo)
> *Amastra (A.) elegantula* Hyatt & Pilsbry, 1911c: 277, pl. 40, fig. 15. Molokai: [no additional details].

fusiformis.
> *Achatinella fusiformis* Pfeiffer, 1855c: 5, pl. 30, fig. 18. Hawaiian Islands: [no additional details].
> > *Remarks.* Synonym of *mucronata* Newcomb, *teste* Hyatt & Pilsbry (1911c: 268).

+georgii. (Mo)
> *Amastra nubilosa georgii* Pilsbry & Cooke, 1914b: 39, pl. 7, figs. 6–8. Molokai: "Pleistocene, in sand dunes at base of the bluff, Moomomi".

gigantea.
> *Achatinella gigantea* Newcomb, 1854a: 10, pl. 22, fig. 17 [1854b: 136, pl. 22, fig. 17]. Maui: "Hale a Ka la".
> > *Remarks.* Synonym of *magna* Adams, *teste* Hyatt & Pilsbry (1911b: 239).

globosa.
> *Achatinella globosa* Hyatt & Pilsbry, 1911c: 293. *Nom. nud.*
>> Remarks. Originally a Gulick manuscript name, listed in the synonymy of *nigra* Newcomb by Hyatt & Pilsbry. Junior primary homonym of *globosa* Pfeiffer, 1855 (in *Achatinella*).

goniops. (M)
> *Amastra goniops* Pilsbry & Cooke, 1914b: 41, pl. 4, figs. 1–5. Maui: "West Maui: upper Olowalu gulch . . . Mt. Lihau and Mt. Helu".

goniostoma.
> *Achatinella* (*Laminella*) *goniostoma* Pfeiffer, 1856b: 203 [not "1855, p. 103", as stated by Hyatt & Pilsbry (1911c: 297)]. Hawaiian Islands ("Sandwich Islands" in publication title): [no additional details].
>> Remarks. Synonym of *affinis* Newcomb, *teste* Hyatt & Pilsbry (1911c: 297).

grayana. (L)
> *Achatinella* (*Laminella*) *grayana* Pfeiffer, 1856b: 204. Hawaiian Islands ("Sandwich Islands" in publication title): [no additional details].

hitchcocki. (Mo)
> *Amastra* (*A.*) *hitchcocki* Cooke, 1917: 243, pl. C, fig. 7. Molokai: "second valley west of Puukapele".

humilis. (Mo)
> *Achatinella humilis* Newcomb, 1855c: 143. Molokai: "Kalai".

humilis.
> *Achatinella humilis* Pfeiffer, 1856c: 207. Molokai: [no additional details].
>> Remarks. Primary junior homonym of *humilis* Newcomb (see comments regarding Pfeiffer and Newcomb in the introduction to the family Achatinellidae). Synonym of *humilis* Newcomb. **N. syn.**

inopinata. (M)
> *Amastra* (*A.*) *inopinata* Cooke, 1933: 23, pl. 2, figs. 11, 12. Maui: "East Maui: Kula, near the division between the lands of Keokea and Kamaole . . . from an extremely rocky pasture extending between the upper and lower roads".

johnsoni. (M)
> *Amastra* (*A.*) *johnsoni* Hyatt & Pilsbry, 1911c: 304, pl. 45, fig. 16. Maui: "West Maui: Wailuku".

+kahakuloensis. (M)
> *Amastra baldwiniana kahakuloensis* Pilsbry & Cooke, 1914b: 43, pl. 8, figs. 5, 6. Maui: "Kahakuloa, W. Maui".

kalamaulensis. (Mo)
> *Amastra kalamaulensis* Pilsbry & Cooke, 1914b: 37, pl. 3, figs. 4–7. Molokai: "Kalamaula".

kaunakakaiensis. (Mo)
> *Amastra kaunakakaiensis* Pilsbry & Cooke, 1914b: 36, pl. 3, figs. 8–10. Molokai: "pipe-line trail in upper Kaunakakai district, above and a short distance below the spring".

+kaupakaluana. (M)
> *Amastra* (*A.*) *affinis kaupakaluana* Hyatt & Pilsbry, 1911c: 301, pl. 46, figs. 1–6. Maui: "East Maui: Kaupakalua".

lahainana. (M)
> *Amastra lahainana* Pilsbry & Cooke, 1914b: 43, pl. 6, figs. 1–10. Maui: "West Maui: Olowalu Gulch, district of Lahaina".

lineolata. (?M, ?H)
> *Achatinella lineolata* Newcomb, 1853: 29 [1854a: 14, pl. 23, fig. 29; 1854b: 140, pl. 23, fig. 29]. Maui: [no additional details].
>> Remarks. Listed as "*Sedis incertae*" within the genus *Amastra* by Hyatt & Pilsbry (1911b: 141); subsequently treated as such by Hyatt & Pilsbry (1911c: 320); but in-

cluded among *Amastra s. str.* by Pilsbry & Cooke (1914b: 44). According to Pilsbry & Cooke (1914b: 44) the shell figured by Newcomb (1854a: pl. 23, fig. 29; 1854b: pl. 23, fig. 29) does not match the description (1853: 29, 1854a: 14; 1854b: 140), but is referable (Pilsbry & Cooke, 1914b: 45) to their new species *Amastra neglecta*. Both Newcomb (1854a: 14; 1854b: 140) and Pfeiffer (1854b: 128) gave "Hawaii" as the locality.

+**longa**. (L)
Amastra longa Sykes, 1896: 128. Lanai: "windward side of Lanai".
Remarks. Variety of *moesta* Newcomb, *teste* Hyatt & Pilsbry (1911c: 247).

+**macerata**. (Mo)
Amastra (A.) nubilosa macerata Hyatt & Pilsbry, 1911c: 260, pl. 27, figs. 5–7, 11, 12. Molokai: [no additional details].

magna. (?Mo, ?M, L)
Achatinella magna Adams, 1851a: 125 [1851b: 41]. Type locality not given.

makawaoensis. (M)
Amastra (A.) makawaoensis Hyatt & Pilsbry, 1911c: 294, pl. 43, figs. 7–9. Maui: "East Maui: Makawao".

malleata. (M)
Amastra malleata Smith *in* Gulick & Smith, 1873: 85, pl. 10, fig. 18. Maui: "Kula, on East Maui".

mastersi. (M)
Achatinella mastersi Newcomb, 1854a: 27, pl. 24, fig. 67 [1854b: 153, pl. 24, fig. 67]. Maui (as "Mani"): [no additional details].

maura.
Amastra simularis var. *maura* Ancey & Sykes *in* Ancey, 1899: 270, pl. 13, fig. 16 [plate by Sykes (1899)]. Molokai: [no additional details].
Remarks. For explanation of authorship, see Remarks under *semicarnea* Ancey & Sykes. Synonym of *simularis* Hartman, *teste* Hyatt & Pilsbry (1911c: 270); *simularis* subsequently synonymized with *mucronata* Newcomb by Pilsbry & Cooke (1914b: 35). Synonym of *mucronata* Newcomb. **N. syn.**

mirabilis. (M)
Amastra (A.) mirabilis Cooke, 1917: 245, pl. B, fig. 9. Maui: "East Maui: Just below Kaupo gap along the trail, about 5,000 feet elevation".

modesta. (?Mo)
Achatinella modesta Adams, 1851a: 128 [1851b: 44]. Hawaiian Islands: [no additional details].

moesta. (L)
Achatinella moesta Newcomb, 1854a: 31, pl. 24, fig. 77 [1854b: 157, pl. 24, fig. 77]. Lanai (as "Ranai"): [no additional details].

montana. (M)
Amastra montana Baldwin, 1906b: 136. Maui: "Mt. Kukui, summit of West Maui, 6,000ft. alt."

+**moomomiensis**. (Mo)
Amastra humilis moomomiensis Pilsbry & Cooke, 1914b: 38, pl. 7, figs. 9–14. Molokai: "Pleistocene of the sand dunes of Moomomi, around base of the bluff".

mucronata. (Mo, ?M)
Achatinella mucronata Newcomb, 1853: 28 [1854a: 20, pl. 23, fig. 49; 1854b: 146, pl. 23, fig. 49]. Molokai: [no additional details].
Remarks. Newcomb (1854a: 20; 1854b: 146) gave "Mani [= Maui]" as the locality. Pilsbry & Cooke (1914b: 35) stated that "probably Mapulehu [Molokai] is the type locality".

nana.
Achatinella (Amastra) nana Baldwin, 1893: 9. *Nom. nud.*

nana. (M)
>   *Achatinella* (*Amastra*) *nana* Baldwin, 1895: 232, pl. 11, figs. 48, 49. Maui: "Makawao".

neglecta. (?M)
>   *Amastra neglecta* Pilsbry & Cooke, 1914b: 45, pl. 7, fig. 20. Type locality not given [Pilsbry & Cooke (1914b: 45) say "Maui?"].
>   *Remarks.* See *lineolata* Newcomb.

nigra. (M)
>   *Achatinella nigra* Newcomb, 1855b: 219. Maui: "E. Maui".

nigra.
>   *Achatinella nigra* Pfeiffer, 1856c: 209. Maui (as "Mani"): [no additional details].
>   *Remarks.* Primary junior homonym of *nigra* Newcomb (see comments regarding Pfeiffer and Newcomb in the introduction to the family Achatinellidae). Synonym of *nigra* Newcomb. **N. syn.**

nubifera. (Mo)
>   *Amastra* (*A.*) *nubifera* Hyatt & Pilsbry, 1911c: 274, pl. 41, figs. 6, 7. Molokai: [no additional details].

nubilosa. (Mo)
>   *Achatinella nubilosa* Mighels, 1845: 20. Oahu [?error = Molokai (Hyatt & Pilsbry, 1911c: 260)]: [no additional details].

nucula. (?L)
>   *Amastra nucula* Smith *in* Gulick & Smith, 1873: 85, pl. 10, fig. 19. Type locality not given ["probably on the island of Lanai" (Smith *in* Gulick & Smith, 1873: 85)].

+obscura. (L)
>   *Achatinella obscura* Newcomb, 1854a: 31, pl. 24, fig. 78 [1854b: 157, pl. 24, fig. 78]. Lanai (as "Ranai"): [no additional details].
>   *Remarks.* "Variety" of *moesta* Newcomb, *teste* Hyatt & Pilsbry (1911c: 247).

+puella. (Mo)
>   *Amastra subobscura puella* Pilsbry & Cooke, 1914b: 36. Molokai: "Ualapue".

pulla.
>   *Achatinella pulla* Pfeiffer, 1856c: 209. Lanai (as "Ranaï"): [no additional details].
>   *Remarks.* Synonym of *pusilla* Newcomb, *teste* Hyatt & Pilsbry (1911b: 235).

pullata.
>   *Achatinella* (*Amastra*) *pullata* Baldwin, 1893: 9. Nom. nud.

pullata. (Mo)
>   *Achatinella* (*Amastra*) *pullata* Baldwin, 1895: 228, pl. 11, figs. 31, 32. Molokai: "Waikolu".

pumila.
>   *Achatinella pumila* Pfeiffer & Clessin, 1879: 313. Nom. nud.
>   *Remarks.* Listed in the synonymy of *modesta* Adams by Pfeiffer & Clessin (1879: 313). Originally a Gulick manuscript name.

pupoidea.
>   *Achatinella pupoidea* Newcomb, 1854a: 18, pl. 23, fig. 42 [1854b: 144, pl. 23, fig. 42]. Maui: "E. Mani [= Maui]".
>   *Remarks.* Synonym of *affinis* Newcomb, *teste* Pilsbry & Cooke (1914b: 43).

pusilla. (L)
>   *Achatinella pusilla* Newcomb, 1855c: 144. Lanai (as "Ranai"): [no additional details].

+roseotincta. (Mo)
>   *Amastra simularis* var. *roseotincta* Sykes, 1896: 130. Molokai: "Molokai Mountains".
>   *Remarks.* Subspecies of *mucronata* Newcomb, *teste* Hyatt & Pilsbry (1911c: 270).

rubristoma. (L)
Amastra rubristoma Baldwin, 1906b: 137. Lanai: [no additional details].

rustica.
Amastra rustica Gulick *in* Gulick & Smith, 1873: 84, pl. 10, fig. 17. Maui: "Kula, on East Maui".
Remarks. Synonym of *affinis* Newcomb, *teste* Hyatt & Pilsbry (1911c: 297–98).

+semicarnea. (Mo)
Amastra simularis var. *semicarnea* Ancey & Sykes *in* Ancey, 1899: 270, pl. 13, fig. 8. Molokai: [no additional details].
Remarks. Subspecies of *mucronata* Newcomb, *teste* Hyatt & Pilsbry (1911c: 272). Because Ancey (1899) and Sykes (1899) were published simultaneously, various combinations of authorship can be attributed to this taxon. Ancey provided the description but in addition cited pl. 13. Sykes alone was responsible for pl. 13 (see Sykes, 1899: 275). As first revisers, we select "Ancey & Sykes" as authors.

seminuda.
Achatinella (Amastra) seminuda Baldwin, 1893: 10. *Nom. nud.*

seminuda. (Mo)
Amastra seminuda Baldwin, 1906b: 137. Molokai: "Waikolu".

+sepulta. (Mo)
Amastra humilis sepulta Pilsbry & Cooke, 1914b: 39, pl. 7, figs. 15–17. Molokai: "Moomomi, with the preceding [i.e., *Amastra humilis moomomiensis*]".

simularis.
Amastra simularis Hartman, 1888c: 252, pl. 13, fig. 7. Molokai: [no additional details].
Remarks. Synonym of *mucronata* Newcomb, *teste* Pilsbry & Cooke (1914b: 35). See also Hyatt & Pilsbry (1911c: 269), who, although considering it "specifically identical with *mucronata*" and its "retention . . . as a varietal term . . . of doubtful utility", did not include it in the synonymy of *mucronata*; we take this to mean that Hyatt & Pilsbry retained it, if doubtfully, as a valid varietal name. Pilsbry & Cooke (1914b: 35) considered Mapulehu as the type locality.

+subassimilis. (M)
Amastra (A.) assimilis subassimilis Hyatt & Pilsbry, 1911c: 307, pl. 46, figs. 7, 8. Maui: "Waialua".
Remarks. "Probably a local race" of *assimilis* Newcomb, *teste* Hyatt & Pilsbry (1911c: 307).

subcrassilabris. (M)
Amastra (A.) subcrassilabris Hyatt & Pilsbry, 1911c: 293, pl. 45, figs. 14, 15. Maui: "East Maui: Kula".

+subnigra. (Mo)
Amastra (A.) pullata var. *subnigra* Hyatt & Pilsbry, 1911c: 263, pl. 27, figs. 19–21. Molokai: [no additional details].

subobscura. (Mo)
Amastra (A.) subobscura Hyatt & Pilsbry, 1911c: 276, pl. 42, fig. 6. Molokai: "Ulapue".

subpulla.
Amastra (A.) affinis subpulla Hyatt & Pilsbry, 1911c: 300, pl. 45, figs. 3, 4. Maui: "Kula, East Maui".
Remarks. Synonym of *affinis* Newcomb, *teste* Pilsbry & Cooke (1914b: 43).

sykesi. (Mo)
Amastra (A.) sykesi Hyatt & Pilsbry, 1911c: 273, pl. 42, figs. 2, 3. Molokai: "Halawa".

tricincta. (Mo)
Amastra (A.) tricincta Hyatt & Pilsbry, 1911c: 277, pl. 39, fig. 15. Molokai: "Kolekole".

umbrosa.
: *Achatinella (Amastra) umbrosa* Baldwin, 1893: 10. *Nom. nud.*

+**umbrosa**. (Mo)
: *Achatinella (Amastra) umbrosa* Baldwin, 1895: 229, pl. 11, figs. 36, 37. Molokai: "Kamalo".
: *Remarks.* Subspecies of *pullata* Baldwin, *teste* Hyatt & Pilsbry (1911c: 263). "Variety" of *pullata* Baldwin, *teste* Pilsbry & Cooke (1914b: 34).

uniplicata. (Mo)
: *Achatinella (Amastra) uniplicata* Hartman, 1888b: 50, pl. 1, fig. 7. Molokai: [no additional details].

+**vetuscula**. (Mo)
: *Amastra (A.) uniplicata* var. *vetuscula* Cooke, 1917: 244, pl. A, fig. 5. Molokai: "shifting sands north of Mauna Loa and directly south of Laina where the pipe line crosses the shifting sands".

violacea. (Mo)
: *Achatinella violacea* Newcomb, 1853: 18 [1854a: 9, pl. 22. fig. 14; 1854b: 135, pl. 22, fig. 14]. Molokai: [no additional details].

+**wailauensis**. (Mo)
: *Amastra (A.) violacea* var. *wailauensis* Hyatt & Pilsbry, 1911c: 258, pl. 27, figs. 1, 2, pl. 39, figs. 13, 14. Molokai: "Wailau".

Subgenus AMASTRELLA Sykes

AMASTRELLA Sykes, 1900: 352. Type species: *Amastra rugulosa* Pease, 1870, by original designation.

abavus. (Mo)
: *Amastra (Amastrella) abavus* Hyatt & Pilsbry, 1911c: 255, pl. 42, figs. 4, 5. Molokai: "Pukoa".
: *Remarks.* Described in section *Amastrella* of subgenus *Amastra.*

+**annosa**. (K)
: *Amastra (Amastrella) rugulosa* var. *annosa* Cooke, 1917: 230, pl. A, fig. 9. Kauai: "Pleistocene deposits of Hanamaulu plains south of Wailua river".

anthonii. (K)
: *Achatinella anthonii* Newcomb, 1861: 93. Kauai: [no additional details].

brevis.
: *Achatinella brevis* Pfeiffer, 1846a: 90. Hawaiian Islands: [no additional details].
: *Remarks.* Synonym of *nucleola* Gould, *teste* Hyatt & Pilsbry (1911b: 154).

+**castanea**. (O)
: *Amastra (Amastrella) rubens castanea* Hyatt & Pilsbry, 1911b: 194, pl. 32, fig. 18. Oahu: "Waianae".

conica. (H)
: *Amastra conica* Baldwin, 1906b: 137. Hawaii: "Hamakua".

+**corneiformis**. (O)
: *Amastra (Amastrella) rubens corneiformis* Hyatt & Pilsbry, 1911b: 194, pl. 31, figs. 17, 18. Oahu: "Waianae".

decorticata. (O)
: *Amastra decorticata* Gulick *in* Gulick & Smith, 1873: 84, pl. 10, fig. 14. Oahu: "Kawailoa . . . Helemano, Opaiula, Waimea, Waialei, and Kahuku".
: *Remarks.* Hyatt & Pilsbry (1911b: 200) considered Kawailoa as the type locality, but no lectotype designation was made.

AMASTRIDAE

elliptica. (O)
> *Amastra elliptica* Gulick *in* Gulick & Smith, 1873: 83, pl. 10, fig. 15. Oahu: "Waialei . . . Kahuku and Hanula [= Hauula], and . . . Kawailoa".
>> *Remarks*. Hyatt & Pilsbry (1911b: 204) considered Waialei the type locality, but no lectotype was designated.

+emortua. (H)
> *Amastra (Amastrella) flavescens* var. *emortua* Cooke, 1917: 231, pl. A, fig. 6. Hawaii: "Huehue in the district of North Kona, on the northwestern slopes of Hualalai about 1,700 feet elevation".

+fastigata. (K)
> *Amastra (Amastrella) rugulosa* var. *fastigata* Cooke, 1917: 229, pl. B, fig. 1. Kauai: "Koloa".
>> *Remarks*. Cooke gave more than 1 locality but his designated holotype (BPBM 41997) came from "Koloa". (The lot contains 3 specimens but 1 of these has "holotype" written on the shell).

flavescens. (H)
> *Achatinella flavescens* Newcomb, 1854a: 25, pl. 24, fig. 62 [1854b: 151, pl. 24, fig. 62]. Hawaii: [no additional details].

fossilis. (H)
> *Amastra fossilis* Baldwin, 1903: 35. Hawaii: "Palihoukapapa on the Hamakua slope of Maunakea, at an elevation of 4,000 feet" [in text describing *Amastra senilis* (see Baldwin, 1903: 35)].

fragosa. (H)
> *Amastra (Amastrella) fragosa* Cooke, 1917: 236, pl. A, fig. 4. Hawaii: "Kapulehu, about three miles north of Huehue and nearly the same elevation".

fuliginosa.
> *Achatinella fuliginosa* Gould, 1845: 28. Hawaiian Islands ("Sandwich Islands" in publication title): [no additional details].
>> *Remarks*. Synonym of *tristis* Férussac, *teste* Hyatt & Pilsbry (1911b: 206–07).

+gentilis. (H)
> *Amastra (Amastrella) conica* var. *gentilis* Cooke, 1917: 233, pl. A, fig. 1. Hawaii: "Waikii station, land of Waikoloa about 6,000 feet elevation".

+gyrans. (H)
> *Amastra (Amastrella) conica* var. *gyrans* Hyatt *in* Hyatt & Pilsbry, 1911c: 314, pl. 47, figs. 3, 4. Hawaii: [no additional details].

hawaiiensis. (H)
> *Amastra (Amastrella) hawaiiensis* Hyatt & Pilsbry, 1911c: 319, pl. 42, figs. 7, 8. Hawaii: "Waimanu, in the northeastern part of the island".

+henshawi. (H)
> *Amastra henshawi* Baldwin, 1903: 34. Hawaii: "South Kona".
>> *Remarks*. Subspecies of *flavescens* Newcomb, *teste* Hyatt & Pilsbry (1911c: 318).

+infelix. (O)
> *Amastra (Amastrella) rubens infelix* Hyatt & Pilsbry, 1911b: 195, pl. 31, fig. 15. Oahu: "Kahana".

inflata. (O)
> *Achatinella (Laminella) inflata* Pfeiffer, 1856b: 203. Hawaiian Islands ("Sandwich Islands" in publication title): [no additional details].
>> *Remarks*. Hyatt & Pilsbry (1911b: 202) considered Wahiawa the type locality.

+janeae. (K)
> *Amastra (Amastrella) rugulosa janeae* Cooke, 1933: 12, pl. 1, fig. 6. Kauai: "Ohia Valley, Anahola watershed west of the Kaneha Reservoir, along ditch trail near bottom of the valley".

+**kahana.** (O)
    Amastra (Amastrella) rubens kahana Hyatt & Pilsbry, 1911b: 195, pl. 31, fig. 16. Oahu: "Kahana, on the north side of the eastern range".

+**kauensis.** (H)
    Amastra melanosis kauensis Pilsbry & Cooke, 1915a: 50, pl. 1, fig. 18. Hawaii: "Waiohinu, Kau, near the southern end of Hawaii".

+**kohalensis.** (H)
    Amastra (Amastrella) conica var. *kohalensis* Hyatt & Pilsbry, 1911c: 314, pl. 49, fig. 13. Hawaii: "Kohala Mts."

luctuosa. (O)
    Achatinella (Laminella) luctuosa Pfeiffer 1856b: 204. Hawaiian Islands ("Sandwich Islands" in publication title): [no additional details].

+**meineckei.** (K)
    Amastra (Amastrella) anthonii meineckei Cooke, 1933: 14, pl. 1, figs. 7, 11–15. Kauai: "Kalalau, on the northern side of the valley, east of double waterfall".

melanosis. (H)
    Achatinella melanosis Newcomb, 1854a: 18, pl. 23, fig. 41 [1854b: 144, pl. 23, fig. 41]. Hawaii: [no additional details].
        Remarks. According to Pilsbry & Cooke (1915a: 49), Newcomb received his shells from Baldwin, labeled "Hamakua", but Newcomb's label says "Mauna Loa". Verification of the exact type locality requires further research.

+**normalis.** (K)
    Amastra (Amastra) rugulosa normalis Hyatt & Pilsbry, 1911b: 153, pl. 15, figs. 8–10, pl. 42, fig. 9. Kauai: [no additional details].

nucleola. (K, ?O)
    Achatinella nucleola Gould, 1845: 28. Hawaiian Islands ("Sandwich Islands" in publication title): [no additional details].

ovatula. (K)
    Amastra (Amastrella) ovatula Cooke, 1933: 17, pl. 2, figs. 5–7. Kauai: "near the western edge of the Haena plain".

pagodula. (H)
    Amastra (Amastrella) pagodula Cooke, 1917: 237, pl. B, fig. 4. Hawaii: "Huehue about 1,800 feet elevation".
        Remarks. In the original publication, Cooke gave 2 localities but did designate a "holotype" and "cotypes" (all BPBM 41974) and "paratypes" (BPBM 41975); the labels associated with all of these specimens say "Huehue".

petricola. (Mo)
    Achatinella petricola Newcomb, 1855c: 143. Molokai: [no additional details].

petricola.
    Achatinella petricola Pfeiffer, 1856c: 208. Molokai (as "Modonai"): [no additional details].
        Remarks. Primary junior homonym of *petricola* Newcomb (see comments regarding Pfeiffer and Newcomb in the introduction to the family Achatinellidae). Synonym of *petricola* Newcomb. **N. syn**.

porcus. (O)
    Amastra (Amastrella) porcus Hyatt & Pilsbry, 1911b: 207, pl. 38, fig. 3. Oahu: "Mokuleia, Waianae mountains".

+**remota.** (K)
    Amastra (Amastrella) remota Cooke, 1917: 228, pl. A, fig. 3. Kauai: "Pleistocene deposits of the southwestern bluff of Kalalau valley about 100 feet elevation".
        Remarks. Subspecies of *anthonii* Newcomb, *teste* Cooke (1933: 14).

rubens. (O)
    Achatinella rubens Gould, 1845: 27. Hawaiian Islands ("Sandwich Islands" in publication title): [no additional details].
        Remarks. Hyatt & Pilsbry (1911b: 193) considered Waianae (Oahu) the type locality.

rubicunda.
    Achatinella (Amastra) rubicunda Baldwin, 1893: 9. *Nom. nud.*

+rubicunda. (O)
    Achatinella (Amastra) rubicunda Baldwin, 1895: 229, pl. 11, fig. 38. Oahu: "Konahuanui Mt."
        Remarks. Subspecies of *tenuilabris* Gulick, *teste* Hyatt & Pilsbry (1911b: 197).

rubida. (O)
    Amastra rubida Gulick *in* Gulick & Smith, 1873: 84, pl. 10, fig. 12. Oahu: "Kahuku".

+rubinia. (O)
    Amastra (Amastrella) rubens var. *rubinia* Hyatt & Pilsbry, 1911b: 193, pl. 32, fig. 16. Oahu: "Kukuiala".

rugulosa. (K)
    Amastra rugulosa Pease, 1870a: 95. Kauai (in publication title): [no additional details].
        Remarks. Also listed by Pease (1870b: 649) with no description.

+saxicola. (H)
    Amastra saxicola Baldwin, 1903: 34. Hawaii: "Kau".
        Remarks. "Var." of *flavescens* Newcomb, *teste* Hyatt & Pilsbry (1911c: 317).

seminigra. (O)
    Amastra (Amastrella) seminigra Hyatt & Pilsbry, 1911b: 195, pl. 32, figs. 8, 9, 11. Oahu: "Waimano".

senilis. (H)
    Amastra senilis Baldwin, 1903: 35. Hawaii: "Palihoukapapa on the Hamakua slope of Maunakea, at an elevation of 4,000 feet".

spicula. (H)
    Amastra (Amastrella) spicula Cooke, 1917: 234, pl. A, fig. 2. Hawaii: "Waikii station in the land of Waikaloa about 6,000 feet elevation".

+subglobosa. (K)
    Amastra (Amastrella) anthonii subglobosa Cooke, 1933: 16, pl. 1, fig. 16. Kauai: "Waimea".

+sulphurea. (O)
    Amastra luctuosa var. *sulphurea* Ancey, 1904a: 121, pl. 7, fig. 9. Oahu: "Waialae".

tenuilabris. (O)
    Amastra tenuilabris Gulick *in* Gulick & Smith, 1873: 83, pl. 10, fig. 16. Oahu: "Pauoa" (see Remarks below).
        Remarks. Gulick stated that "the specimens are . . . reported by Dr. Frick to be from Pauoa, on Oahu; but there is some reason to believe that they could not have been found on Oahu". Gulick's rationale for this statement apparently was because he considered it was allied to a species from the Island of Hawaii. Hyatt & Pilsbry (1911b: 196) considered Pauoa the type locality, as Gulick's type material corresponded closely to shells from Nuuanu valley, immediately west of Pauoa.

tristis.
    Helix (Cochlogena) tristis Férussac, 1821c: 60. *Nom. nud.*

tristis. (O)
    Helix tristis Férussac *in* Quoy & Gaimard, 1825d: 482, pl. 68, figs. 6, 7. Hawaiian Islands: [no additional details].

viriosa. (H)
> *Amastra (Amastrella) viriosa* Cooke, 1917: 235, pl. C, fig. 6. Hawaii: "Kahauloa and Kealakekua elevation 4,360 feet".
>> *Remarks.* Cooke (1917: 236) gave more than 1 locality but did designate a "holotype" (BPBM 41963), "cotypes" and "paratypes" (all BPBM 39661). The lot (BPBM 39661) containing the "cotypes" (each cotype marked with a cross on the shell) and from which the "holotype" was taken, came from "Kahauloa and Kealakekua el. 4360 ft.", according to the catalog ledger entry. The catalog entry for an additional lot (BPBM 194560), not mentioned by Cooke (1917) but labelled as "paratypes", says "Kahauloa = 39661. el. 4360 ft." The labels of the lots containing the "cotypes" (39661) and "paratypes" (194560) say "Kahauloa" only. The label and catalog entry for the "holotype" (41963) simply say "Hawaii".

whitei. (H)
> *Amastra (Amastrella) whitei* Cooke, 1917: 232, pl. C, fig. 4. Hawaii: "Kahauloa, elevation 3,250 feet".
>> *Remarks.* Cooke's designated holotype measurements (Cooke, 1917: 233) do not match the measurements of the specimen so marked as holotype (BPBM 41964). This leads to confusion as to the identity of the actual type specimen. The syntypes (BPBM 41964, 39642) all come from the locality indicated above.

Subgenus ARMIELLA Hyatt & Pilsbry

ARMIELLA Hyatt & Pilsbry, 1911b: 145 (as *Amastra* (*Kauaia*) sect.). Type species: *Achatinella* (*Amastra*) *knudsenii* Baldwin, 1895, by monotypy.

Treated as a section of subgenus *Kauaia* in the *Manual of Conchology* but as a subgenus of genus *Kauaia* by Vaught (1989). We here treat both *Kauaia* and *Armiella* as subgenera of *Amastra*.

+armillata. (K)
> *Amastra (Kauaia) (Armiella) ricei* var. *armillata* Cooke, 1917: 223, pl. A, fig. 8. Kauai: "Milolii".

knudsenii. (K)
> *Achatinella (Amastra) knudsenii* Baldwin, 1895: 234, pl. 11, figs. 43, 44. Kauai: "Halemanu".

ricei. (K)
> *Amastra (Kauaia) (Armiella) ricei* Cooke, 1917: 221, pl. C, fig. 1. Kauai: "In the Kauaiula branch of Milolii valley".

Subgenus CYCLAMASTRA Pilsbry & Vanatta

CYCLAMASTRA Pilsbry & Vanatta, 1905: 570 (as *Amastra* sect.). Type species: *Achatinella cyclostoma* Baldwin, 1895, by original designation.

agglutinans. (M)
> *Achatinella obesa* var. *agglutinans* Newcomb, 1854a: 17 [1854b: 143]. Type locality not given.
>> *Remarks.* Treated as a species by Hyatt & Pilsbry (1911c: 283).

antiqua.
> *Achatinella (Amastra) antiqua* Baldwin, 1893: 8. Nom. nud.

antiqua. (O)
> *Achatinella (Amastra) antiqua* Baldwin, 1895: 233, pl. 11, fig. 47. Oahu: "Ewa".

+arenarum. (Mo)
> *Amastra (Cyclamastra) umbilicata arenarum* Pilsbry & Cooke, 1914b: 23, pl. 2, figs. 1–4. Molokai: "Pleistocene sand dunes of Moomoni".

+aurora. (M)
> *Amastra (Cyclamastra) obesa aurora* Pilsbry & Cooke, 1914b: 18, pl. 4, figs. 9–12. Maui: "East Maui: Auwahi, at about 4200 ft. elevation.

AMASTRIDAE

+bembicodes. (O)
>   *Amastra (Cyclamastra) thurstoni bembicodes* Cooke, 1933: 8, pl. 1, fig. 4. Oahu: "Manoa, fossil in road cuttings near the corner of Ferdinand Street and Adolphe Street".

carinata.
>   *Amastra carinata* Gulick *in* Gulick & Smith, 1873: 83. Maui: "Wailuku, West Maui".
>   >   Remarks. Synonym of *agglutinans* Newcomb, *teste* Gulick & Smith (1873: 83) and Hyatt & Pilsbry (1911c: 283).

cyclostoma.
>   *Achatinella (Amastra) cyclostoma* Baldwin, 1893: 8. Nom. nud.

cyclostoma. (K)
>   *Achatinella (Amastra) cyclostoma* Baldwin, 1895: 234, pl. 11, fig. 53. Kauai: "Makaweli".

+debilis. (M)
>   *Amastra (Cyclamastra) metamorpha debilis* Pilsbry & Cooke, 1914b: 20, pl. 5, figs. 6, 7. Maui: "Olowalu Gulch, at a greater elevation than *A. metamorpha*".

delicata. (K)
>   *Amastra (Cyclamastra) delicata* Cooke, 1933: 7, pl. 1, fig. 2. Kauai: "Nonou Mountains in high valley just west of peak, 700–900 feet elevation".

elephantina. (O)
>   *Amastra (Cyclamastra) elephantina* Cooke, 1917: 226, pl. B, fig. 3. Oahu: "Waimano gulch".

extincta. (?K, ?O, ?H)
>   *Achatinella (Laminella) extincta* Pfeiffer, 1856b: 204. Oahu: [no additional details].
>   >   Remarks. Pease (1870a: 87) included *extincta* Pfeiffer in his list of species of Kauai. Pilsbry & Cooke (1914b: 30) suggested Kona [Hawaii] as a possible locality.

fragilis. (Mo)
>   *Amastra (Cyclamastra) fragilis* Pilsbry & Cooke, 1914b: 24, pl. 2, figs. 11, 12. Molokai: "Pipe-line trail in upper Kaunakakai, above and below the spring".

globosa. (O)
>   *Amastra (Cyclamastra) globosa* Cooke, 1933: 7, pl. 1, fig. 3. Oahu: "Kahuku, in limestone bank above the sand beach, slightly northeast of wireless station".

gouveii. (O)
>   *Amastra (Cyclamastra) gouveii* Cooke, 1917: 225, pl. C, fig. 3. Oahu: "Wailupe, east side a little more than half way up the ridge".

+gregoryi. (K)
>   *Amastra (Cyclamastra) cyclostoma gregoryi* Cooke, 1933: 5, pl. 1, fig. 1. Kauai: "limestone bluff about 0.5 mile east of Aweoweonui Bay, in Mahaulepu, 50 to 75 feet above sea level".

hartmani.
>   *Leptachatina hartmani* Newcomb *in* Hartman, 1888b: 54, pl. 1, fig. 12 [unnecessary n.n. for *extincta* Pfeiffer]. Oahu [?error = Kauai (Pilsbry & Cooke, 1914b: 21)]: [no additional details].
>   >   Remarks. Hartman (1888b: 54) explicitly stated that Newcomb provided the change of name. The type specimen of *hartmani* is automatically that of *extincta*. Hartman's designation of the specimen in pl. 1, fig. 12 as the "type" and Hyatt & Pilsbry's (1911b: 160) following of this type specimen and treatment of *hartmani* as separate from *extincta* are thus incorrect. A new name is needed for the specimens treated as *hartmani* by Hyatt & Pilsbry. Also p. 56 as *hartmanii*. We select *hartmani* as the correct original spelling.

hartmanii.
: *Leptachatina hartmanii* Hartman, 1888b: 56. Incorrect original spelling of *hartmani* Newcomb.

juddii. (K)
: *Amastra (Cyclamastra) juddii* Cooke, 1917: 223, pl. B, fig. 5. Kauai: [no additional details].

+kawaihapaiensis. (O)
: *Amastra (Cyclamastra) antiqua kawaihapaiensis* Pilsbry & Cooke, 1914b: 22, pl. 2, fig. 13. Oahu: "Kawaihapai; at the western end of the Waianae Mountains, in soil of a field, perhaps 20–30 ft. above sea level".

metamorpha. (M)
: *Amastra (Cyclamastra) metamorpha* Pilsbry & Cooke, 1914b: 19, pl. 5, figs. 1–5. Maui: "West Maui: Olowalu Gulch".

modicella. (H)
: *Amastra (Cyclamastra) modicella* Cooke, 1917: 227, pl. A, fig. 7. Hawaii: "Waikii station, in the land of Waikoloa, about 6,000 feet elevation".

morticina. (M)
: *Amastra (Cyclamastra) morticina* Hyatt & Pilsbry, 1911c: 280, pl. 36, figs. 8, 9. Maui: "West Maui: Kanaio".
  : Remarks. Pilsbry & Cooke (1914b: 25) were "inclined to reduce *morticina* to a subspecies of *A. umbilicata*", but did not explicitly do so.

obesa. (M)
: *Achatinella obesa* Newcomb, 1853: 24 [1854a: 17, pl. 23, figs. 39, 39a; 1854b: 143, pl. 23, figs. 39, 39a]. Maui: "Hale-a-ka-la".

+pluscula. (H)
: *Amastra (Cyclamastra) umbilicata* var. *pluscula* Cooke, 1917: 228, pl. C, fig. 2. Hawaii: "Kapulehu, in the district of North Kona, about 18,000 feet elevation" [elevation incorrect; no point on the island of Hawaii is that high].

problematica. (O)
: *Amastra (Cyclamastra) problematica* Cooke, 1933: 9, pl. 1, fig. 5. Oahu: "Kawailoa, 0.3 mile northeast of railway station in shallow railroad cutting".
  : Remarks. Cooke gave more than 1 locality but his designated "type" (BPBM 10772) is listed in the Bishop Museum catalog ledger with the locality above.

similaris. (K)
: *Amastra rugulosa* var. *similaris* Pease, 1870a: 96. Kauai (in publication title): [no additional details].
  : Remarks. Raised to species status by Baldwin (1893: 10), which was followed by Hyatt & Pilsbry (1911b: 150).

sola. (O)
: *Amastra (Cyclamastra) sola* Hyatt & Pilsbry, 1911b: 158, pl. 38, figs. 6, 9, 10. Oahu: "Wahiawa".

sphaerica. (K)
: *Amastra sphaerica* Pease, 1870a: 94. Kauai (in publication title): [no additional details].
  : Remarks. Also listed by Pease (1870b: 649) with no description.

thurstoni. (O)
: *Amastra (Cyclamastra) thurstoni* Cooke, 1917: 224, pl. B, fig. 2. Oahu: "Manoa, in excavation for Mr. R. Mist's house".

ultima. (H)
: *Amastra (Cyclamastra) ultima* Pilsbry & Cooke, 1914b: 25, pl. 2, figs. 9, 10. Hawaii: "Kahuku, Kau, under lava slabs on a nearly naked flow".

AMASTRIDAE [103]

umbilicata. (O, Mo)
> *Achatinella* (*Laminella*) *umbilicata* Pfeiffer, 1856b: 205. Hawaiian Islands ("Sandwich Islands in publication title): [no additional details].
>> *Remarks.* Pilsbry & Cooke (1914b: 23) suggested "that the Kahuku deposit be considered type locality".

Subgenus HETERAMASTRA Pilsbry

HETERAMASTRA Pilsbry *in* Hyatt & Pilsbry, 1911b: 137 (in key), 141 (heading a list) [1911c: 283 (description)] (as *Amastra* (*Amastra*) sect.). Type species: *Helicter hutchinsonii* Pease, 1862 (as "*A. hutchinsonii*"), by original designation.

acuta.
> *Achatinella acuta* Newcomb, 1854a: 16, pl. 23, fig. 36 [1854b: 142, pl. 23, fig. 36]. Oahu: "Lehui".
>> *Remarks.* Primary junior homonym of *acuta* Swainson (now in subgenus *Paramastra*). Synonym of *elongata* Newcomb, *teste* Newcomb (1858: 328). See also Hyatt & Pilsbry (1911b: 231).

+auwahiensis. (M)
> *Amastra* (*Heteramastra*) *subsoror auwahiensis* Pilsbry & Cooke, 1914b: 48, pl. 5, figs. 8–10. Maui: "East Maui: Auwahi, at about 4200 ft."

dwightii. (M)
> *Amastra* (*Heteramastra*) *dwightii* Cooke, 1933: 24, pl. 2, fig. 14. Maui: "East Maui: Hana".

elongata. (?O, ?Mo, ?M)
> *Achatinella elongata* Newcomb, 1853: 26. Oahu: [no additional details].
>> *Remarks.* Although Hyatt & Pilsbry (1911b: 230–31) questioned the presence of this species on Oahu, 1 of us (CCC) considers that specimens recently collected along the Waianae coast of Oahu may be referable to it.

flemingi. (M)
> *Amastra* (*Heteramastra*) *flemingi* Cooke, 1917: 247, pl. B, fig. 7. Maui: "East Maui: Kanaio, about 2,000 feet elevation, two miles east of Ulupalakua".

fraterna. (L)
> *Amastra fraterna* Sykes, 1896: 129. Lanai: "Mountains of Lanai, behind Koele".

hutchinsonii. (Mo, M)
> *Helicter hutchinsonii* Pease, 1862: 7. Maui: [no additional details].

implicata. (M)
> *Amastra* (*Heteramastra*) *implicata* Cooke, 1933: 25, pl. 2, fig. 15. Maui: "East Maui: Kipahulu, ridge on south side of valley".

+interjecta. (M)
> *Amastra* (*Heteramastra*) *soror* var. *interjecta* Hyatt & Pilsbry, 1911c: 287, pl. 48, figs. 9–11. Maui: "Lahaina".

laeva. (M)
> *Amastra* (*Laminella*) *laeva* Baldwin, 1906b: 138. Maui: "Haleakala Mt., East Maui".

+laticeps. (M)
> *Amastra* (*Heteramastra*) *soror* var. *laticeps* Hyatt & Pilsbry, 1911c: 287, pl. 48, fig. 8. Maui: [no additional details].

nannodes. (M)
> *Amastra* (*Heteramastra*) *nannodes* Cooke, 1933: 26, pl. 2, fig. 16. Maui: "East Maui: Kula, near the division between the lands of Keokea and Kamaole".

nubigena. (M)
>*Amastra* (*Heteramastra*) *nubigena* Pilsbry & Cooke, 1914b: 48, pl. 3, fig. 15, pl. 5, figs. 11, 12 [1915a: 49]. Maui: "West Maui: gulch to the right of Maunahoomaha, above Lahaina . . . Mt. Helu, Honokawai gulch, Moomuku and Honokohau" [1915a: 49].

perversa. (Mo)
>*Amastra* (*Heteramastra*) *perversa* Hyatt & Pilsbry, 1911c: 278, pl. 49, fig. 5. Molokai: "Halawa, near the eastern end of the island".

pilsbryi. (M)
>*Amastra pilsbryi* Cooke, 1913: 68. Maui: "Mt. Helu, West Maui".

sinistrorsa. (H)
>*Amastra* (*Laminella*) *sinistrorsa* Baldwin, 1906b: 138. Hawaii: "Hamakua".

soror. (M)
>*Achatinella soror* Newcomb, 1854a: 17, pl. 23, fig. 38 [1854b: 143, pl. 23, fig. 38]. Maui (as "Mani"): [no additional details].

sororem.
>*Achatinella sororem* Pfeiffer, 1868: 179. *Nom. nud.*
>>Remarks. Listed by Pfeiffer as a synonym of *elongata* Newcomb. Originally a Newcomb manuscript name.

subsoror. (M)
>*Amastra* (*Heteramastra*) *subsoror* Hyatt & Pilsbry, 1911c: 287, pl. 48, figs. 6, 7. Maui: "West Maui: Lahaina".

villosa.
>*Amastra villosa* Sykes, 1896: 129. Molokai: [no additional details].
>>Remarks. Synonym of *hutchinsonii* Pease, *teste* Hyatt & Pilsbry (1911c: 289).

Subgenus KAUAIA Sykes

CARINELLA Pfeiffer, 1875a: 116 (as "eine neue Gruppe"). Type species: *Achatinella kauaiensis* Newcomb, 1860, by monotypy. [Preoccupied, Sowerby, 1839].

KAUAIA Sykes, 1900: 355. Type species: *Achatinella kauaiensis* Newcomb, 1860, automatic. [n.n. for *Carinella* Pfeiffer, 1875].

kauaiensis. (K)
>*Achatinella kauaiensis* Newcomb, 1860: 145. Kauai: [no additional details].

Subgenus METAMASTRA Hyatt & Pilsbry

METAMASTRA Hyatt & Pilsbry, 1911b: 162. Type species: *Achatinella reticulata* Newcomb, 1854, by original designation.

In the *Manual of Conchology*, Hyatt & Pilsbry (1911b) established *Metamastra* on p. 139, listing the included species and designating *Achatinella reticulata* Newcomb, 1854 as the type species. However, when they treated the individual species of *Metamastra*, beginning on p. 162, they indicated *Achatinella variegata* Pfeiffer, 1849 (as "*variegata* Nc." [= Newcomb]) as the type species (p. 162, 163). Nevertheless, the last line of p. 162 refers to the species under discussion in *Metamastra* as the "*reticulata* series", in agreement with p. 139. The key on p. 163, although titled as a key to the "*variegata* series", contains the species listed on p. 139 as the *reticulata* series. That list (p. 139) does not contain *variegata*, which was listed on p. 139 and treated on p. 229 in *Paramastra*. Thus, *variegata* is not eligible as the type species of *Metamastra* because it was not included in the subgenus (*Code* Art. 67(g)). It is clear that *reticulata* was intended as the type species of *Metamastra*, as designated on p. 139, and that the use of *variegata* was a *lapsus* (P.K. Tubbs, *in litt.* to RHC,

22 October 1993). Hyatt & Pilsbry (1911c: 357) noted this mistake, and indeed in the copy of the *Manual* in the ANSP Pilsbry wrote *reticulata* in place of *variegata* on p. 162 (G. Rosenberg, *in litt.* to RHC, 28 September 1993). We accept the type species of *Metamastra* to be *reticulata* Newcomb, 1854.

+acuminata. (O)
    Amastra (Metamastra) subrostrata acuminata Cooke, 1933: 21, pl. 2, fig. 10. Oahu: "Waianae Mountains, Lualualei, southeast side of valley, high up the sides in damp glen".

aemulator. (O)
    Amastra (Metamastra) aemulator Hyatt & Pilsbry, 1911b: 190, pl. 38, fig. 7. Oahu: "Kahauiki".

albolabris. (O)
    Achatinella albolabris Newcomb, 1854a: 23, pl. 24, fig. 56 [1854b: 149, pl. 24, fig. 56]. Oahu: "Waianoe [= Waianae]".

+anaglypta. (O)
    Amastra (Metamastra) sericea var. *anaglypta* Cooke, 1917: 239, pl. C, fig. 9. Oahu: "Punaluu on the trail to Kaliuwaa just below the summit of the ridge".

badia. (O)
    Achatinella (Amastra) badia Baldwin, 1895: 230, pl. 11, fig. 40. Oahu: "Ewa".

breviata. (O)
    Amastra breviata Baldwin, 1895: 231, pl. 11, figs. 45, 46. Oahu: "Palolo and Halawa".

+bryani. (O)
    Amastra (Metamastra) transversalis bryani Pilsbry & Cooke, 1914b: 26, pl. 2, fig. 16. Oahu: "Outside of Punchbowl, Honolulu, about 300 ft. below the summit".

caputadamantis. (O)
    Amastra (Metamastra) caputadamantis Hyatt & Pilsbry, 1911b: 184, pl. 30, figs. 19, 20. Oahu: "Diamond Head, near Honolulu, in a Pleistocene or Holocene deposit".
        Remarks. "Probably to be regarded as a subspecies of *A. transversalis*" (Pilsbry & Cooke, 1914b: 27).

+conspersa. (O)
    Achatinella conspersa Pfeiffer, 1855c: 7, pl. 30, fig. 26. Hawaiian Islands: [no additional details].
        Remarks. Subspecies of *reticulata* Newcomb, *teste* Hyatt & Pilsbry (1911b: 180).

cookei.
    Amastra (Metamastra) cookei Hyatt & Pilsbry, 1911b: 182, pl. 38, figs. 4, 5. Oahu: "Moanalua".
        Remarks. Synonym of *textilis* Férussac, *teste* Pilsbry & Cooke (1914b: 28).

cornea. (O)
    Achatinella cornea Newcomb, 1854a: 15, pl. 23, fig. 32 [1854b: 141, pl. 23, fig. 32]. Type locality not given.

crassilabrum. (O)
    Achatinella crassilabrum Newcomb, 1854a: 15, pl. 23, fig. 31 [1854b: 141, pl. 23, fig. 31]. Oahu: "Waianoe [= Waianae]".

davisiana. (O)
    Amastra davisiana Cooke, 1908b: 215, fig. 1. Oahu: "About a mile from the summit of Konahuanui".

+dichroma. (O)
>   *Amastra (Metamastra) gulickiana dichroma* Cooke, 1933: 19, pl. 2, fig. 8. Oahu: "on both sides near the crest of the ridge between Punaluu and Kaluanui . . . on division ridge between Helemano, Waikane, and Kahana".
>>   Remarks. Although Cooke (1933: 19) designated a "type" (BPBM 10748), the locality ("Kuliuwaa") on the label and in the catalog ledger entry for this specimen is not the same as those published. Identification of the type locality requires further research.

+dispersa. (O)
>   *Amastra (Metamastra) reticulata dispersa* Hyatt & Pilsbry, 1911b: 180, pl. 29, figs. 5–7. Oahu: "Waianae".

ellipsoidea.
>   *Achatinella ellipsoidea* Gould, 1847b: 200. Maui [?error = Oahu (Hyatt & Pilsbry, 1911b: 167)]: [no additional details].
>>   Remarks. Synonym of *textilis* Férussac, *teste* Hyatt & Pilsbry (1911b: 167).

eos. (O)
>   *Amastra (Metamastra) eos* Pilsbry & Cooke, 1914b: 26, pl. 2, figs. 14, 15. Oahu: "Keawaawaa".

+errans. (O)
>   *Amastra (Metamastra) reticulata errans* Hyatt & Pilsbry, 1911b: 182, pl. 29, figs. 12, 13. Oahu: "Waimano".

forbesi. (O)
>   *Amastra (Metamastra) forbesi* Cooke, 1917: 242, pl. B, fig. 8. Oahu: "Makua in the Waianae mountains".

gulickiana. (O)
>   *Amastra (Metamastra) gulickiana* Hyatt & Pilsbry, 1911b: 168, pl. 38, fig. 8. Oahu: "Opaiula".

irwiniana. (O)
>   *Amastra irwiniana* Cooke, 1908b: 213, fig. 3. Oahu: "Summit of Lanihuli, at 2700 ft."

+kaipaupauensis. (O)
>   *Amastra (Metamastra) textilis kaipaupauensis* Hyatt & Pilsbry, 1911b: 168, pl. 38, fig. 12. Oahu: "Kaipaupau".

+media. (O)
>   *Amastra (Metamastra) textilis media* Hyatt & Pilsbry, 1911b: 167, pl. 30, figs. 11, 12. Oahu: "Aeia [= Aiea] (a short valley splitting the ridge northwest of Halawa)".

microstoma.
>   *Achatinella microstoma* Gould, 1845: 28. Hawaiian Islands ("Sandwich Islands" in publication title): [no additional details].
>>   Remarks. Synonym of *textilis* Férussac, *teste* Hyatt & Pilsbry (1911b: 165–67).

montagui. (O)
>   *Amastra montagui* Pilsbry, 1913: 39. Oahu: "Waiahole".

montivaga. (O)
>   *Amastra (Metamastra) montivaga* Cooke, 1917: 240, pl. C, fig. 5. Oahu: "Top of Maunakope, the peak as [sic] the head of the western Kalihi ridge".

+orientalis. (O)
>   *Amastra (Metamastra) reticulata orientalis* Hyatt & Pilsbry, 1911b: 181, pl. 29, figs. 8–11. Oahu: "Wahiawa".

oswaldi. (O)
>   *Amastra (Metamastra) oswaldi* Cooke, 1933: 22, pl. 2, fig. 13. Oahu: "close to the crest of Kualoa ridge east of Puu Kanehoalani".

AMASTRIDAE [107]

paulula. (O)
    Amastra (Metamastra) paulula Cooke, 1917: 240, pl. B, fig. 6. Oahu: "Malaikahana".
pellucida.
    Achatinella (Amastra) pellucida Baldwin, 1893: 9. *Nom. nud.*
pellucida. (O)
    Achatinella (Amastra) pellucida Baldwin, 1895: 231, pl. 11, figs. 41, 42. Oahu: "Waianae Valley".
praeopima. (O)
    Amastra (Metamastra) praeopima Cooke, 1917: 241, pl. C, fig. 8. Oahu: "Waiahole at the crest of the Koolau range where the trail crosses the ridge".
reticulata. (O)
    Achatinella reticulata Newcomb, 1854a: 22, pl. 24, fig. 54 [1854b: 148, pl. 24, fig. 54]. Oahu: "Waianoe [= Waianae]".
sericea. (?O)
    Achatinella (Laminella) sericea Pfeiffer, 1859c: 31. Hawaiian Islands: [no additional details].
        Remarks. Recorded "with doubt" from Waialua, Oahu by Baldwin, but otherwise apparently known only from Pfeiffer's description (Hyatt & Pilsbry, 1911b: 175).
spaldingi. (O)
    Amastra spaldingi Cooke, 1908: 214, fig. 2. Oahu: "Summit of Konahuanui, at 3300 ft."
subcornea. (O)
    Amastra (Metamastra) subcornea Hyatt & Pilsbry, 1911b: 189, pl. 31, fig. 11. Oahu: "near the base of Roundtop, towards Rocky Hill, where Manoa road enters the valley back of Punahou".
subrostrata. (O)
    Achatinella (Laminella) subrostrata Pfeiffer, 1859c: 31. Hawaiian Islands: [no additional details].
textilis.
    Helix (Cochlogena) textilis Férussac, 1821c: 60. *Nom. nud.*
textilis. (O)
    Helix textilis Férussac *in* Quoy & Gaimard, 1825d: 482. Hawaiian Islands: [no additional details].
thaanumi. (O)
    Amastra (Metamastra) thaanumi Hyatt & Pilsbry, 1911b: 177, pl. 38, figs. 1, 2. Oahu: "Kaaawa".
transversalis. (O)
    Achatinella (Laminella) transversalis Pfeiffer, 1856b: 204. Hawaiian Islands ("Sandwich Islands" in publication title): [no additional details].
        Remarks. Hyatt & Pilsbry (1911b: 184) considered Keawaawaa the type locality, but no lectotype designation was made.
undata.
    Achatinella (Amastra) undata Baldwin, 1893: 10. *Nom. nud.*
undata. (O)
    Achatinella (Amastra) undata Baldwin, 1895: 230, pl. 11, fig. 39. Oahu: "Nuuanu".
+vespertina. (O)
    Amastra (Metamastra) reticulata vespertina Pilsbry & Cooke, 1914b: 30, pl. 2, fig. 17. Oahu: "Kawaihapai, in soil of a plowed field between the railroad and the bluff".
vetusta.
    Achatinella (Amastra) vetusta Baldwin, 1893: 10. *Nom. nud.*

vetusta. (O)
    Achatinella (Amastra) vetusta Baldwin, 1895: 233, pl. 11, fig. 50. Oahu: "near Honolulu . . . near the base of Punchbowl Hill, at an altitude of twenty or thirty feet above sea level".

Subgenus PARAMASTRA Hyatt & Pilsbry

PARAMASTRA Hyatt & Pilsbry, 1911b: 137, 139, 208 (as *Amastra (Amastra)* sect.). Type species: *Helix spirizona* Férussac, 1825 (as "*A. spirizona*"), by original designation.

acuta.
    Achatinella acuta Swainson, 1828: 84. "Islands of the Pacific Ocean" (in introduction to genus): [no additional details].
        Remarks. Synonym of *spirizona* Férussac, *teste* Pilsbry & Cooke (1914b: 32), see also Hyatt & Pilsbry (1911b: 217–18).

+aiea. (O)
    Amastra (Paramastra) turritella var. *aiea* Hyatt & Pilsbry, 1911b: 215, pl. 35, figs. 5, 9. Oahu: "Aiea, a short valley dividing the ridge west of Halawa".

albida.
    Achatinella (Laminella) albida Pfeiffer, 1856b: 203. Hawaiian Islands ("Sandwich Islands" in publication title): [no additional details].
        Remarks. Synonym of *chlorotica* Pfeiffer, *teste* Hyatt & Pilsbry (1911b: 220).

baetica.
    Achatinella baetica Jay, 1850: 214. Unavailable name; proposed in synonymy with *spirizona* Férussac; not validated before 1961.
        Remarks. Originally a Mighels name in the Cuming collection. Pilsbry & Cooke (1914b: 32) added the synonym "*boetica*" to *spirizona*, apparently having overlooked the inclusion of *baetica* in the synonymy of *spirizona* by Hyatt & Pilsbry (1911b: 216).

+chlorotica. (O)
    Achatinella (Newcombia) chlorotica Pfeiffer, 1856b: 203. Hawaiian Islands ("Sandwich Islands" in publication title): [no additional details].
        Remarks. Subspecies of *spirizona* Férussac, *teste* Hyatt & Pilsbry (1911b: 220).

conicospira.
    Amastra conicospira Smith *in* Gulick & Smith, 1873: 86, pl. 10, fig. 10. Hawaiian Islands: [no additional details].
        Remarks. Considered a synonym of *intermedia* Newcomb by Hyatt & Pilsbry (1911b: 222–24); *intermedia* Newcomb subsequently treated as a junior synonym of *cylindrica* Newcomb by Pilsbry & Cooke (1914b: 33). Synonym of *cylindrica* Newcomb. **N. syn.**

cylindrica. (O)
    Achatinella cylindrica Newcomb, 1854a: 8, pl. 22, fig. 11 [1854b: 134, pl. 22, fig. 11]. Oahu: "Waianoe [= Waianae]".

decepta.
    Achatinella decepta Adams, 1851a: 127 [1851b: 43]. Type locality not given.
        Remarks. Synonym of *variegata* Pfeiffer, *teste* Hyatt & Pilsbry (1911b: 229).

erecta.
    Helicter (Laminella) erecta Pease, 1869a: 174. Maui [?error = Oahu, see Remarks]: [no additional details].
        Remarks. Hyatt & Pilsbry (1911c: 305) quoted Cooke (*in litt.*) as considering that "*erecta* is identical with *micans*" of Oahu. Pilsbry & Cooke (1914b: 32) synonymized *erecta* with *micans* Pfeiffer, saying that it should be "deleted from the list of Mauian species".

frosti.
    Amastra frosti Ancey, 1892: 719. Oahu: [no additional details].
        Remarks. Synonym of *micans* Pfeiffer, *teste* Pilsbry & Cooke (1914b: 32).

grossa.
>*Achatinella (Laminella) grossa* Pfeiffer, 1856b: 204. Hawaiian Islands ("Sandwich Islands" in publication title): [no additional details].
>>Remarks. Considered a synonym of *porphyrea* Newcomb by Hyatt & Pilsbry (1911b: 226); *porphyrea* Newcomb subsequently treated as a junior synonym of *cylindrica* Newcomb by Pilsbry & Cooke (1914b: 33). Synonym of *cylindrica* Newcomb. **N. syn.**

inornata.
>*Achatinella inornata* Mighels, 1845: 21. Oahu: [no additional details].
>>Remarks. Synonym of *turritella* Férussac, *teste* Hyatt & Pilsbry (1911b: 214).

intermedia.
>*Achatinella intermedia* Newcomb, 1854a: 9, pl. 22, fig. 13 [1854b: 135, pl 22, fig. 13]. Oahu: "Waianoe [= Waianae]".
>>Remarks. Synonym of *cylindrica* Newcomb, *teste* Pilsbry & Cooke (1914b: 33).

micans. (O)
>*Achatinella (Laminella) micans* Pfeiffer, 1859c: 31. Hawaiian Islands: [no additional details].

+nigrolabris. (O)
>*Amastra nigrolabris* Smith *in* Gulick & Smith, 1873: 85, pl. 10, fig. 9. Oahu: "Wahiawa"; "all the valleys from Kalaikoa to Waimea".
>>Remarks. "Pattern" of *spirizona* Férussac, *teste* Pilsbry & Cooke (1914b: 32). The earlier treatment of *nigrolabris* by Hyatt & Pilsbry (1911b: 216–19) is not clear, these authors treating it variously as a "form", "pattern", part of a "hybrid race", yet dealing with it under a separate heading, as was their usual practice for subspecies. Nowhere did they formally synonymize it. We retain it here as an infraspecific taxon.

oahuensis.
>*Achatina oahuensis* Green, 1827: 49, pl. 4, fig. 5. Hawaiian Islands: [no additional details].
>>Remarks. Synonym of *turritella* Férussac, *teste* Hyatt & Pilsbry (1911b: 213–14).

porphyrea.
>*Achatinella porphyrea* Newcomb, 1854a: 10, pl. 22, fig. 16 [1854b: 136, pl. 22, fig. 16]. Oahu: "Waianoe [= Waianae]".
>>Remarks. Synonym of *cylindrica* Newcomb, *teste* Pilsbry & Cooke (1914b: 33).

porphyrostoma. (O)
>*Helicter (Amastra) porphyrostoma* Pease, 1869a: 172. Oahu: [no additional details].

+rudis. (O)
>*Achatinella rudis* Pfeiffer, 1855c: 5, pl. 30, fig. 17. Hawaiian Islands: [no additional details].
>>Remarks. Subspecies of *spirizona* Férussac, *teste* Hyatt & Pilsbry (1911b: 219).

spirizona.
>*Helix (Cochlogena) spirizona* Férussac, 1821c: 60. *Nom. nud.*

spirizona. (O)
>*Helix spirizona* Férussac *in* Quoy & Gaimard, 1825d: 480. ?Hawaiian Islands ("Elle habite probablement les îles Sandwich [= it probably lives in the Sandwich Islands]"): [no additional details].
>>Remarks. Hyatt & Pilsbry (1911b: 217) considered "Waianae valley on Mt. Kaala" the type locality, and Pilsbry & Cooke (1914b: 32) considered "Helemano or in that immediate neighborhood" the type locality.

tenuispira.
>*Achatinella (Amastra) tenuispira* Baldwin, 1893: 10. *Nom. nud.*

tenuispira. (O)
>*Achatinella (Amastra) tenuispira* Baldwin, 1895: 232, pl. 11, fig. 51. Oahu: "Kaala Mt."

turritella. (O)
>> *Helix (Cochlogena) turritella* Férussac, 1821c: 60. Hawaiian Islands: [no additional details].
>>> *Remarks.* Name validated by being the only member of a Férussac species-group distinguished from his other "*Helicteres*" species as having a "Coquille turriculée". Hyatt & Pilsbry (1911b: 214) said "Nuuanu valley may be taken as the type locality".

unicolor.
>> *Amastra frosti* var. *unicolor* Ancey, 1899: 269, pl. 12, fig. 11. Oahu: "Waianae Mountains".
>>> *Remarks.* Synonym of *micans* Pfeiffer, *teste* Hyatt & Pilsbry (1911b: 210).

variegata. (O)
>> *Achatinella variegata* Pfeiffer, 1849c: 90. Hawaiian Islands: [no additional details].
>>> *Remarks.* See comments under subgenus *Metamastra*.

+waiawa. (O)
>> *Amastra (Paramastra) turritella* var. *waiawa* Hyatt & Pilsbry, 1911b: 215, pl. 35, figs. 7, 8. Oahu: "Waiawa".

Incertae Sedis in Genus AMASTRA Adams & Adams

amicta. (Hawaiian Islands)
>> *Amastra amicta* Smith *in* Gulick & Smith, 1873: 86, pl. 10, fig. 20. Hawaiian Islands: [no additional details].
>>> *Remarks.* Listed as "*Sedis incertae*" by Hyatt & Pilsbry (1911b: 141). Listed under *Achatinella* by Paetel (1873: 105).

breviana.
>> *Achatinella (Amastra) breviana* Baldwin, 1893: 8. *Nom. nud.*
>>> *Remarks.* Probably a spelling mistake intended for *breviata* (now in subg. *Metamastra*), published by Baldwin (1895: 231).

farcimen. (M)
>> *Achatinella (Laminella) farcimen* Pfeiffer, 1857a: 334. Maui (as "Mani"): [no additional details].
>>> *Remarks.* See Hyatt & Pilsbry (1911c: 291) and Pilsbry & Cooke (1914b: 46) for discussion of its possible affinities.

ferruginea.
>> *Achatinella (Amastra) ferruginea* Baldwin, 1893: 9. *Nom. nud.*

luteola. (?H)
>> *Helix luteola* Férussac *in* Quoy & Gaimard, 1825d: 480. "probable qu'elle vient des îles Mariannes [= probably came from the Mariana Islands]".
>>> *Remarks.* Newcomb (1858: 308) and Hyatt & Pilsbry (1911c: 321) considered the Marianas locality incorrect. Hyatt & Pilsbry (1911b: 138) listed it under subg. *Amastrella* but also (p. 141) as "*Sedis incertae*" and subsequently (1911c: 321) appeared to treat it as of unknown status.

peasei. (?O, ?L)
>> *Amastra peasei* Smith *in* Gulick & Smith, 1873: 86, pl. 10, fig. 13. Hawaiian Islands: [no additional details].
>>> *Remarks.* Listed as "*Sedis incertae*" by Hyatt & Pilsbry (1911b: 141) and as of unknown status (1911c: 322).

solida.
>> *Helicter (Amastra) solida* Pease, 1869a: 173. Oahu: [no additional details].
>>> *Remarks.* According to Pilsbry & Cooke (1914b: 31), *solida* Pease is a synonym of either *substrata* Pfeiffer (subg. *Metamastra*) or *decorticata* Gulick (subg. *Amastrella*). Its true affinity must await further revision and examination of the type(s).

testudinea.
>> *Achatinella (Amastra) testudinea* Baldwin, 1893: 10. *Nom. nud.*

AMASTRIDAE [111]

Genus CARELIA Adams & Adams

CARELIA Adams & Adams, 1855c: 132 (as *Achatina* subg.). Type species: *Achatina bicolor* Jay, 1839, by subsequent designation of Martens (1860: 208).

The most recent work dealing with the complete genus, as far as it was then known, was by Cooke (1931), superseding the *Manual of Conchology* (Hyatt & Pilsbry, 1911a). Subsequently, Cooke & Kondo (1952) described a number of new taxa. The genus *Carelia* is endemic to the Hawaiian Islands of Niihau and Kauai. We exclude *C. glutinosa* Ancey, 1893 as not being Hawaiian. It was originally described in *Carelia*, but apparently not from Hawaiian material, and seems perhaps to belong in the subulinid genus *Homorus* (cf. Hyatt & Pilsbry, 1911a: 118; Cooke, 1931: 5).

adusta.
    Achatina adusta Gould, 1845: 26. Hawaiian Islands ("Sandwich Islands" in publication title): [no additional details].
        Remarks. Synonym of *bicolor* Jay, *teste* Cooke (1931: 37).

anceophila. (K)
    Carelia anceophila Cooke, 1931: 31, pl. 3, figs. 9, 10. Kauai: "along the Olokele trail".

+angulata. (K)
    Carelia adusta var. *angulata* Pease, 1871a: 403. Polynesia (in publication title): [no additional details].
        Remarks. Subspecies of *bicolor* Jay, *teste* Cooke (1931: 42).

azona.
    Carelia turricula var. *azona* Ancey, 1904a: 121. Kauai: [no additional details].
        Remarks. Synonym of *turricula* Mighels, *teste* Cooke (1931: 60).

+baldwini. (K)
    Carelia olivacea baldwini Cooke, 1931: 71, pl. 13, fig. 8. Kauai: [no additional details].

bicolor. (K)
    Achatina bicolor Jay, 1839: 119, pl. 6, fig. 3. Type locality not given.

cochlea. (K)
    Achatina cochlea Reeve, 1849a: pl. 1, fig. 5. Peru [?error = Kauai (Hyatt & Pilsbry, 1911a: 109; Cooke, 1931: 55)]: [no additional details].

cumingiana. (K)
    Spiraxis cumingiana Pfeiffer, 1855h: 106, pl. 32, fig. 1. Kauai (as "Kanai"): [no additional details].

dolei. (K)
    Carelia dolei Ancey, 1893: 328. Kauai: "Hanalei".

evelynae. (K)
    Carelia evelynae Cooke & Kondo, 1952: 331, figs. 2a–f. Kauai: "Polihale, base of Polihale Ridge: 500 ft. inland from ocean, 150 ft. alt."

extincta.
    Carelia extincta Hyatt & Pilsbry, 1911a: 117, pl. 16, fig. 8. Type locality not given.
        Remarks. Originally a Newcomb manuscript name under which he distributed specimens of *sinclairi* Ancey; validated by Hyatt & Pilsbry (1911a: 117) but never formally synonymized with *sinclairi* Ancey (see also Cooke, 1931: 22). Synonym of *sinclairi* Ancey. **N. syn.**

fuliginea.
    Achatina fuliginea Pfeiffer, 1854a: 66. Type locality not given.
        Remarks. Synonym of *bicolor* Jay, *teste* Cooke (1931: 38).

glossema. (K)
> *Carelia glossema* Cooke, 1931: 80, pl. 15, figs. 1–3. Kauai: "Olokele, hillside . . . 10 mi. from shore, el. 1400'".
> *Remarks.* Cooke gave more than 1 locality but his designated "type" (BPBM 86083) has a Bishop Museum catalog ledger entry indicating the locality above.

hyattiana. (K)
> *Carelia hyattiana* Pilsbry *in* Hyatt & Pilsbry, 1911a: 108, pl. 21, figs. 1, 2. Hawaiian Islands: [no additional information].

hyperleuca.
> *Carelia bicolor* var. *hyperleuca* Hyatt & Pilsbry, 1911a: 114, pl. 20, figs. 5, 6. Type locality not given.
> *Remarks.* Synonym of *bicolor* Jay, *teste* Cooke (1931: 38).

+infrequens. (K)
> *Carelia olivacea* var. *infrequens* Cooke, 1931: 74, pl. 14, fig. 4. Kauai: "Anahola, in plowed field just south and near the mouth of the stream."

+isenbergi. (K)
> *Carelia dolei isenbergi* Cooke, 1931: 53, pl. 6, figs. 6–12. Kauai: "Haena plain, a few hundred yards east of the Manoa stream".

kalalauensis. (K)
> *Carelia kalalauensis* Cooke, 1931: 32, pl. 3, figs. 1–8, 11–17. Kauai: "Kalalau . . . north side of valley east of double waterfall".

knudseni. (K)
> *Carelia knudseni* Cooke, 1931: 82, pl. 15, fig. 4. Kauai: "between the Waimea and Hanapepe drainage basins".

kobelti.
> *Carelia cumingiana* var. *kobelti* Borcherding, 1910: 239, pl. 20, figs. 1, 2. Kauai (in publication title): [no additional details].
> *Remarks.* Synonym of *dolei* Ancey, *teste* Cooke (1931: 49).

lirata. (K)
> *Carelia lirata* Cooke, 1931: 78, pl. 14, figs. 9–12. Kauai: "limestone bluff about one-half mile east of Aweoweonui Bay in Mahaulepu, from 50 to 150 feet above sea level".

lymani. (K)
> *Carelia lymani* Cooke, 1931: 74, pl. 14, fig. 5. Kauai: [no additional details].

+magnapustulata. (K)
> *Carelia paradoxa magnapustulata* Cooke & Kondo, 1952: 341, figs. 5a–c. Kauai: "Lepeuli: On beach".

+meineckei. (K)
> *Carelia cumingiana meineckei* Cooke, 1931: 46, pl. 5, figs. 9–13. Kauai: "ridge between Lumahai and Wainiha".

minor.
> *Carelia adusta* var. *minor* Borcherding, 1910: 244, pl. 20, figs. 17, 18. Kauai (in publication title): [no additional details].
> *Remarks.* Synonym of *bicolor* Jay, *teste* Cooke (1931: 38).

mirabilis. (K)
> *Carelia mirabilis* Cooke, 1931: 29, pl. 1, figs. 9–11. Kauai: "limestone bluff about 0.5 of a mile east of Aweoweonui Bay, in Mahaulepu, from 50–150 feet above sea level".

+moloaaensis. (K)
> *Carelia olivacea moloaaensis* Cooke & Kondo, 1952: 338, figs. 4a–f. Kauai: "Lepeuli Beach".

necra. (K)
    Carelia necra Cooke, 1931: 85, pl. 16, figs. 1–3. Kauai: "Hanamalu Flats, about 1 mile south of the Wailua Stream".

newcombi.
    Achatina newcombi Pfeiffer, 1853a: 262. Hawaiian Islands: [no additional details].
        Remarks. Synonym of *turricula* Mighels, *teste* Cooke (1931: 60).

obeliscus.
    Achatina obeliscus Reeve, 1850a: pl. 23, fig. 129. Type locality not given.
        Remarks. Synonym of *turricula* Mighels, *teste* Cooke (1931: 59).

olivacea. (K)
    Carelia olivacea Pease, 1866: 293. Hawaiian Islands: [no additional details].

paradoxa. (K)
    Spiraxis paradoxa Pfeiffer, 1854d: 128. Kauai: [no additional details].

periscelis. (K)
    Carelia periscelis Cooke, 1931: 83, pl. 15, figs. 5, 6. Kauai: "Hanalei Valley, Kaapoko branch, at 2000 feet elevation".

pilsbryi. (K)
    Carelia pilsbryi Sykes, 1909: 204, unnumbered fig. Kauai: [no additional details].

+priggei. (K)
    Carelia olivacea var. *priggei* Cooke, 1931: 73, pl. 14, figs. 1–3. Kauai: "southern side of the Anahola ridge at Kiokala".

+propinquella. (K)
    Carelia olivacea propinquella Cooke, 1931: 72, pl. 13, figs. 9, 10. Kauai: "at about 1000 feet elevation, Wailua, just within the native forest in the valley on the south branch of the north fork of the Wailua River".

rigida.
    Carelia rigida Hyatt *in* Hyatt & Pilsbry, 1911a: 111, pl. 21, figs. 8, 13. Kauai: [no additional details].
        Remarks. Synonym of *cochlea* Reeve, *teste* Cooke (1931: 54).

sinclairi. (N)
    Carelia sinclairi Ancey, 1892: 720. Niihau: [no additional details].

+spaldingi. (K)
    Carelia necra spaldingi Cooke, 1931: 88, pl. 16, figs. 4–7. Kauai: "northern half of Waipouli race track".

suturalis.
    Carelia fuliginea var. *suturalis* Ancey, 1904a: 122. Kauai: [no additional details].
        Remarks. Synonym of *bicolor* Jay, *teste* Cooke (1931: 38).

tenebrosa. (K)
    Carelia tenebrosa Cooke, 1931: 75, pl. 14, figs. 6–8. Kauai: "on the upper slopes and top of Haupu, the highest peak of the Kipu range, 1500–2200 feet elevation".

+thaanumi. (K)
    Carelia paradoxa thaanumi Cooke, 1931: 92, pl. 16, fig. 8. Kauai: "'south side of the [Wailua] river in a cane field south of the gate' near the base of the Kalepa range."

+tsunami. (K)
    Carelia pilsbryi tsunami Cooke & Kondo, 1952: 335, figs. 3a–d. Kauai: "Lepeuli Beach".

turricula. (K)
    Achatina turricula Mighels, 1845: 20. Oahu [?error = Kauai (Hyatt & Pilsbry, 1911a: 103; Cooke, 1931: 59)]: [no additional details].

variabilis.
> *Carelia variabilis* Pease, 1871a: 402. Kauai: [no additional details].
> *Remarks.* Synonym of *olivacea* Pease, *teste* Cooke (1931: 67).

viridans.
> *Carelia variabilis* var. *viridans* Pease, 1871b: 473. *Nom. nud.*
> *Remarks.* Considered an error for *viridis* Pease by Hyatt & Pilsbry (1911a: 107).

viridis.
> *Carelia variabilis* var. *viridis* Pease, 1871a: 402. Kauai: [no additional details].
> *Remarks.* Synonym of *olivacea* Pease, *teste* Cooke (1931: 67).

+**waipouliensis.** (K)
> *Carelia paradoxa waipouliensis* Cooke & Kondo, 1952: 340. Kauai: "Olohena, northwest corner of Waipouli race track, in freshly plowed ground".

zonata.
> *Carelia adusta* var. *zonata* Borcherding, 1910: 244, pl. 20, figs. 13, 14. Kauai (in publication title): [no additional details].
> *Remarks.* Synonym of *bicolor* Jay, *teste* Cooke (1931: 38).

zonata.
> *Carelia adusta* var. *minor*, forma *zonata* Borcherding, 1910: 244, pl. 20, figs, 15, 16. Unavailable name; infrasubspecific (*Code* Art. 45).

Genus LAMINELLA Pfeiffer

LAMINELLA Pfeiffer, 1854b: 126 (as *Achatinella* subg.). Type species: *Helix gravida* Férussac, 1825 (as "*Achatinella*"), by subsequent designation of Martens (1860: 250).

alexandri. (M)
> *Achatinella alexandri* Newcomb, 1865: 182. Maui: [no additional details].

aspera. (M)
> *Laminella aspera* Baldwin, 1908: 68. Maui: "Wailuku valley, West Maui".

+**aurantium.** (O)
> *Laminella gravida aurantium* Pilsbry & Cooke, 1915a: 54, pl. 1, figs. 3–5. Oahu: "main ridge, above Waiahole . . . Head of Waiawa . . . Eastern ravines of Kaliuwaa".

bulbosa. (M)
> *Achatinella bulbosa* Gulick, 1858: 253, pl. 8, fig. 71. Maui: "Honuaula, E. Maui. Kula, E. Maui".

+**circumcincta.** (L)
> *Laminella concinna* var. *circumcincta* Hyatt & Pilsbry, 1911c: 337, pl. 54, figs. 12, 13. Lanai: [no additional details].
> *Remarks.* Originally a Dall collection name.

citrina. (Mo)
> *Achatinella citrina* Pfeiffer, 1848b: 234. Hawaiian Islands: [no additional details].
> *Remarks.* Originally a Mighels name in the Cuming collection.

concinna. (L)
> *Achatinella concinna* Newcomb, 1854a: 31, pl. 24, fig. 79 [1854b: 157, pl. 24, fig. 79]. Lanai (as "Ranai"): [no additional details].

concolor.
> *Achatinella gravida* var. *concolor* Martens, 1860: 250. Impermissible alternative name for *suffusa* Reeve; available by bibliographic reference.
> *Remarks.* Synonym of *suffusa* Reeve, *teste* Martens (1860: 250).

depicta.
> *Achatinella* (*Laminella*) *depicta* Baldwin, 1893: 7. *Nom. nud.*

depicta. (Mo)
>*Achatinella* (*Laminella*) *depicta* Baldwin, 1895: 228, pl. 11, figs. 33–35. Molokai: "Kamalo".
>>Remarks. Reduced to a subspecies of *alexandri* Newcomb by Hyatt & Pilsbry (1911c: 345) but restored to species status by Pilsbry & Cooke (1915a: 56).

+dimondi. (O)
>*Achatinella dimondi* Adams, 1851a: 126 [1851b: 42]. Hawaiian Islands: [no additional details].
>>Remarks. Subspecies of *gravida* Férussac, *teste* Pilsbry & Cooke (1915a: 52).

+duoplicata. (M)
>*Laminella duoplicata* Baldwin, 1908: 68. Maui: "Waichu [= Waiehu] Valley, West Maui".
>>Remarks. Subspecies of *alexandri* Newcomb, *teste* Hyatt & Pilsbry (1911c: 342).

ferussaci.
>*Achatinella* (*Laminella*) *ferussaci* Pfeiffer, 1856b: 203 ["*Fernsfaci*" in some copies, according to Pfeiffer (see Hyatt & Pilsbry, 1911c: 330)]. Oahu: [no additional details].
>>Remarks. Synonym of *sanguinea* Newcomb, *teste* Hyatt & Pilsbry (1911c: 330–32).

+gracilior. (L)
>*Laminella tetrao* var. *gracilior* Hyatt & Pilsbry, 1911c: 335, pl. 54, fig. 8. Lanai: [no additional details].

gracilis.
>*Helix gracilis* Férussac *in* Quoy & Gaimard, 1825d: 478. *Nom. nud.*
>>Remarks. Synonym of *gravida* Férussac, *teste* Hyatt & Pilsbry (1911c: 328).

gravida. (O)
>*Helix gravida* Férussac *in* Quoy & Gaimard, 1825d: 478. Hawaiian Islands: [no additional details].
>>Remarks. Hyatt & Pilsbry (1911c: 327) said that "Palolo valley [Oahu] may be considered the type locality".

helvina.
>*Achatinella* (*Laminella*) *helvina* Baldwin, 1893: 7. *Nom. nud.*

+helvina. (Mo)
>*Achatinella* (*Laminella*) *helvina* Baldwin, 1895: 227, pl. 11, fig. 30. Molokai: "Ohia valley, near Kaluaaha".
>>Remarks. Subspecies of *citrina* Pfeiffer, *teste* Hyatt & Pilsbry (1911c: 352).

+kalihiensis. (O)
>*Laminella gravida kalihiensis* Pilsbry & Cooke, 1915a: 54, pl. 1, fig. 6. Oahu: "Kalihi".

+kamaloensis. (Mo)
>*Laminella depicta kamaloensis* Pilsbry & Cooke, 1915a: 56. Molokai: "northwestern Kamalo above the amphitheatre, along the old Kamalo ditch".

kuhnsi. (M)
>*Amastra* (*Laminella*) *kuhnsi* Cooke, 1908c: 217, unnumbered fig. Maui: "West Maui: Kahakuloa".

lata.
>*Achatinella dimondi* var. *lata* Adams, 1851a: 127 [1851b: 43]. Hawaiian Islands: [no additional details].
>>Remarks. Synonym of *dimondi* Adams, *teste* Hyatt & Pilsbry (1911c: 328) and Pilsbry & Cooke (1915a: 53).

+leucoderma. (O)
>*Laminella sanguinea* var. *leucoderma* Pilsbry & Cooke, 1915a: 55. Oahu: "Near the middle of the western ridge of Popouwela, Waianae Mts."

+muscaria. (Mo)
    Laminella venusta var. *muscaria* Hyatt & Pilsbry, 1911c: 349, pl. 51, figs. 13–16. Molokai: [no additional details].

+orientalis. (Mo)
    Laminella venusta var. *orientalis* Hyatt & Pilsbry, 1911c: 350, pl. 51, fig. 11. Molokai: "Puukaeha (on the central ridge near the east end of Molokai)".

picta. (?O, M)
    Achatinella picta Mighels, 1845: 21. Oahu [?error = Maui (Hyatt & Pilsbry, 1911c: 339)]: [no additional details].

picta.
    Achatinella picta Pfeiffer, 1846a: 90. Hawaiian Islands: [no additional details].
        Remarks. Primary junior homonym of *picta* Mighels. Synonym of *picta* Mighels, *teste* Hyatt & Pilsbry (1911c: 338).

remyi. (L)
    Achatinella remyi Newcomb, 1855c: 146. Lanai (as "Ranai"): [no additional details].

remyi.
    Achatinella remyi Pfeiffer, 1856c: 207. Hawaii: [no additional details].
        Remarks. Primary junior homonym of *remyi* Newcomb (see comments regarding Pfeiffer and Newcomb in the introduction to the family Achatinellidae). Synonym of *remyi* Newcomb. **N. syn.**

sanguinea. (O)
    Achatinella sanguinea Newcomb, 1854a: 9, pl. 22, fig. 15 [1854b: 135, pl. 22, fig. 15]. Oahu: "Lehui".

semivenulata.
    Laminella semivenulata Borcherding, 1906: 92, pl. 8, figs. 23, 24. Molokai: "Manawai".
        Remarks. "Certainly not a subspecies, merely a mutation" of *citrina* Pfeiffer, *teste* Pilsbry & Cooke (1915a: 56).

+semivestita. (Mo)
    Laminella venusta color-var. *semivestita* Hyatt & Pilsbry, 1911c: 349, pl. 51, figs. 6–10, 12. Molokai: [no additional details].

straminea. (O)
    Achatinella straminea Reeve, 1850c: pl. 5, fig. 38. Hawaiian Islands: [no additional details].

+suffusa. (O)
    Achatinella suffusa Reeve, 1850c: pl. 2, fig. 11. Type locality not given.
        Remarks. Subspecies of *gravida* Férussac, *teste* Pilsbry & Cooke (1915a: 53).

tetrao. (L)
    Achatinella tetrao Newcomb, 1855a: 311 (animal only, shell not described) [1855b: 219 (description of shell)]. Lanai (as "Ranai"): [no additional details].

tetrao.
    Achatinella tetrao Pfeiffer, 1856c: 207. Lanai (as "Ranaï"): [no additional details].
        Remarks. Primary junior homonym of *tetrao* Newcomb (see comments regarding Pfeiffer and Newcomb in the introduction to the family Achatinellidae). Synonym of *tetrao* Newcomb. **N. syn.**

venusta. (?O, Mo)
    Achatinella venusta Mighels, 1845: 21. Oahu: [no additional details].
        Remarks. Correct identification of the type locality requires further research since Hyatt & Pilsbry (1911c: 348) gave only Molokai localities and considered that "no such shell occurs on Oahu".

+waianaensis. (O)
    Laminella gravida waianaensis Pilsbry & Cooke, 1915a: 54, pl. 1, figs. 7, 8. Oahu: "Haleauau, Waianae Mountains".

AMASTRIDAE   [117]

Genus PLANAMASTRA Pilsbry

PLANAMASTRA Pilsbry *in* Hyatt & Pilsbry, 1911b: 129. Type species: *Patula digonophora* Ancey, 1889, by original designation.

Two species (*depressiformis* Pease, 1865 and *prostrata* Pease, 1865), tentatively included as Hawaiian *Planamastra* by Pilsbry (*in* Hyatt & Pilsbry, 1911b: 131–32), were subsequently correctly identified by Pilsbry & Cooke (1922: 17) in the genera *Trochomorpha* and *Planorbis*, respectively, neither of which are Hawaiian. We exclude them from this catalog.

digonophora. (O)
> *Patula digonophora* Ancey, 1889a: 171. Oahu: [no additional details].

+koolauensis. (O)
> *Planamastra spaldingi koolauensis* Cooke, 1933: 4, pl. 2, figs. 3, 4. Oahu: "Kahuku, in deposit on limestone bluff above pumping station 1.2 miles northeast of Kahuku Mill".

peaseana. (Hawaiian Islands)
> *Planamastra peaseana* Pilsbry *in* Hyatt & Pilsbry, 1911b: 130, pl. 25, figs. 8–10. Hawaiian Islands: [no additional details].

spaldingi. (O)
> *Planamastra spaldingi* Cooke, 1933: 3, pl. 2, figs. 1, 2. Oahu: "Waianae Mountains: Pukaloa".

Genus TROPIDOPTERA Ancey

TROPIDOPTERA Ancey, 1889a: 191. Type species: *Helix alata* Pfeiffer, 1856 [misidentification, = *Endodonta wesleyi* Sykes, 1896], by monotypy.
PTERODISCUS Pilsbry, 1893: 36 (as *Endodonta* sect.). Type species: *Helix alata* Pfeiffer, 1856 [misidentification, = *Endodonta wesleyi* Sykes, 1896], automatic. [Unnecessary n.n. for *Tropidoptera* Ancey, 1889].
HELICAMASTRA Pilsbry & Vanatta, 1905: 570 (as *Amastra* sect.). Type species: *Amastra discus* Pilsbry & Vanatta, 1905, by original designation.

Most authors (e.g., Hyatt & Pilsbry, 1911a: 119–20; Thiele, 1931: 502) seem to have discounted Ancey's genus *Tropidoptera*, apparently following Pilsbry (1893: 36) in considering, incorrectly, that it is preoccupied by *Tropidopterus* Agassiz (as "Blanch.") (Coleoptera). *Tropidoptera* Ancey has priority over *Pterodiscus* Pilsbry, as indicated by Zilch (1959a: 143), so all species-group names, except *alata* which was the basis of Ancey's genus, now fall in new combination with *Tropidoptera*. An application by us to the ICZN (no. S.2903) to clarify the misidentification of the type species of *Tropidoptera* has been submitted in accordance with *Code* Art. 70(b).

alata. (L)
> *Helix alata* Pfeiffer, 1856e: 33. Hawaiian Islands: [no additional details].

cookei.
> *Pterodiscus cookei* Hyatt & Pilsbry, 1911a: 127, pl. 23, figs. 6–8. Oahu: "Nuuanu valley". **N. comb**.
>> *Remarks*. Synonym of *rex* Sykes, *teste* Pilsbry & Cooke (1914b: 17).

discus. (O)
> *Amastra (Helicamastra) discus* Pilsbry & Vanatta, 1905: 571, pl. 38, figs. 1–3. Oahu: "Waianae". **N. comb**.

+ewaensis. (O)
    Pterodiscus wesleyi ewaensis Pilsbry *in* Hyatt & Pilsbry, 1911a: 125, pl. 23, fig. 10. Oahu: "Ewa". **N. comb.**
heliciformis. (O)
    Amastra heliciformis Ancey, 1890: 340. Oahu: "Waianae". **N. comb.**
+lita. (L)
    Pterodiscus alatus litus Pilsbry *in* Hyatt & Pilsbry, 1911a: 122, pl. 22, figs. 4–6. Lanai: [no additional details]. **N. comb.**
rex. (O)
    Amastra (*Kauaia*) *rex* Sykes, 1904b: 159, 2 unnumbered figs. Oahu: "Summit of Konahuanui". **N. comb.**
thaanumi.
    Pterodiscus thaanumi Pilsbry *in* Hyatt & Pilsbry, 1911a: 125, pl. 24, figs. 1, 2. Oahu: "Kukaeiole, near Kaaawa, on the northeastern coast". **N. comb.**
        Remarks. Synonym of *rex* Sykes, *teste* Pilsbry & Cooke (1914b: 17).
wesleyi. (O)
    Endodonta (*Pterodiscus*) *wesleyi* Sykes, 1896: 127. Hawaiian Islands (by bibliographic reference to Pilsbry [1893: 36]): [no additional details]. **N. comb.**

Subfamily LEPTACHATININAE

Genus ARMSIA Pilsbry

ARMSIA Pilsbry *in* Hyatt & Pilsbry, 1911b: 132. Type species: *Pterodiscus petasus* Ancey, 1899 (as "*A. petasus*"), by original designation.

petasus. (O)
    Pterodiscus petasus Ancey, 1899: 268, pl. 12, fig. 4. Oahu: "Waianae Mountains".

Genus LEPTACHATINA Gould

LEPTACHATINA Gould, 1847b: 201 (as "group" of *Achatinella*). Type species: *Achatinella acuminata* Gould, 1847, by monotypy.

Subgenus ANGULIDENS Pilsbry & Cooke

ANGULIDENS Pilsbry & Cooke, 1914b: 8 (as *Leptachatina* sect.). Type species: *Leptachatina subcylindracea* Cooke, 1910, by original designation.

anceyana. (H)
    Leptachatina (*L.*) *anceyana* Cooke *in* Hyatt & Pilsbry, 1910: 39, pl. 1, figs. 18, 19. Hawaii: "Mana".
cookei. (O)
    Leptachatina cookei Pilsbry, 1914: 61. Oahu: "Kawaihapai, on a steep wooded bluff about 500 ft. above the coastal plain, and perhaps 3/4 mile from the sea".
fossilis. (K)
    Leptachatina (*L.*) *fossilis* Cooke *in* Hyatt & Pilsbry, 1910: 61, pl. 8, figs. 58, 59. Kauai: [no additional details].

AMASTRIDAE [119]

hyperodon. (M)
    Leptachatina (Angulidens) hyperodon Pilsbry & Cooke, 1914b: 12, pl. 11, figs. 6, 7. Maui: "East Maui".

microdon. (O)
    Leptachatina (Angulidens) microdon Pilsbry & Cooke, 1914b: 10, pl. 9, fig. 3. Oahu: "western ridge of Popouwela, Waianae Mountains".

subcylindracea. (O, Mo, Kah)
    Leptachatina (L.) oryza var. *subcylindracea* Cooke *in* Hyatt & Pilsbry, 1910: 28, pl. 3, figs. 60, 61. Oahu: [no additional details].
        Remarks. Raised to species status by Pilsbry & Cooke (1914b: 11).

Subgenus ILIKALA Cooke

ILIKALA Cooke *in* Hyatt & Pilsbry, 1911a: 89 (as *Leptachatina* subg.). Type species: *Achatinella fusca* Newcomb, 1853 (as "*L. fusca*"), by original designation.

fraterna. (K)
    Leptachatina (Ilikala) fraterna Cooke *in* Hyatt & Pilsbry, 1911a: 91, pl. 12, figs. 8, 11. Kauai: [no additional details].

fusca. (O)
    Achatinella fusca Newcomb, 1853: 28 [1854a: 19, pl. 23, fig. 44; 1854b: 145, pl. 23, fig. 44]. Oahu: ["Manoa" (Newcomb, 1854a: 19; 1854b: 145)].

irregularis.
    Achatinella (Amastra) irregularis Pfeiffer, 1856a: 164. *Nom. nud.*

irregularis. (?O)
    Achatinella (Amastra) irregularis Pfeiffer, 1856b: 205. Hawaiian Islands ("Sandwich Islands" in publication title): [no additional details].
        Remarks. Placed here following Hyatt & Pilsbry (1911c: 356) who considered it "almost certainly identical with *L. fusca* Newc. . . . which has priority"; although we consider this statement not to have formally synonymized them.

nematoglypta. (O)
    Leptachatina nematoglypta Pilsbry & Cooke, 1914b: 14, pl. 9, figs. 9, 10. Oahu: "Halawa".

petila. (O)
    Achatinella petila Gulick, 1856: 189, pl. 6, fig. 17. Oahu: "Koko on the eastern end of Oahu".

+striatella. (O)
    Achatinella striatella Gulick, 1856: 178, pl. 6, fig. 6. Oahu: "On the mountain ridge of Keawaawa".
        Remarks. Subspecies of *fusca* Newcomb, *teste* Cooke (*in* Hyatt & Pilsbry, 1911a: 91).

Subgenus LABIELLA Pfeiffer

LABIELLA Pfeiffer, 1854b: 142 (as *Achatinella* subg.). Type species: *Achatinella labiata* Newcomb, 1853, by monotypy.

callosa. (O)
    Achatinella (Labiella) callosa Pfeiffer, 1857a: 334. Oahu: [no additional details].

dentata.
    Achatinella dentata Pfeiffer, 1855b: 4 [1855c: 7]. Hawaiian Islands: [no additional details].
        Remarks. Synonym of *labiata* Newcomb, *teste* Cooke (*in* Hyatt & Pilsbry, 1911a: 77).

labiata. (O)
    Achatinella labiata Newcomb, 1853: 27 [1854a: 15, pl. 23, fig. 33; 1854b: 141, pl. 23, fig. 33]. Oahu: ["Lehui" (Newcomb, 1854a: 15; 1854b: 141)].

lagena. (O)
>Achatinella lagena Gulick, 1856: 175, pl. 6, fig. 3. Oahu: "Helemanu, Wahiawa, and Kalaikoa".

lenta. (M)
>Leptachatina (Labiella) lenta Cooke in Hyatt & Pilsbry, 1911a: 79, pl. 2, figs. 23, 24. Maui: "West Maui: Maunahoomaha, Wahakuli".

Subgenus LEPTACHATINA Gould

LEPTACHATINA Gould, 1847b: 201 (as "group" of Achatinella). Type species: Achatinella acuminata Gould, 1847, by monotypy.

accincta.
>Achatina accincta Gould, 1852: 88. Unjustified emendation of accineta Mighels, 1845.
>>Remarks. Objective synonym of accineta Mighels. Gould explicitly considered "accineta" a typographical error; see also Johnson (1949: 220).

accineta. (O)
>Achatina accineta Mighels, 1845: 20. Oahu: "Waianai".
>>Remarks. Numerous subsequent authors have implicitly assumed this to be a misspelling of "accincta" (e.g., Pfeiffer, 1848c: 271; Reeve, 1849f: pl. 19, fig. 101; Pease, 1870b: 650; Cooke in Hyatt & Pilsbry, 1910: 25). Original spelling of "accineta" valid by Code Art. 32(c)(ii).

acuminata. (K)
>Achatinella acuminata Gould, 1847b: 200. Kauai: [no additional details].

antiqua. (K)
>Leptachatina antiqua Pease, 1870a: 94. Kauai (in publication title): [no additional details].
>>Remarks. Also p. 87 as "antiquata". We select antiqua as the correct original spelling. Also listed by Pease (1870b: 651) with no description.

antiquata.
>Leptachatina antiquata Pease, 1870a: 87. Incorrect original spelling of antiqua Pease.

approximans. (O)
>Leptachatina approximans Ancey, 1897b: 222. Oahu: "Waianae".

arborea. (H)
>Leptachatina arborea Sykes, 1900: 357, pl. 11, fig. 21. Hawaii: "Kona at 4000 feet; Olaa, Hilo".

attenuata. (K)
>Leptachatina (L.) attenuata Cooke in Hyatt & Pilsbry, 1911a: 69, pl. 7, figs. 45, 46. Kauai: "Haleieie, Makaweli, Waiakoali, Ekaula; Hanalei".

+avus. (Mo)
>Leptachatina oryza avus Pilsbry & Cooke, 1914b: 5, pl. 10, figs. 12–16. Molokai: "Sand dunes of Moomomi".

baldwini. (M)
>Leptachatina (L.) baldwini Cooke in Hyatt & Pilsbry, 1910: 12, pl. 2, figs. 33, 41. Maui: "West Maui Maunahoomaha; Mt. Lihau, Honokawai Gulch and Akauka-imu; Lahaina. East Maui: Kaliili."

balteata. (K)
>Leptachatina balteata Pease, 1870a: 91. Kauai (in publication title): [no additional details].
>>Remarks. Also listed by Pease (1870b: 651) with no description.

brevicula. (K)
>Helicter (Leptachatina) brevicula Pease, 1869a: 169. Kauai: [no additional details].

AMASTRIDAE [121]

+brevis. (K)
    Leptachatina (L.) pachystoma var. *brevis* Cooke *in* Hyatt & Pilsbry, 1910: 52, pl. 8, fig. 53. Kauai: "Haleieie".

captiosa. (O)
    Leptachatina (L.) captiosa Cooke *in* Hyatt & Pilsbry, 1910: 29, pl. 11, fig. 12. Oahu: "Waianae Mts., back of Leilehua".

cerealis. (O)
    Achatinella cerealis Gould, 1847b: 201. Oahu: "Waianai [= Waianae]".

cingula. (O)
    Achatinella cingula Mighels, 1845: 21. Oahu: [no additional details].

clara.
    Achatinella clara Pfeiffer, 1846a: 90. Hawaiian Islands: [no additional details].
        Remarks. Synonym of *striatula* Gould, *teste* Cooke (*in* Hyatt & Pilsbry, 1911a: 74).

compacta. (M)
    Helicter (Labiella) compacta Pease, 1869a: 172. Maui: [no additional details].

concolor. (Mo)
    Leptachatina (L.) concolor Cooke *in* Hyatt & Pilsbry, 1910: 31, pl. 6, figs. 9, 10. Molokai: "Kamalo".

conicoides. (Mo)
    Leptachatina conicoides Sykes, 1900: 359, pl. 11, fig. 26. Molokai: [no additional details].

conspicienda. (M)
    Leptachatina (L.) conspicienda Cooke *in* Hyatt & Pilsbry, 1910: 56, pl. 11, fig. 10. Maui: "East Maui: Mt. Kukui" [= West Maui].
        Remarks. Both label and catalog entry for the holotype (BPBM 15160), designated by Cooke (*in* Hyatt & Pilsbry, 1910: 56), correctly indicate that the type locality is in West Maui.

convexiuscula. (O)
    Leptachatina convexiuscula Sykes, 1900: 360, pl. 11, fig. 11. Oahu: "Waiolani".

corneola. (O)
    Achatinella corneola Pfeiffer, 1846a: 90. Hawaiian Islands: [no additional details].

coruscans. (Mo)
    Achatinella (Leptachatina) coruscans Hartman, 1888b: 52, pl. 1, fig. 16. Molokai: [no additional details].

costulata. (O)
    Achatinella costulata Gulick, 1856: 177, pl. 6, fig. 5. Oahu: "Pupukea, Waimea, and Kawailoa".

costulosa. (K)
    Leptachatina costulosa Pease, 1870a: 90. Kauai (in publication title): [no additional details].
        Remarks. Also listed by Pease (1870b: 651) with no description.

crystallina. (O)
    Achatinella crystallina Gulick, 1856: 186, pl. 6, fig. 14. Oahu: "Kamoo, Waialua".

cuneata. (K)
    Leptachatina (L.) cuneata Cooke *in* Hyatt & Pilsbry, 1910: 6, pl. 10, figs. 1, 2. Kauai: "Kapaa".

cylindrata. (K)
    Helicter (Leptachatina) cylindrata Pease, 1869a: 168. Kauai: [no additional details].

+cylindrella. (K)
>   *Leptachatina* (*L.*) *pachystoma* var. *cylindrella* Cooke *in* Hyatt & Pilsbry, 1910: 51, pl. 8, fig. 49. Kauai: [type locality not restricted by the statement that "this variety was especially abundant at Haleieie"].

deceptor. (K)
>   *Leptachatina deceptor* Cockerell, 1927: 117. Kauai: "Haena sand-hills".

defuncta. (H)
>   *Leptachatina* (*L.*) *defuncta* Cooke *in* Hyatt & Pilsbry, 1910: 39, pl. 1, fig. 16. Hawaii: "Mana".

dimidiata. (O)
>   *Achatinella* (*L.*) *dimidiata* Pfeiffer, 1856b: 205. Hawaiian Islands ("Sandwich Islands" in publication title): [no additional details].

+dissimilis. (Mo)
>   *Leptachatina* (*L.*) *coruscans* var. *dissimilis* Cooke *in* Hyatt & Pilsbry, 1910: 42, pl. 6, fig. 8. Molokai: "near Waikolu".

dormitor. (Mo)
>   *Leptachatina dormitor* Pilsbry & Cooke, 1914b: 6, pl. 11, fig. 3. Molokai: "near the top of Mauna Loa, at about 1350 ft. elevation, on the south side of the 'crater'".

elevata.
>   *Achatinella elevata* Pfeiffer, 1856c: 209. Hawaiian Islands ("Sandwich Islands" in publication title): [no additional details].
>   *Remarks.* Synonym of *gracilis* Pfeiffer, *teste* Cooke (*in* Hyatt & Pilsbry, 1910: 16).

emerita. (Mo)
>   *Leptachatina emerita* Sykes, 1900: 361, pl. 11, fig. 10. Molokai: "Kalamaula, and at 4000 feet".

exilis. (O)
>   *Achatinella exilis* Gulick, 1856: 188, pl. 6, fig. 16. Oahu: "Keawaawa".

exoptabilis. (O)
>   *Leptachatina* (*L.*) *exoptabilis* Cooke *in* Hyatt & Pilsbry, 1910: 21, pl. 10, figs. 5, 6. Oahu: "Diamond Head; Waianae Mts. back of Leilehua".

extensa. (K)
>   *Leptachatina extensa* Pease, 1870a: 92. Kauai (in publication title): [no additional details].
>   *Remarks.* Also listed by Pease (1870b: 651) with no description.

fragilis.
>   *Achatinella fragilis* Gulick, 1856: 183, pl. 6, fig. 11. Oahu: "Helemanu".
>   *Remarks.* Synonym of *gummea* Gulick, *teste* Cooke (*in* Hyatt & Pilsbry, 1910: 27).

fulgida. (M)
>   *Leptachatina* (*L.*) *fulgida* Cooke *in* Hyatt & Pilsbry, 1910: 12, pl. 2, figs. 39, 40. Maui: "West Maui: Mt. Lihau, Mt. Kukui, Akau-ka-imu, Ahoa".

fumida. (O)
>   *Achatinella fumida* Gulick, 1856: 181, pl. 6, fig. 9. Oahu: "Waialei, Pupukea, Waimea, Kawailoa, and Helemanu".

fumosa.
>   *Achatinella fumosa* Newcomb, 1854a: 14, pl. 23, fig. 28 [1854b: 140, pl. 23, fig. 28]. Oahu: "Manoa".
>   *Remarks.* Synonym of *cingula* Mighels, *teste* Cooke (*in* Hyatt & Pilsbry, 1910: 53).

gayi. (K)
>   *Leptachatina* (*L.*) *gayi* Cooke *in* Hyatt & Pilsbry, 1911a: 72, pl. 7, figs. 39, 40. Kauai: "Makaweli".

AMASTRIDAE [123]

glutinosa. (O)
    Achatinella (Laminella) glutinosa Pfeiffer, 1856b: 204. Hawaiian Islands ("Sandwich Islands" in publication title): [no additional details].
gracilis. (O)
    Achatinella gracilis Pfeiffer, 1855c: 6, pl. 30, fig. 22 [1855g: 68]. Hawaiian Islands: [no additional details].
grana. (M)
    Achatinella grana Newcomb, 1853: 29 [1854a: 20, pl. 23, fig. 46; 1854b: 146, pl. 23, fig. 46]. Maui: "E. Maui".
granifera.
    Achatinella granifera Gulick, 1856: 185, pl. 6, fig. 13. Oahu: "Keawaawa".
        Remarks. Synonym of *accineta* Mighels (*"accincta"* of most authors), *teste* Cooke (*in* Hyatt & Pilsbry, 1910: 25).
gummea. (O)
    Achatinella gummea Gulick, 1856: 182, pl. 6, fig. 10. Oahu: "Mokuleia and Lihue".
guttula. (M)
    Achatinella guttula Gould, 1847b: 201. Maui: [no additional details].
haenensis. (K)
    Leptachatina haenensis Cockerell, 1927: 117. Kauai: "Haena sand-hills".
+hesperia. (O)
    Leptachatina oryza hesperia Pilsbry & Cooke, 1914b: 5, pl. 10, figs. 10, 11. Oahu: "Kawaihapai, in soil of a plowed field between the railroad and the bluff, and on the latter. Haleiwa, on the golf links, in superficial deposits of calcareous sand".
illimis. (O)
    Leptachatina (L.) illimis Cooke *in* Hyatt & Pilsbry, 1910: 10, pl. 10, fig. 3. Oahu: "Palehuna in the Waianae Mts."
imitatrix. (H)
    Leptachatina imitatrix Sykes, 1900: 364, pl. 11, fig. 9. Hawaii: "Mauna Loa at 4000 feet".
impressa. (L)
    Leptachatina impressa Sykes, 1896: 127. Lanai: "Mountains of Lanai, behind Koele".
isthmica. (M)
    Leptachatina isthmica Ancey & Sykes *in* Ancey, 1899: 270, pl. 13, fig. 20 [plate by Sykes (1899)]. Maui: "Sand Hills, between East and West Maui".
        Remarks. For explanation of authorship, see Remarks under *semicarnea* Ancey & Sykes.
knudseni. (K)
    Leptachatina (L.) knudseni Cooke in Hyatt & Pilsbry, 1910: 8, pl. 9, figs. 11, 12. Kauai: "Waipo, near Halemanu, at an altitude of 3500 ft., Ekaula, alt. 1900 ft."
konaensis. (H)
    Leptachatina konaensis Sykes, 1900: 364, pl. 11, fig. 13. Hawaii: "Kona at 4000 feet".
kuhnsi. (M)
    Leptachatina (L.) kuhnsi Cooke *in* Hyatt & Pilsbry, 1910: 48, pl. 11, fig. 3. Maui: "West Maui: Abau-ka-imu, Maunahoomaha".
lacrima.
    Achatinella lacrima Gulick, 1856: 176, pl. 6, fig. 4. Oahu: "Lihue, Oahu . . . Kalaikoa, Wahiawa, Helemanu and Peula".
        Remarks. Synonym of *glutinosa* Pfeiffer, *teste* Cooke (*in* Hyatt & Pilsbry, 1910: 46).

laevigata. (Mo)
>*Leptachatina (L.) laevigata* Cooke *in* Hyatt & Pilsbry, 1910: 11, pl. 6, figs. 4, 5. Molokai: "Mapulehu Ridge".

laevis. (K)
>*Leptachatina laevis* Pease, 1870a: 91. Kauai (in publication title): [no additional details].
>>*Remarks.* Also listed by Pease (1870b: 651) with no description.

lanaiensis. (L)
>*Leptachatina (L.) lanaiensis* Cooke *in* Hyatt & Pilsbry, 1911a: 67, pl. 12, figs. 2, 3. Lanai: [no additional details].

lanceolata. (Mo)
>*Leptachatina (L.) lanceolata* Cooke *in* Hyatt & Pilsbry, 1911a: 65, pl. 6, figs. 12, 13. Molokai: "Kamalo, near Waikolu".

leiahiensis. (O)
>*Leptachatina (L.) leiahiensis* Cooke *in* Hyatt & Pilsbry, 1910: 22, pl. 10, figs. 9, 10. Oahu: "Diamond Head".

lepida. (H)
>*Leptachatina (L.) lepida* Cooke *in* Hyatt & Pilsbry, 1910: 40, pl. 1, figs. 12, 13. Hawaii: "Mana".

leucochila. (K)
>*Achatinella leucochila* Gulick, 1856: 173, pl. 6, fig. 1. Kauai: [no additional details].

longiuscula. (L)
>*Leptachatina (L.) longiuscula* Cooke *in* Hyatt & Pilsbry, 1910: 57, pl. 11, fig. 11. Lanai: [no additional details].

lucida. (K)
>*Leptachatina lucida* Pease, 1870a: 93. Kauai (in publication title): [no additional details].
>>*Remarks.* Also listed by Pease (1870b: 650) with no description.

+manana. (O)
>*Leptachatina opipara manana* Pilsbry & Cooke, 1914b: 7. Oahu: "North side of the summit of the peak at intersection of the Waimano-Manana ridge and the main range".

maniensis. (M)
>*Achatina maniensis* Pfeiffer, 1855d: 126. Maui (as "Mani"): [no additional details].
>>*Remarks.* Implicitly considered a spelling mistake for "*mauiensis*" by most authors, including Hyatt & Pilsbry (1911c: 355).

manoaensis.
>*Achatinella manoaensis* Pfeiffer, 1859d: 545. *Nom. nud.*
>>*Remarks.* Originally a Newcomb manuscript name, listed as a synonym of *melampoides* by Pfeiffer (1859d: 545). Synonym of *ventulus* Férussac, *teste* Hyatt & Pilsbry (1911c: 356).

margarita.
>*Achatinella (Leptachatina) margarita* Pfeiffer, 1856b: 206. Hawaiian Islands ("Sandwich Islands" in publication title): [no additional details].
>>*Remarks.* Synonym of *accineta* Mighels ("*accincta*" of most authors), *teste* Cooke (*in* Hyatt & Pilsbry, 1910: 25).

marginata. (O)
>*Achatinella marginata* Gulick, 1856: 179, pl. 6, fig. 7. Oahu: "Kalaikoa".

mcgregori. (M)
>*Leptachatina mcgregori* Pilsbry & Cooke, 1914b: 8, pl. 11, fig. 8. Maui: "West Maui: near Lahaina, at 1000 ft. elevation".

melampoides.
 Achatinella melampoides Pfeiffer, 1853a: 262. Hawaiian Islands: [no additional details].
  Remarks. Synonym of *ventulus* Férussac, *teste* Cooke (*in* Hyatt & Pilsbry, 1910: 54).
+**micra**. (K)
 Leptachatina (*L.*) *brevicula* var. *micra* Cooke *in* Hyatt & Pilsbry, 1910: 24, pl. 8, fig. 55. Kauai: "Haleieie at 1700 ft. and Miloii at 1500 ft."
molokaiensis. (Mo)
 Leptachatina (*L.*) *molokaiensis* Cooke *in* Hyatt & Pilsbry, 1910: 22, pl. 10, figs. 11, 12. Molokai: "Mapulehu Ridge, Kaluaaha and Wailau Pali".
nitida. (?O, ?Mo, M)
 Achatinella nitida Newcomb, 1853: 29 [1854a: 14, pl. 23, fig. 30; 1854b: 140, pl. 23, fig. 30]. Maui: "E. Maui".
obclavata.
 Achatinella (*Leptachatina*) *obclavata* Pfeiffer, 1855f: 98 [1855g: 70]. Hawaiian Islands: [no additional details].
  Remarks. Synonym of *sandwicensis* Pfeiffer, *teste* Cooke (*in* Hyatt & Pilsbry, 1910: 9).
obsoleta. (M)
 Spiraxis obsoleta Pfeiffer, 1857a: 335. Hawaiian Islands: [no additional details].
obtusa. (O)
 Achatinella obtusa Pfeiffer, 1856c: 209. Hawaiian Islands ("Sandwich Islands" in publication title): [no additional details].
+**occidentalis**. (M)
 Leptachatina (*L.*) *nitida* var. *occidentalis* Cooke *in* Hyatt & Pilsbry, 1910: 43, pl. 2, fig. 22. Maui: "West Maui: Maunahoomaha; Lahaina".
octavula.
 Achatinella octavula Paetel, 1873: 106. *Nom. nud.*
  Remarks. Paetel (1873: 106) listed "*octavula* Pfr." in the genus *Achatinella*. Pfeiffer & Clessin (1879: 316) included "*octavula* Pät.?" in the synonymy of *obclavata* in the subgenus *Leptachatina* of the genus *Achatinella*. Cooke (*in* Hyatt & Pilsbry, 1910: 9) included "*Achatinella* (*Leptachatina*) *octavula* Paetel, CLESSIN, Nomen. Helic. Viv., 1881, p. 316" in the synonymy of *Leptachatina sandwicensis* (Pfeiffer). And Pilsbry & Cooke (1914a: 369) listed "*Achatinella octavula* Pfr." (Paetel, 1873: 106) among their "unrecognized species", suggesting that it was an "error for *obclavata*?" We place it in *Leptachatina s. str.* solely on the basis of these tentative assignments.
octogyrata. (O)
 Achatinella octogyrata Gulick, 1856: 190, pl. 6, fig. 18. Oahu: "Palolo valley".
+**olaaensis**. (H)
 Leptachatina (*L.*) *konaensis* var. *olaaensis* Cooke *in* Hyatt & Pilsbry, 1910: 45, pl. 1, fig. 4. Hawaii: "Olaa".
opipara. (O)
 Leptachatina (*L.*) *opipara* Cooke *in* Hyatt & Pilsbry, 1910: 30, pl. 12, fig. 1. Oahu: "Apex of mountain range back of Palolo Valley".
oryza. (O)
 Achatinella (*Leptachatina*) *oryza* Pfeiffer, 1856b: 206. Oahu: [no additional details].
ovata. (M)
 Leptachatina (*L.*) *ovata* Cooke *in* Hyatt & Pilsbry, 1910: 33, pl. 2, fig. 30. Maui: "West Maui: Maunahoomaha, Honokahau Gulch, Honokawai Gulch, Iao Valley, Kauaula, Honolua".
pachystoma. (K)
 Helicter (*Labiella*) *pachystoma* Pease, 1869a: 171. Kauai: [no additional details].

+parvula. (?M)
    Achatinella parvula Gulick, 1856: 195, pl. 6, fig. 24. Hawaiian Islands: [no additional details].
        Remarks. "Variety" of *vitreola* Gulick, *teste* Cooke (*in* Hyatt & Pilsbry, 1910: 35–36).

perkinsi. (L)
    Leptachatina perkinsi Sykes, 1896: 128. Lanai: "Mountains of Lanai, behind Koele".

persubtilis. (O)
    Leptachatina (*L.*) *persubtilis* Cooke *in* Hyatt & Pilsbry, 1910: 15, pl. 10, fig. 4. Oahu: "Waianae Mts. back of Waialua".

pilsbryi. (O)
    Leptachatina (*L.*) *pilsbryi* Cooke *in* Hyatt & Pilsbry, 1910: 55, pl. 11, figs. 5, 6. Oahu: "Kukaeiole in Kaaawa".

popouwelensis. (O)
    Leptachatina popouwelensis Pilsbry & Cooke, 1914b: 1, pl. 9, fig. 4. Oahu: "western ridge of Popouwela".

praestabilis. (M)
    Leptachatina (*L.*) *praestabilis* Cooke *in* Hyatt & Pilsbry, 1910: 43, pl. 2, figs. 37, 38. Maui: "West Maui: Lahaina".

pulchra. (O)
    Leptachatina (*L.*) *pulchra* Cooke *in* Hyatt & Pilsbry, 1910: 29, pl. 10, figs. 7, 8. Oahu: "Waianae Mts. back of Leilehua".

pumicata. (?K)
    Bulimus pumicatus Mighels, 1845: 19. Oahu: [no additional details].
        Remarks. Lectotype selected by Johnson (1949: 228), who, referring to Cooke, indicated that it was "Undoubtedly from Kauai" (see also Johnson, 1949: 217). Placed here with *Leptachatina* on the basis of Pilsbry & Cooke (1916: 271).

pupoidea. (K)
    Leptachatina (*L.*) *pupoidea* Cooke *in* Hyatt & Pilsbry, 1911a: 74, pl. 7, figs. 43, 44. Kauai: "Milolii, at 1500 ft.".

pyramis. (O, ?K)
    Achatinella pyramis Pfeiffer, 1846a: 90. Hawaiian Islands: [no additional details].

resinula. (O)
    Achatinella resinula Gulick, 1856: 174, pl. 6, fig. 2. Oahu: "Kawailoa, Waimea, Pupukea, Waialei, and Punaluu".

saccula. (Hawaiian Islands)
    Achatinella (*Leptachatina*) *saccula* Hartman, 1888b: 55, pl. 1, fig. 15. Hawaiian Islands: [no additional details].

sagittata. (Mo)
    Leptachatina sagittata Pilsbry & Cooke, 1914b: 2, pl. 11, fig. 9. Molokai: "Pipeline trail, upper Kaunakakai".

sandwicensis. (O, ?Mo)
    Achatina sandwicensis Pfeiffer, 1846b: 32. Hawaiian Islands: [no additional details].

saxatilis. (O)
    Achatinella saxatilis Gulick, 1856: 187, pl. 6, fig. 15. Oahu: "Mokuleia".

sculpta. (O)
    Achatina sculpta Pfeiffer, 1856d: 211. Oahu: [no additional details].

scutilus. (O)
    Bulimus scutilus Mighels, 1845: 20. Oahu: [no additional details].

semipicta. (L)
    Leptachatina semipicta Sykes, 1896: 128. Lanai: "Mountains of Lanai, behind Koele".

simplex. (H)
    Helicter (Leptachatina) simplex Pease, 1869a: 170. Hawaii: [no additional details].

smithi. (L)
    Leptachatina smithi Sykes, 1896: 128. Lanai: "Mountains of Lanai, above Koele".

somniator. (Mo)
    Leptachatina somniator Pilsbry & Cooke, 1914b: 7, pl. 11, figs. 4, 5. Molokai: "near the top of Mauna Loa, at about 1350 ft. elevation".

stiria. (O)
    Achatinella stiria Gulick, 1856: 194, pl. 6, fig. 22. Oahu: "Helemanu, Peula, and Kawailoa".

striata. (K)
    Tornatellina striata Newcomb, 1861: 93. Kauai: [no additional details].

striatula. (K)
    Achatinella striatula Gould, 1845: 28. Hawaiian Islands ("Sandwich Islands" in publication title): [no additional details].

subovata. (L)
    Leptachatina (L.) subovata Cooke *in* Hyatt & Pilsbry, 1910: 37, pl. 11, fig. 2. Lanai: [no additional details].

subula. (O)
    Achatinella subula Gulick, 1856: 191, pl. 6, fig. 19. Oahu: "Palolo valley".

succincta. (O)
    Achatinella succincta Newcomb, 1855b: 220. Oahu: "Ewa".

succincta.
    Achatinella succincta Pfeiffer, 1856c: 209. Hawaiian Islands ("Wahai"; locality not known).
        Remarks. Primary junior homonym of *succincta* Newcomb (see comments regarding Pfeiffer and Newcomb in the introduction to the family Achatinellidae). Synonym of *succincta* Newcomb. **N. syn.**

supracostata. (L)
    Leptachatina supracostata Sykes, 1900: 370, pl. 11, fig. 22. Lanai: "Mts. behind Koele".

tenebrosa. (K)
    Leptachatina tenebrosa Pease, 1870a: 92. Kauai (in publication title): [no additional details].
        Remarks. Also listed by Pease (1870b: 651) with no description.

tenuicostata. (H)
    Helicter (Leptachatina) tenuicostata Pease, 1869a: 170. Hawaii: [no additional details].

terebralis. (O)
    Achatinella terebralis Gulick, 1856: 193, pl. 6, fig. 21. Oahu: "Kawailoa".

teres. (?O)
    Achatinella (Leptachatina) teres Pfeiffer, 1856b: 206. Hawaiian Islands ("Sandwich Islands" in publication title): [no additional details].

triticea. (O)
    Achatinella triticea Gulick, 1856: 184, pl. 6, fig. 12. Oahu: "Keawaawa".
        Remarks. Treated as a synonym of *oryza* Pfeiffer by Cooke (*in* Hyatt & Pilsbry, 1910: 28), who misspelled it "*tritacea*", but regarded as a distinct species by Pilsbry & Cooke (1914b: 2).

+turgidula. (K)
> *Leptachatina turgidula* Pease, 1870a: 89. Kauai: [no additional details].
> *Remarks.* "Variety" of *pachystoma* Pease, *teste* Cooke (*in* Hyatt & Pilsbry, 1910: 51). Also listed by Pease (1870b: 651) with no description.

turrita. (O)
> *Achatinella turrita* Gulick, 1856: 192, pl. 6, fig. 20. Oahu: "Mountain ravines of Lihue".

vana. (O)
> *Leptachatina vana* Sykes, 1900: 372, pl. 11, fig. 27. Oahu: "Mt. Kaala".

varia. (Mo)
> *Leptachatina* (*L.*) *varia* Cooke *in* Hyatt & Pilsbry, 1910: 32, pl. 11, fig. 1. Molokai: "Pali-ko-i in Halawa and . . . the mouth of Halawa Valley".

ventulus.
> *Helix* (*Cochlogena*) *ventulus* Férussac, 1821c: 60. *Nom. nud.*

ventulus. (O)
> *Helix ventulus* Férussac *in* Quoy & Gaimard, 1825d: 481. Guam [?error = Oahu (Hyatt & Pilsbry, 1911a: 54)]: [no additional details].

vitrea.
> *Achatinella vitrea* Newcomb, 1854a: 16, pl. 23, fig. 34 [1854b: 142, pl. 23, fig. 34]. Oahu: "Manoa".
> *Remarks.* Synonym of *cingula* Mighels, *teste* Cooke (*in* Hyatt & Pilsbry, 1910: 53).

vitreola. (M)
> *Achatinella vitreola* Gulick, 1856: 194, pl. 6, fig. 23. Hawaiian Islands: [no additional details].

Subgenus THAANUMIA Ancey

THAANUMIA Ancey, 1899: 269 (as genus). Type species: *Thaanumia omphalodes* Ancey, 1899, by monotypy.

dulcis. (M)
> *Leptachatina* (*Thaanumia*) *dulcis* Cooke *in* Hyatt & Pilsbry, 1911a: 85, pl. 13, figs. 8, 10. Maui: "East Maui: Ulapalakua, Makawao".

fuscula. (O)
> *Achatinella fuscula* Gulick, 1856: 180, pl. 6, fig. 8. Oahu: "Mountain forests of Mokuleia".

henshawi. (H)
> *Leptachatina henshawi* Sykes, 1903: 1, unnumbered fig. Hawaii: "Bucholtz, Kona, 1,800 feet".

morbida. (Mo)
> *Leptachatina* (*Thaanumia*) *morbida* Cooke *in* Hyatt & Pilsbry, 1911a: 87, pl. 13, fig. 12. Molokai: "Puu Kolekole".

omphalodes. (O)
> *Thaanumia omphalodes* Ancey, 1899: 269, pl. 12, fig. 8. Oahu: "Waianae Mountains".

optabilis. (O)
> *Leptachatina* (*Thaanumia*) *optabilis* Cooke *in* Hyatt & Pilsbry, 1911a: 84, pl. 13, fig. 9. Oahu: "Waianae Mts., back of Leilehua".

perforata. (K)
> *Leptachatina* (*Thaanumia*) *perforata* Cooke *in* Hyatt & Pilsbry, 1911a: 88, pl. 7, fig. 32. Kauai: "Puukapele".

thaanumi. (Mo)
> *Leptachatina* (*Thaanumia*) *thaanumi* Cooke *in* Hyatt & Pilsbry, 1911a: 88, pl. 6, figs. 16, 17. Molokai: "Mapulehu ridge".

Genus PAUAHIA Cooke

PAUAHIA Cooke *in* Hyatt & Pilsbry, 1911a: 80 (as *Leptachatina* subg.). Type species: *Leptachatina artata* Cooke, 1911, by original designation.

Raised to generic status by Pilsbry & Cooke (1914b: 15).

artata. (O)
    Leptachatina (*Pauahia*) *artata* Cooke *in* Hyatt & Pilsbry, 1911a: 80, pl. 13, figs. 1–4. Oahu: "Halawa, 1,500 ft.; Mt. Tantalus 2,000 ft."

chrysallis. (O)
    Achatina chrysallis Pfeiffer, 1855f: 99. Hawaiian Islands: [no additional details].

columna.
    Leptachatina columna Ancey, 1889b: 266. Oahu: [no additional details].
        Remarks. Synonym of *chrysallis* Pfeiffer, *teste* Cooke (*in* Hyatt & Pilsbry, 1911a: 82).

tantilla. (O)
    Leptachatina (*Pauahia*) *tantilla* Cooke *in* Hyatt & Pilsbry, 1911a: 81, pl. 13, figs. 5–7. Oahu: "Waianae Mts. back of Leilehua".

Questionably Included AMASTRIDAE in the Hawaiian Fauna

semicostata. (?not Hawaiian)
    Achatinella (*Leptachatina*) *semicostata* Pfeiffer, 1856b: 206. Hawaiian Islands ("Sandwich Islands" in publication title): [no additional details].
        Remarks. Placed in *Pauahia* by Pilsbry & Cooke (1914b: 16). C.M. Cooke, Jr. (unpublished notes) subsequently considered it not Hawaiian but from the Galapagos.

Family PUPILLIDAE

The pupillids are 1 of the major groups of land snails of the Pacific islands, although not endemic. The family-level classifications of different authors vary considerably (e.g., Boss, 1982; Solem, 1989, 1991; Tillier, 1989; Vaught, 1989; Zilch, 1959a), as do assignments of genera to families and subfamilies. For family-group classifications we take the conservative approach adopted by Solem (1989, 1991) and follow the *Manual of Conchology* (Pilsbry, 1935: vii–xii) which included in the Pupillidae the following subfamilies and genera represented in the Hawaiian fauna: Gastrocoptinae (*Gastrocopta*), Nesopupinae (*Lyropupa, Nesopupa, Pronesopupa*), Pupillinae (*Pupoidopsis*) and Vertigininae (*Columella*). Other authors have tended to split the Pupillidae (*sensu* Pilsbry), raising some subfamilies to family level; and, in addition, have placed certain genera in a number of different families/ subfamilies. The *Manual of Conchology* (Pilsbry, 1916–1918, 1920–1921, 1922–1926, 1927–1935; Pilsbry & Cooke, 1918–1920) subdivided the various genera into "sections"; we treat all these sections as subgenera.

If *cubana* Dall is excluded as not being definitively Hawaiian, the native Hawaiian fauna consists of 56 species (22 in *Lyropupa*, 20 in *Nesopupa*, 10 in *Pronesopupa* [all Nesopupinae], 1 in *Pupoidopsis* [Pupillinae], 3 in *Columella* [Vertigininae]) and 30 infraspecific taxa (15 in *Lyropupa*, 13 in *Nesopupa* (including *seminulum* Boettger, a junior homonym for which we do not provide a replacement name), 2 in *Pronesopupa*). One introduced species and 2 introduced infraspecific taxa, have also been described from the Hawaiian Islands, all in *Gastrocopta* (Gastrocoptinae).

Subfamily GASTROCOPTINAE

Genus GASTROCOPTA Wollaston

GASTROCOPTA Wollaston, 1878: 515 (as *Pupa* subg.). Type species: *Pupa acarus* Benson, 1856 (as "*G. acarus*") [not Hawaiian], by subsequent designation of Pilsbry (1916: 7) [see also ICZN (1957a: 167; Direction 72)].

Subgenus GASTROCOPTA Wollaston

GASTROCOPTA Wollaston, 1878: 515 (as *Pupa* subg.). Type species: *Pupa acarus* Benson, 1856 (as "*G. acarus*") [not Hawaiian], by subsequent designation of Pilsbry (1916: 7) [see also ICZN (1957a: 167; Direction 72)].

+**kailuana**. (O; introduced)
    Gastrocopta lyonsiana form *kailuana* Pilsbry, 1917a: 143, pl. 24, figs. 5, 6. Oahu: "On the north (Koolau or windward) side of Oahu, at Kaelepulu, Kailua".

lyonsiana.
    Pupa lyonsiana Ancey, 1892: 713. Oahu: "Punahou".
    Remarks. Synonym of *servilis* Gould, 1843 [not Hawaiian], *teste* Pilsbry (1934: 159).

servilis. (Midway, Pearl & Hermes, Laysan, O; introduced)
: *Pupa servilis* Gould, 1843 [not Hawaiian].
: *Remarks.* Included here only because *lyonsiana* Ancey, described from the Hawaiian Islands, is now considered a junior synonym of *servilis* Gould. Northwestern Hawaiian Islands localities from Conant *et al.* (1984).

Subgenus SINALBINULA Pilsbry

SINALBINULA Pilsbry, 1916: 11. Type species: *Pupa armigerella* Reinhardt, 1877 (as "*G. armigerella*") [not Hawaiian], by original designation.

+**nacca**. (O, H; introduced)
: *Vertigo nacca* Gould, 1862: 280. Hawaii: [no additional details].
: *Remarks.* Pilsbry (1917a: 147) included *nacca* in the synonymy of *pediculus* Shuttleworth, 1852 [not Hawaiian (Christensen & Kirch, 1981)] and (p. 148) included Oahu in the distribution of *pediculus*. Subsequently (p. 149–50) it seems that he treated *nacca* as the only "race" in the Hawaiian Islands, giving localities on both the islands of Oahu and Hawaii.

Subfamily NESOPUPINAE

Genus LYROPUPA Pilsbry

LYROPUPA Pilsbry, 1900: 432 (as *Nesopupa* sect.). Type species: *Pupa lyrata* Gould, 1843 (as "*N. lyrata*"), by original designation.

Subgenus LYROPUPA Pilsbry

LYROPUPA Pilsbry, 1900: 432 (as *Nesopupa* sect.). Type species: *Pupa lyrata* Gould, 1843, (as "*N. lyrata*"), by original designation.

+**baldwiniana**. (M)
: *Lyropupa rhabdota baldwiniana* Cooke *in* Pilsbry & Cooke, 1920: 241, pl. 20, figs. 7, 8. Maui: "West Maui: Iao".

carbonaria.
: *Lyropupa carbonaria* Ancey, 1904a: 125, pl. 7, fig. 21. Oahu: "In valle Nuuanu, prope Honolulu [= in Nuuanu Valley, near Honolulu]".
: *Remarks.* Synonym of *lyrata* Gould, *teste* Pilsbry & Cooke (1920: 236). We take "*carbonifera*" in Cooke (1907: 12) to be an incorrect spelling.

clathratula. (H)
: *Lyropupa clathratula* Ancey, 1904a: 125, pl. 7, fig. 19. Hawaii: "Olaa".

+**fossilis**. (O)
: *Lyropupa lyrata fossilis* Cooke & Pilsbry *in* Pilsbry & Cooke, 1920: 237, pl. 19, figs. 7, 11. Oahu: "Manoa . . . Waimanalu".

+**gouldi**. (O)
: *Lyropupa lyrata* form *gouldi* Pilsbry & Cooke, 1920: 235, pl. 19, figs. 8, 9. Type locality not given.

+**lanaiensis**. (L)
: *Lyropupa rhabdota lanaiensis* Cooke *in* Pilsbry & Cooke, 1920: 241, pl. 20, fig. 6. Lanai: [no additional details].

lyrata. (O)
: *Pupa lyrata* Gould, 1843: 139. Hawaiian Islands: [no additional details].

magdalenae.
>*Pupa magdalenae* Ancey, 1892: 716. Hawaiian Islands ("Iles Sandwich" in publication title): [no additional details].
>*Remarks.* Synonym of *lyrata* Gould, *teste* Pilsbry & Cooke (1920: 235).

microthauma. (O)
>*Lyropupa microthauma* Ancey, 1904a: 126, pl. 7, fig. 20. Oahu: "in valle Nuuanu [= in Nuuanu Valley]".

+pluris. (Mo)
>*Lyropupa rhabdota pluris* Pilsbry & Cooke, 1920: 240, pl. 20, figs. 3–5. Molokai: "along the pipe-line trail, upper Kaunakakai".

prisca. (H)
>*Lyropupa magdalenae* var. *prisca* Ancey, 1904b: 68, pl. 5, fig. 19. Hawaii: "Palihoukapapa, on the Hamakua slope of Mauna Kea, Kawaii [= Hawaii], at an elevation of 4,000 feet." [?error = Mana (Pilsbry & Cooke, 1920: 244)].
>*Remarks.* Raised to species status by Pilsbry & Cooke (1920: 243).

rhabdota. (Mo)
>*Lyropupa rhabdota* Cooke & Pilsbry *in* Pilsbry & Cooke, 1920: 239, pl. 20, fig. 2. Molokai: "Pelekunu".

striatula. (H)
>*Vertigo striatula* Pease, 1871b: 461. Hawaii: [no additional details].
>*Remarks.* Possibly "identical with *L. clathratula*" (Pilsbry & Cooke, 1920: 246). "Very close to *L. clathratula* . . . [but] . . . we allow both to stand as species" (Pilsbry & Cooke *in* Pilsbry, 1926: 223).

thaanumi. (M)
>*Lyropupa thaanumi* Cooke & Pilsbry *in* Pilsbry & Cooke, 1920: 242, pl. 20, figs. 12, 13. Maui: "East Maui: Auwahi".

truncata. (H)
>*Lyropupa truncata* Cooke, 1908a: 211, unnumbered fig. Hawaii: "Kohala Mts."

+uncifera. (O)
>*Lyropupa lyrata uncifera* Cooke & Pilsbry *in* Pilsbry & Cooke, 1920: 236, pl. 19, figs. 12, 13. Oahu: "coral bluff 1 1/2 miles west of Kahuku".

Subgenus LYROPUPILLA Pilsbry & Cooke

LYROPUPILLA Pilsbry & Cooke, 1920: 247 (as *Lyropupa* sect.). Type species: *Lyropupa spaldingi* Pilsbry & Cooke, 1920, by original designation.

anceyana. (H)
>*Lyropupa anceyana* Cooke & Pilsbry *in* Pilsbry & Cooke, 1920: 253, pl. 26, figs. 3, 6. Hawaii: "Olaa".

antiqua. (O)
>*Lyropupa antiqua* Cooke & Pilsbry *in* Pilsbry & Cooke, 1920: 250, pl. 21, figs. 8, 9, 11. Oahu: "Manoa, in pleistocene deposits along the Upper Manoa Road".

hawaiiensis. (H)
>*Lyropupa mirabilis* var. *hawaiiensis* Ancey, 1904b: 68, pl. 5, fig. 19. Hawaii: "Palihoukapapa, on the Hamakua slope of Mauna Kea, Kawaii [= Hawaii], at an elevation of 4,000 feet."
>*Remarks.* Raised to species status by Pilsbry & Cooke (1920: 251).

mirabilis. (O)
>*Pupa mirabilis* Ancey, 1890: 339. Oahu: [no additional details].

scabra. (M)
>*Lyropupa scabra* Pilsbry and Cooke, 1920: 254, pl. 26, figs. 1, 2. Maui: "East Maui: Ukulele".

PUPILLIDAE   [133]

+**sinulifera**. (Mo)
>   *Lyropupa sparna sinulifera* Pilsbry & Cooke, 1920: 253, pl. 22, fig. 13. Molokai: "Western ravine of Kamalo; also . . . along the pipe-line trail".

spaldingi. (O)
>   *Lyropupa spaldingi* Pilsbry & Cooke, 1920: 248, pl. 21, figs. 10, 12, 13. Oahu: "Puu Kaua".

sparna. (Mo, L)
>   *Lyropupa sparna* Cooke & Pilsbry *in* Pilsbry & Cooke, 1920: 252, pl. 22, figs. 6, 7, 10, 11. Molokai: "Kalihi".

Subgenus MIRAPUPA Cooke & Pilsbry

MIRAPUPA Cooke & Pilsbry *in* Pilsbry & Cooke, 1920: 255 (as *Lyropupa* sect.). Type species: *Vertigo perlonga* Pease, 1871 (as "*Lyropupa*"), by original designation.

costata. (Mo, L, Kah, H)
>   *Vertigo costata* Pease, 1871b: 461. Hawaii: [no additional details].
>   >   *Remarks.* Pilsbry & Cooke (1920: 273) considered *costata* of "uncertain genus", but later Pilsbry & Cooke (*in* Pilsbry, 1926: 223) referred it to *Lyropupa*. Being dextral, it falls in the subgenus *Mirapupa*, although Pilsbry & Cooke (*in* Pilsbry, 1926: 223) did not place it here explicitly.

cubana. (?Hawaiian Islands)
>   *Vertigo cubana* Dall, 1890: 1, figs. 1, 2. Cuba [?error = Hawaiian Islands (Pilsbry & Cooke, 1920: 269)]: [no additional details].

+**cylindrata**. (O)
>   *Lyropupa perlonga* form *cylindrata* Pilsbry & Cooke, 1920: 261, pl. 23, fig. 8. Oahu: "Makua".

cyrta. (H)
>   *Lyropupa cyrta* Cooke & Pilsbry *in* Pilsbry & Cooke, 1920: 268, pl. 23, figs. 9, 10. Hawaii: "Mana".

+**filicostata**. (N, K)
>   *Lyropupa perlonga filicostata* Cooke & Pilsbry *in* Pilsbry & Cooke, 1920: 262, pl. 23, fig. 12. Kauai: "Limahuli".

+**interrupta**. (O)
>   *Lyropupa perlonga interrupta* Pilsbry & Cooke, 1920: 261, pl. 22, fig. 5, pl. 25, figs. 1–4, 10. Oahu: "debris of the 'coral bluff' 1 1/2 miles west of Kahuku . . . Maleakahana [= Malaekahana]".

kahoolavensis.
>   *Lyropupa kahoolavensis* Pilsbry & Cooke, 1920: 256, pl. 22, figs. 1–4, 8, 9. Kahoolawe: "Hakioawa".
>   >   *Remarks.* Synonym of *costata* Pease, *teste* Pilsbry & Cooke (*in* Pilsbry, 1926: 224).

+**kona**. (Mo, ?M, H)
>   *Lyropupa ovatula kona* Pilsbry & Cooke, 1920: 266, pl. 26, figs. 10, 11, 14, 15. Hawaii: "North Kona at Huehue".

+**maunaloae**. (Mo)
>   *Lyropupa micra maunaloae* Pilsbry & Cooke, 1920: 264, pl. 25, figs. 8, 9. Molokai: "summit of Mauna Loa, and at Moomomi on the north shore, near sea level . . . near the shifting sands, Mauna Loa".

micra. (O)
>   *Lyropupa micra* Cooke & Pilsbry *in* Pilsbry & Cooke, 1920: 263, pl. 23, fig. 7, pl. 25, figs. 5–7. Oahu: "Kaelepulu, Kailua, on a lime-rock bench about 1/4 mile from the north shore".

+**moomomiensis.** (Mo)
>*Lyropupa ovatula moomomiensis* Pilsbry & Cooke *in* Pilsbry, 1926: 225, pl. 28, figs. 3–5. Molokai: "Moomomi".

ovatula. (O)
>*Lyropupa ovatula* Cooke & Pilsbry *in* Pilsbry & Cooke, 1920: 265, pl. 23, fig. 11, pl. 24, figs. 1–5. Oahu: "Kaelepulu, Kailua, in crevices and along the base of a low lime-rock bluff about 1/4 mile from the shore".

+**percostata.** (O)
>*Lyropupa micra percostata* Pilsbry & Cooke, 1920: 264, pl. 25, figs. 11, 12. Oahu: "Kaelepulu, Kailua, on a lime-rock bluff about a quarter mile from the shore".

perlonga. (O)
>*Vertigo perlonga* Pease, 1871b: 462. Oahu: [no additional details].

plagioptyx. (O)
>*Lyropupa plagioptyx* Pilsbry & Cooke, 1920: 267, pl. 24, figs. 8, 11, 12. Oahu: "Kawaihapai, on a steep, wooded hillside about 500 ft. above the plain and perhaps 3/4 mile from the sea".

+**puukolekolensis.** (Mo)
>*Lyropupa kahoolavensis puukolekolensis* Pilsbry & Cooke, 1920: 258, pl. 26, figs. 9, 12. Molokai: "Puukolekole".

thaumasia. (K)
>*Lyropupa thaumasia* Cooke & Pilsbry *in* Pilsbry & Cooke, 1920: 270, pl. 24, figs. 13–15, pl. 25, fig. 14. Kauai: "Hanakapiai".

Genus NESOPUPA Pilsbry

PTYCHOCHILUS Boettger, 1881: 47 (as *Vertigo* subsect.). Type species: *Pupa (Vertigo) tantilla* Gould, 1847 [not Hawaiian], by original designation.

PTYCHOCHYLUS: Incorrect original spelling of *Ptychochilus* Boettger (Boettger, 1881: 48).

NESOPUPA Pilsbry, 1900: 431. Type species: *Pupa (Vertigo) tantilla* Gould, 1847 [not Hawaiian], automatic. [Unnecessary n.n. for *Ptychochilus* Boettger, 1881].

Ptychochilus Boettger has priority over *Nesopupa* Pilsbry. However, the latter name has been in common usage since its proposal, while the former appears never to have been used. We have applied to the ICZN (Application no. S.2904, Cowie *et al.*, 1994) to conserve *Nesopupa* Pilsbry and to reject *Ptychochilus* Boettger (and its alternative original spelling). *Ptychochilus* is considered the correct original spelling.

Subgenus INFRANESOPUPA Cooke & Pilsbry

INFRANESOPUPA Cooke & Pilsbry *in* Pilsbry & Cooke, 1920: 289 (as *Nesopupa* sect.). Type species: *Nesopupa (Infranesopupa) limatula* Cooke & Pilsbry, 1920, by original designation.

anceyana. (H)
>*Nesopupa (Infranesopupa) anceyana* Cooke & Pilsbry *in* Pilsbry & Cooke, 1920: 293, pl. 28, figs. 2, 3. Hawaii: "Humuula".

bishopi. (M)
>*Nesopupa (Infranesopupa) bishopi* Cooke & Pilsbry *in* Pilsbry & Cooke, 1920: 296, pl. 28, fig. 4. Maui: "E. Maui: Haleakala Crater, near Crystal Cave".

dubitabilis. (Mo)
>*Nesopupa (Infranesopupa) dubitabilis* Cooke & Pilsbry *in* Pilsbry & Cooke, 1920: 291, pl. 28, fig. 9. Molokai: "Poholua".

PUPILLIDAE [135]

forbesi. (H)
> *Nesopupa (Infranesopupa) forbesi* Cooke & Pilsbry *in* Pilsbry & Cooke, 1920: 297, pl. 28, fig. 5. Hawaii: "Huumula . . . in a large *kipuka* in the 1855 Flow, about half way between Halealoha and Ainahou, at about 5,000 ft. elevation".

infrequens. (K)
> *Nesopupa (Infranesopupa) infrequens* Cooke & Pilsbry *in* Pilsbry & Cooke, 1920: 298, pl. 28, fig. 7. Kauai: "Halemanu".

+kaalaensis. (O)
> *Nesopupa (Infranesopupa) dubitabilis kaalaensis* Cooke & Pilsbry *in* Pilsbry & Cooke, 1920: 292, pl. 28, fig. 13. Oahu: "Kaala, eastern spur, about 2,500 ft. elevation".
>> *Remarks.* Cooke & Pilsbry did not designate a single type locality, but they did designate a "type" (BPBM 11069), the label of which says "Kaala".

limatula. (M)
> *Nesopupa (Infranesopupa) limatula* Cooke & Pilsbry *in* Pilsbry & Cooke, 1920: 290, pl. 28, figs. 6, 10. Maui: "Ainahou".
>> *Remarks.* Cooke & Pilsbry did not designate a single type locality, but they did designate a "type" (BPBM 11067), the label of which says "Ainahou".

subcentralis. (H)
> *Nesopupa (Infranesopupa) subcentralis* Cooke & Pilsbry *in* Pilsbry & Cooke, 1920: 294, pl. 28, fig. 8. Hawaii: "Palihoukapapa".

Subgenus LIMBATIPUPA Cooke & Pilsbry

LIMBATIPUPA Cooke & Pilsbry *in* Pilsbry & Cooke, 1920: 306 (as *Nesopupa* sect.). Type species: *Pupa newcombi* Pfeiffer, 1853 (as "*N. newcombi*"), by original designation.

alloia. (K)
> *Nesopupa (Limbatipupa) alloia* Cooke & Pilsbry *in* Pilsbry & Cooke, 1920: 321, pl. 29, fig. 10. Kauai: "Hanapepe falls".

+angusta. (K)
> *Nesopupa (Limbatipupa) newcombi* form *angusta* Cooke & Pilsbry *in* Pilsbry & Cooke, 1920: 315, text-fig. 4 (p. 309). Kauai: "Kipu" (p. 308).

costulosa.
> *Vertigo costulosa* Pease, 1871b: 462. Hawaii: [no additional details].
>> *Remarks.* Synonym of *newcombi* Pfeiffer, *teste* Cooke & Pilsbry (*in* Pilsbry & Cooke, 1920: 309).

+disjuncta. (O)
> *Nesopupa (Limbatipupa) newcombi* form *disjuncta* Pilsbry & Cooke, 1920: 317, text-fig. 13 (p. 309). Oahu: "Mokuleia" (p. 308).

+gnampta. (?K, O)
> *Nesopupa (Limbatipupa) newcombi gnampta* Cooke & Pilsbry *in* Pilsbry & Cooke, 1920: 317; text-figs. 14, 15 (p. 309). Oahu: "Luakaha, Nuuanu" (p. 308).

+interrupta. (H)
> *Nesopupa (Limbatipupa) newcombi interrupta* Cooke & Pilsbry *in* Pilsbry & Cooke, 1920: 315, text-figs. 4a–6a (p. 309). Hawaii: "Waiaha" (p. 308).

kauaiensis. (K)
> *Nesopupa kauaiensis* Ancey, 1904a: 124, pl. 7, fig. 17. Kauai: "Kipu".

+multidentata. (O)
> *Nesopupa (Limbatipupa) newcombi* form *multidentata* Cooke & Pilsbry *in* Pilsbry & Cooke, 1920: 315, text-figs. 3, 3a (p. 309). Oahu: "Glen Ada, Nuuanu".

newcombi. (O, Mo, L, H)
> *Pupa newcombi* Pfeiffer, 1853d: 530 [1854a: 69]. Hawaiian Islands: [no additional details].

oahuensis. (O)
> *Nesopupa* (*Limbatipupa*) *oahuensis* Cooke & Pilsbry *in* Pilsbry & Cooke, 1920: 317, pl. 29, figs. 11, 12. Oahu: "Nuuanu Valley at Luakaha falls".

seminulum. (K, O, Mo, M, H)
> *Pupa newcombi* var. *seminulum* Boettger, 1881: 58, pl. 12, fig. 14. Hawaii: [no additional details].
>> Remarks. Primary junior homonym of *Pupa seminulum* Lowe, 1852 [not Hawaiian]. No new name proposed here, pending further revision (cf. Cooke & Pilsbry *in* Pilsbry & Cooke, 1920: 314).

singularis. (O, M)
> *Nesopupa* (*Limbatipupa*) *singularis* Cooke & Pilsbry *in* Pilsbry & Cooke, 1920: 320, pl. 29, fig. 8. Oahu: "Kaliuwaa".
>> Remarks. Cooke & Pilsbry did not designate a single type locality, but they did designate a "type" (= holotype, BPBM 11077), the label of which says "Kaliuwaa".

Subgenus NESODAGYS Cooke & Pilsbry

NESODAGYS Cooke & Pilsbry *in* Pilsbry & Cooke, 1920: 299 (as *Nesopupa* sect.). Type species: *Nesopupa wesleyana* Ancey, 1904, by subsequent designation of Zilch (1959a: 152).

+gouveiae. (H)
> *Nesopupa* (*Nesodagys*) *wesleyana* form *gouveiae* Cooke & Pilsbry *in* Pilsbry & Cooke, 1920: 301, pl. 29, fig. 4. Hawaii: "Hookena".
>> Remarks. Cooke & Pilsbry did not designate a single type locality, but they did designate a "type" (= holotype, BPBM 11081), the label of which says "Hookena".

kamaloensis.
> *Nesopupa* (*Nesodagys*) *wesleyana rhadina* form *kamaloensis* Cooke & Pilsbry *in* Pilsbry & Cooke, 1920: 303, pl. 29, fig. 9. Unavailable name; infrasubspecific.

+rhadina. (K, O, Mo, M, L)
> *Nesopupa* (*Nesodagys*) *wesleyana rhadina* Cooke & Pilsbry *in* Pilsbry & Cooke, 1920: 301, pl. 29, fig. 13. Molokai: "Poholua".

thaanumi. (O, M, L, H)
> *Nesopupa thaanumi* Ancey, 1904a: 123. Hawaii: "Olaa".

+tryphera. (K, O, Mo)
> *Nesopupa* (*Nesodagys*) *wesleyana* form *tryphera* Cooke & Pilsbry *in* Pilsbry & Cooke, 1920: 301, pl. 29, fig. 3. Oahu: "Palolo".

wesleyana. (O, M, Kah, H)
> *Nesopupa wesleyana* Ancey, 1904a: 123, pl. 7, fig. 16. Hawaii: "Hilo, 4 miles Olaa road".
>> Remarks. Ancey (1904a: 123) gave more than 1 locality. Cooke & Pilsbry (*in* Pilsbry & Cooke, 1920: 300) designated a lectotype from the locality above.

Subgenus NESOPUPILLA Pilsbry & Cooke

NESOPUPILLA Pilsbry & Cooke, 1920: 278 (as *Nesopupa* sect.). Type species: *Nesopupa* (*Nesopupilla*) *waianaeensis* Cooke & Pilsbry, 1920, by original designation.

bacca. (H)
> *Vertigo bacca* Pease, 1871b: 462. Hawaii: "Kalapana".
>> Remarks. Cooke & Pilsbry (*in* Pilsbry & Cooke, 1920: 279) considered this an "unidentified" species but nevertheless placed it in *Nesopupa* (*Nesopupilla*).

PUPILLIDAE [137]

baldwini. (Mo, M, L, H)
> *Nesopupa baldwini* Ancey, 1904a: 122, pl. 7, fig. 13. Maui [see Remarks]: "Kaupakalua".
>> *Remarks.* Ancey (1904a: 122) gave "Molokai" and "Kaupakalua, Maui" as localities. Cooke & Pilsbry (*in* Pilsbry & Cooke, 1920: 287) designated a "holotype" [= lectotype] (BPBM 18698), of which the label and the catalog ledger entry say "Maui"; labels on paralectotype lots also say "Kaupakalua".

+centralis. (H)
> *Nesopupa baldwini* var. *centralis* Ancey, 1904a: 122. Hawaii: "Olaa".

dispersa. (O, Mo, M, L, Kah, H)
> *Nesopupa (Nesopupilla) dispersa* Cooke & Pilsbry *in* Pilsbry & Cooke, 1920: 284, pl. 27, figs. 7, 8. Oahu: "Makua".

+lanaiensis. (L)
> *Nesopupa (Nesopupilla) baldwini lanaiensis* Pilsbry & Cooke, 1920: 289, pl. 27, figs. 13–15. Lanai: [no additional details].

litoralis. (O)
> *Nesopupa (Nesopupilla) litoralis* Cooke & Pilsbry *in* Pilsbry & Cooke, 1920: 283, pl. 28, fig. 1. Oahu: "Ewa".

plicifera. (O)
> *Nesopupa plicifera* Ancey, 1904a: 122, pl. 7, fig. 14. Oahu: "in valle Nuuanu [= in Nuuanu valley]".

+subcostata. (Mo)
> *Nesopupa (Nesopupilla) baldwini subcostata* Pilsbry & Cooke, 1920: 288, pl. 27, figs. 11, 12. Molokai: "upper Kaunakakai, along the pipe-line trail".

waianaensis. (O)
> *Nesopupa (Nesopupilla) waianaensis* Cooke & Pilsbry *in* Pilsbry & Cooke, 1920: 281, pl. 27, figs. 4–6. Oahu: "Waianae Mts. at Pukaloa, in the open valley under stones near the 'Hunter's Cabin'".

Genus PRONESOPUPA Iredale

PRONESOPUPA Iredale, 1913: 384. Type species: *Pronesopupa senex* Iredale, 1913 [not Hawaiian], by monotypy.

Subgenus EDENTULOPUPA Cooke & Pilsbry

EDENTULOPUPA Cooke & Pilsbry *in* Pilsbry, 1920: 11 (as *Pronesopupa* sect.). Type species: *Pupa admodesta* Mighels, 1845, by original designation.

admodesta. (K, O, Mo, H)
> *Pupa admodesta* Mighels, 1845: 19. Oahu: [no additional details].

Subgenus PRONESOPUPA Iredale

PRONESOPUPA Iredale, 1913: 384 (as genus). Type species: *Pronesopupa senex* Iredale, 1913 [not Hawaiian], by monotypy.

acanthinula. (O, Mo, M, H)
> *Pupa acanthinula* Ancey, 1892: 709. Oahu: "Makiki".

boettgeri. (K, O, Mo, M, L, H)
> *Pronesopupa (Pronesopupa) boettgeri* Cooke & Pilsbry *in* Pilsbry, 1920: 8, pl. 1, fig. 17. Oahu: "Tantalus".

hystricella. (K, O, Mo, M, L, H)
>*Pronesopupa (Pronesopupa) hystricella* Cooke & Pilsbry *in* Pilsbry, 1920: 7, pl. 1, fig. 12. Hawaii: "Hilo, Reed's Island".

+spinigera. (K, O, Mo, M, H)
>*Pronesopupa (Pronesopupa) boettgeri spinigera* Cooke & Pilsbry *in* Pilsbry, 1920: 10, pl. 1, fig. 11. Oahu: "Nuuanu, Luukaha".
>>*Remarks.* Cooke & Pilsbry (*in* Pilsbry, 1920: 10) published only very broad locality information, but did designate a holotype (BPBM 11031), the label of which says "Oahu, Nuuanu, Luukaha".

Subgenus SERICIPUPA Cooke & Pilsbry

SERICIPUPA Cooke & Pilsbry *in* Pilsbry, 1920: 13 (as *Pronesopupa* sect.). Type species: *Pronesopupa (Sericipupa) frondicola* Cooke & Pilsbry, 1920, by original designation.

+corticicola. (M)
>*Pronesopupa (Sericipupa) frondicola corticicola* Cooke & Pilsbry *in* Pilsbry, 1920: 14, pl. 1, fig. 3. Maui: "E. Maui: Puunianiau at 7,000 ft."

frondicola. (M)
>*Pronesopupa (Sericipupa) frondicola* Cooke & Pilsbry *in* Pilsbry, 1920: 13, pl. 1, fig. 4. Maui: "E. Maui: Ainahou, at the head of Keanae Gap, Haleakala".

incerta. (K)
>*Pronesopupa (Sericipupa) incerta* Cooke & Pilsbry *in* Pilsbry, 1920: 16, pl. 1, fig. 6. Kauai: "Halemanu".

lymaniana. (H)
>*Pronesopupa (Sericipupa) lymaniana* Cooke & Pilsbry *in* Pilsbry, 1920: 18, pl. 1, fig. 2. Hawaii: "28 1/2 miles Olaa road".

molokaiensis. (Mo)
>*Pronesopupa (Sericipupa) molokaiensis* Cooke & Pilsbry *in* Pilsbry, 1920: 15, pl. 1, fig. 5. Molokai: "Kawela, at about 3,500 ft."

orycta. (H)
>*Pronesopupa (Sericipupa) orycta* Cooke & Pilsbry *in* Pilsbry, 1920: 18, pl. 1, fig. 10. Hawaii: "Palihoukapapa".

sericata. (H)
>*Pronesopupa (Sericipupa) sericata* Cooke & Pilsbry *in* Pilsbry, 1920: 17, pl. 1, fig. 1. Hawaii: "Piihonua, (a hill) in the flow of 1855, about 5,000 feet elevation".

Subfamily PUPILLINAE

Genus PUPOIDOPSIS Pilsbry & Cooke

PUPOIDOPSIS Pilsbry & Cooke *in* Pilsbry, 1921a: 106. Type species: *Pupoidopsis hawaiensis* Pilsbry & Cooke, 1921, by original designation.

hawaiensis. (O, Mo, M)
>*Pupoidopsis hawaiensis* Pilsbry & Cooke *in* Pilsbry 1921a: 107, pl. 17, fig. 2. Oahu: "Kaelepulu, Kailua".
>>*Remarks.* This is 1 of only very few indigenous Hawaiian land snail species known also to occur outside the Hawaiian Islands (Christensen & Kirch, 1986: 59; Cooke & Neal, 1928: 28, 29). The others are *Striatura pugetensis* (Dall, 1895) (Gastrodontidae) and perhaps *Vitrina tenella* Gould, 1846 (Zonitidae), *Succinea tahitensis* Pfeiffer, 1847 (Succineidae), and a number of Ellobiidae.

Subfamily VERTIGININAE

Genus COLUMELLA Westerlund

SPHYRADIUM: authors, not Charpentier, 1837, misidentification.
COLUMELLA Westerlund, 1878: 193. Type species: *Pupa inornata* Michaud, 1831 [= *Pupa edentula* Draparnaud, 1805] [not Hawaiian], by subsequent designation of Kennard & Woodward (1926: 129).

The designation of *Pupa inornata* Michaud, 1831 as type of *Columella* by Kennard & Woodward (1926: 129) was published on 27 March 1926. The designation of *Pupa edentula* Draparnaud, 1805 by Pilsbry (1926: 233) was published on 1 April 1926. *Pupa edentula* Draparnaud, 1805 was included as a synonym of *inornata* by Westerlund (1878: 193) and so qualifies as an originally included nominal species (*Code* Art. 69(a)(i)). We are not aware of an earlier type species designation.

In the *Manual of Conchology*, Pilsbry (1926: 232) referred the Hawaiian species only tentatively to *Columella*.

alexanderi. (M)
> *Sphyradium alexanderi* Cooke & Pilsbry *in* Pilsbry & Cooke, 1906: 216, fig. 3. Maui: "West Maui, at the top of Mt. Kukui, elevation about 6,000 feet."

olaaensis. (H)
> *Columella olaaensis* Pilsbry, 1926: 248, pl. 30, fig. 7. Hawaii: "Olaa".

sharpi. (H)
> *Sphyradium sharpi* Pilsbry & Cooke, 1906: 215, figs. 1, 2. Hawaii: "Crest of the Kilauea crater, about a half mile south of the hotel."

Family FERUSSACIIDAE

The Ferussaciidae are not native to the Hawaiian Islands but a single taxon has been described from Hawaiian material.

Genus CECILIOIDES Férussac

CECILIOIDES Férussac, 1814: 48. Type species: *Buccinum acicula* Müller, 1774 [not Hawaiian], by monotypy [see ICZN (1955: 48, 56; Opinion 335), Warén & Gittenberger (1993: 107)].

CAECILIANELLA Bourguignat, 1856: 378. Type species: *Buccinum acicula* Müller, 1774 [not Hawaiian], automatic. [Unnecessary n.n. for *Cecilioides* Férussac, 1814].

baldwini. (O, H; introduced)
    Caecilianella baldwini Ancey, 1892: 718. Oahu: "Manoa".
        Remarks. Probably a synonym of the widely distributed synanthropic *aperta* Swainson, 1840, *teste* Pilsbry (1906a: 46) and F. Naggs (*in litt.* to RHC, 26 October 1993).

Family SUBULINIDAE

A number of subulinid species have been dispersed by human agency throughout the tropics (e.g., Christensen & Kirch, 1981, 1986; Solem, 1959, 1978, 1989). All the taxa known from the Hawaiian Islands have been artificially introduced to the archipelago. We list those taxa that, although introduced, have been described from Hawaiian material. Some of these are now considered junior synonyms of extralimital taxa, which are therefore also listed.

Following Naggs (1994), we treat *Allopeas* as a genus, not as a subgenus of *Lamellaxis* Strebel & Pfeiffer). Also following Naggs (1994), we treat *Paropeas* Pilsbry as a genus, not as a subgenus of *Prosopeas* Mörch, placing *achatinaceum* Pfeiffer and its synonym *henshawi* Sykes in *Paropeas*.

Subfamily SUBULININAE

Genus ALLOPEAS Baker

ALLOPEAS Baker, 1935: 84 (as *Lamellaxis* subg.). Type species: *Bulimus gracilis* Hutton, 1834 (as "*Lamellaxis*") [not Hawaiian], by original designation.

It appears that *Allopeas* Baker, 1935 is a junior subjective synonym of the non-Hawaiian *Tortaxis* Pilsbry, 1906 (Naggs, 1992: 259). However, we retain *Allopeas* in its current usage, following Naggs' (1992) application to the ICZN and pending the decision of the ICZN. See also Naggs (1994).

gracile. (O, Mo, H ["All the Islands" (see *junceus*; Baldwin, 1893: 17)]; introduced)
    Bulimus gracilis Hutton, 1834 [not Hawaiian].
        Remarks. Included here only because *junceus* Gould, described from the Hawaiian Islands, is now considered a junior synonym of *gracile* Hutton. Placed in *Allopeas* following Baker (1935: 84, 1945: 88), Pilsbry (1946: 175), Solem (1959: 119) and Naggs (1994).

+hawaiiense. (H; introduced)
    Opeas prestoni var. *hawaiiensis* Sykes, 1904a: 113, Fig. 3. Hawaii: "Kawailoa, Mauna Loa at 1,500 feet . . . Hilo".
        Remarks. Subspecies of *clavulinum* Potiez & Michaud, 1838 [not Hawaiian], *teste* Pilsbry (1906b: 136). Placed in *Allopeas*, following Smith (1992: 309) and Naggs (1994).

junceus.
    Bulimus junceus Gould, 1846f: 191. Society Islands, Hawaiian Islands: [no additional details].
        Remarks. Considered a synonym of *oparanus* Pfeiffer [not Hawaiian] by Caum (1928: 61); *oparanus* Pfeiffer subsequently treated as a synonym of *gracile* Hutton [not Hawaiian] by, e.g., Solem (1978: 43). Synonym of *gracile* Hutton. **N. syn.**

pyrgiscus.
> *Bulimus pyrgiscus* Pfeiffer, 1861: 24. Hawaiian Islands: [no additional details].
> *Remarks.* Considered a synonym of *oparanus* Pfeiffer [not Hawaiian] by Pilsbry (1906c: 184); *oparanus* Pfeiffer subsequently synonymized with *gracile* Hutton [not Hawaiian] by, e.g., Solem (1978: 43). Synonym of *gracile* Hutton. **N. syn**.

Genus OPEAS Albers

OPEAS Albers, 1850: 175. Type species: *Helix goodallii* Miller, 1822 (as "*Stenogyra goodalli*") [= *Bulimus pumilus* Pfeiffer, *teste* Zilch (1959b: 351)] [not Hawaiian], by subsequent designation of Martens (1860: 265).

opella. (O, H; introduced)
> *Opeas opella* Pilsbry & Vanatta, 1906: 785, fig. 1. Oahu: "Honolulu"; Hawaii: "Hilo".
> *Remarks.* Possibly not correctly placed in *Opeas* but, as we are aware of no subsequent treatment referring it to another genus, we retain it in *Opeas* for the present.

Genus PAROPEAS Pilsbry

PAROPEAS Pilsbry, 1906a: 14 (as *Prosopeas* subg.). Type species: *Bulimus acutissimum* Mousson, 1857 (as "*P. acutissimum*") [not Hawaiian], by original designation.

achatinaceum. (K, O, H; introduced)
> *Bulimus achatinaceus* Pfeiffer, 1846 [not Hawaiian].
> *Remarks.* Included here only because *henshawi* Sykes, described from the Hawaiian Islands, is now considered a junior synonym of *achatinaceum* Pfeiffer.

henshawi.
> *Opeas henshawi* Sykes, 1904a: 112, fig. 2. Hawaii: "Hilo".
> *Remarks.* Synonym of *achatinaceum* Pfeiffer, *teste* Jutting (1952: 387) and Naggs (1994).

Family ENDODONTIDAE

As treated by Solem (1976), the most recent reviser of the group, the Endodontidae are endemic to the Pacific Basin. Solem (1976) considered the endodontoids of the Pacific (Endodontidae, Charopidae, Punctidae) the most diverse land snail group in the region, surpassing in numbers of species both the Achatinellidae and Amastridae. Endodontids are now rarely seen in the Hawaiian Islands and many species may well be extinct; predation by introduced ants perhaps having been a major factor (Solem, 1974, 1976, 1983; Tillier & Clarke, 1983). As with much of the Pacific land snail fauna, little is known of the habitat or habits of endodontids (summarized by Solem, 1976: 100–01).

Solem (1976) included 31 Hawaiian species of Endodontidae (17 in *Cookeconcha*, 10 in *Endodonta* and 4 in *Nesophila*). Solem (1977) added 2 species (1 in *Cookeconcha* and 1 in his new genus *Protoendodonta*) and Christensen (1982) an additional 1 (in *Endodonta*), giving a total of 34 species. There is, in addition, a single infraspecific taxon (*albina* Ancey). We follow Solem's synonymies, placements of species in genera, etc. Distribution data are from Solem (1976) and the original descriptions. Solem (1976: 3) estimated that 199–205 Hawaiian endodontoid species (including 5–9 Punctidae—see below) are represented in the Bishop Museum (Honolulu) collections, most undescribed.

Genus COOKECONCHA Solem

COOKECONCHA Solem, 1976: 207. Type species: *Helix hystrix* Pfeiffer, 1846, by original designation.

Solem (1976) treated his genera *Kondoconcha* [not Hawaiian] and *Cookeconcha* as masculine and *Rhysoconcha* [not Hawaiian] as feminine. We follow Brown (1956: 530) in treating the term "concha", and hence *Cookeconcha*, as feminine.

antiqua. (Midway)
>*Cookeconcha antiquus* Solem, 1977: 905, pl. 1, figs. 1–6, pl. 2, fig. 1. Midway: "Sand Island core (S-22) at 42.3-m depth in coarse foraminiferal sand."

contorta. (O)
>*Helix contorta* Férussac *in* Quoy & Gaimard, 1825d: 469. Hawaiian Islands: [no additional details].
>>*Remarks.* Validated by Férussac *in* Quoy & Gaimard (1825d: 469) by the provision of distinguishing characters in the text, although the plate referred to in that work (pl. 51a) was not published until 1832. "Probably Waianae Mountains, Oahu" (Solem, 1976: 215).

cookei. (O)
>*Endodonta (Thaumatodon) cookei* Cockerell, 1933: 58. Oahu: "Mt. Tantalus".

decussatula. (Mo, M)
>*Helix decussatula* Pease, 1866: 291. Hawaiian Islands: [no additional details].

elisae. (?Hawaiian Islands)
> *Pitys elisae* Ancey, 1889a: 180. Hawaiian Islands: [no additional details].
> *Remarks.* Although Caum (1928: 650) explicitly considered this species not Hawaiian, its status remains uncertain (Solem, 1976: 216) and it is retained as a Hawaiian species for the present.

filicostata.
> *Pitys filicostata* Pease, 1871b: 454. Kauai: [no additional details].
> *Remarks.* Synonym of *paucicostata* Pease, *teste* Solem (1976: 218).

henshawi. (H)
> *Endodonta* (*Thaumatodon*) *henshawi* Ancey, 1904b: 66, pl. 5, figs. 15, 16. Hawaii: "Palihoukapapa, on the Hamakua slope of Mauna Kea, Kawaii [= Hawaii], at an elevation of 4,000 feet" (in the introduction to the paper).

hystricella. (?K, O, ?M)
> *Helix hystricella* Pfeiffer, 1859a: 25. Hawaiian Islands: [no additional details].

hystrix. (O)
> *Helix hystrix* Pfeiffer, 1846e: 67. Hawaiian Islands: [no additional details].
> *Remarks.* Probably originally a Mighels manuscript name.

intercarinata.
> *Helix intercarinata* Mighels, 1845: 18. Oahu: [no additional details].
> *Remarks.* Synonym of *contorta* Férussac, *teste* Solem (1976: 215).

jugosa. (K, ?O)
> *Helix jugosa* Mighels, 1845: 19. Kauai: "Waioli" (Kauai not explicitly stated by Mighels).

lanaiensis. (?K, L, ?H)
> *Endodonta* (*Nesophila*) *lanaiensis* Sykes, 1896: 127. Lanai: "Mountains of Lanai, behind Koele".

luctifera. (Mo)
> *Endodonta* (*Thaumatodon*) *luctifera* Pilsbry & Vanatta, 1905: 575, pl. 39, figs. 4–6. Hawaiian Islands: [no additional details].

nuda. (H)
> *Endodonta* (*Thaumatodon*) *nuda* Ancey, 1899: 268, pl. 12, fig. 1. Hawaii: "Olaa, Central Hawaii".

paucicostata. (K)
> *Pithys paucicostata* Pease, 1871a: 395. Kauai: [no additional details].

paucilamellata. (H)
> *Endodonta hystricella* var. *paucilamellata* Ancey, 1904b: 67, pl. 5, fig. 17. Hawaii: "Palihoukapapa, on the Hamakua slope of Mauna Kea, Kawaii [= Hawaii], at an elevation of 4,000 feet" (in the introduction to the paper).
> *Remarks.* Raised to species status by Solem (1976: 219).

ringens. (?Mo, L)
> *Endodonta* (*Thaumatodon*) *ringens* Sykes, 1896: 126. Lanai: "Mountains of Lanai, behind Koele".

rubiginosa.
> *Helix rubiginosa* Gould, 1846a: 173. Kauai: [no additional details].
> *Remarks.* Primary junior homonym of *rubiginosa* Rossmässler, 1838 [not Hawaiian]. We follow Solem (1976: 222) in treating *rubiginosa* as a synonym of *jugosa*. See Solem (1976: 220, 222) for discussion of problems of synonymy.

setigera.
> *Helix setigera* Gould, 1844: 174. Hawaiian Islands ("Sandwich Islands" in the introduction to the paper): [no additional details].
> *Remarks.* Primary junior homonym of *setiger* Sowerby, 1841 [not Hawaiian]. Synonym of *hystrix* Pfeiffer, *teste* Solem (1976: 220).

stellula. (M)
>   *Helix stellula* Gould, 1844: 174. Hawaiian Islands [= Maui (Gould, 1852: 57; Johnson, 1964: 152; Solem, 1976: 218)]: [no additional details].

thaanumi. (Mo, M, H)
>   *Endodonta* (*Nesophila*) *thaanumi* Pilsbry & Vanatta, 1905: 574, pl. 39, figs. 1–3. Hawaii: "Kaiwiki, near Hilo".

thwingi. (H)
>   *Endodonta thwingi* Ancey, 1904b: 66. Hawaii: "in an extinct crater of the Kona coast".

Genus ENDODONTA Albers

ENDODONTA Albers, 1850: 89. Type species: *Helix lamellosa* Férussac, 1825, by subsequent designation of Martens (1860: 90).

apiculata. (K)
>   *Endodonta apiculata* Ancey, 1889a: 188. Kauai: [no additional details].

binaria. (K)
>   *Helix binaria* Pfeiffer, 1856e: 33. Hawaiian Islands: [no additional details].

concentrata. (L)
>   *Endodonta concentrata* Pilsbry & Vanatta, 1906: 785, pl. 43, figs. 5, 6. Lanai: [no additional details].

ekahanuiensis. (O)
>   *Endodonta ekahanuiensis* Solem, 1976: 375, figs. 166a–c. Oahu: "Loc. 3 (= W410C–6), north branch, south Ekahanui Gulch, Waianae Mountains".

fricki. (O)
>   *Helix fricki* Pfeiffer, 1858: 21, pl. 40, fig. 3. Hawaiian Islands: [no additional details].
>   *Remarks.* Selected here as the correct original spelling; see also "*frickii*".

frickii.
>   *Helix frickii* Pfeiffer, 1858: pl. 40, fig. 3. Incorrect original spelling of *fricki* Pfeiffer [in the legend to pl. 40].

kalaeloana. (O)
>   *Endodonta kalaeloana* Christensen, 1982: 135, figs. 1–3. Oahu: "Ewa District, ahupua'a (land division) of Honouliuli, Site 50-Oa-B6-78 (sinkhole, distance ca. 1.4 km, bearing 341° from Barbers Point Light) Sample 8 (Layer II, depth 20–30 cm below surface)".

kamehameha. (Mo)
>   *Endodonta kamehameha* Pilsbry & Vanatta, 1906: 784, pl. 43, figs. 3, 4. Molokai: "Wailau Pali, Mapulehu".

lamellosa.
>   *Helix* (*Helicodonta*) *lamellosa* Férussac, 1821c: 38. *Nom nud.*

lamellosa. (?K, O, ?M)
>   *Helix lamellosa* Férussac *in* Quoy & Gaimard, 1825d: 469. "Port Jackson et des îles du grand Océan [= Port Jackson and the islands of the great Ocean]".
>   *Remarks.* Validated by Férussac *in* Quoy & Gaimard (1825d: 469) by the distinguishing characters in the text directly below the names *Helix lamellosa* and *Helix contorta*, although the plate referred to in that work was not published until 1832. "I am fairly certain that this specimen [the "type"] came from the slopes of Konahuanui [= Oahu]" (Cooke, 1928: 16). Kauai and Maui given as additional localities by Pilsbry & Vanatta (1906: 783), discussed by Cooke (1928: 15–16), but not mentioned by Solem (1976: 379–80).

laminata. (K)
>   *Helix laminata* Pease, 1866: 292. Hawaiian Islands: [no additional details].

marsupialis. (O)
>   *Endodonta marsupialis* Pilsbry & Vanatta, 1906: 784, pl. 43, figs. 1, 2. Oahu: [no additional details].

rugata. (M)
>   *Helix rugata* Pease, 1866: 291. Hawaiian Islands: [no additional details].

Genus NESOPHILA Pilsbry

NESOPHILA Pilsbry, 1893: 27 (as *Endodonta* sect.). Type species: *Helix tiara* Mighels, 1845, by original designation.

+albina. (Hawaiian Islands)
>   *Charopa baldwini* var. *albina* Ancey, 1889a: 176. Hawaiian Islands: [no additional details].
>   *Remarks.* Sykes (1900) and Solem (1976) did not mention *albina*.

baldwini. (?K)
>   *Charopa baldwini* Ancey, 1889a: 176. Hawaiian Islands: [no additional details].

capillata. (K)
>   *Helix capillata* Pease, 1866: 292. Hawaiian Islands: [no additional details].
>   *Remarks.* Not a homonym of *capillata* Lightfoot [see Kay (1965a) for discussion of authorship] [not Hawaiian] which is a *nom. nud.*

distans. (K)
>   *Helix distans* Pease, 1866: 290. Hawaiian Islands: [no additional details].

tiara. (K)
>   *Helix tiara* Mighels, 1845: 19. Kauai: [no additional details].

Genus PROTOENDODONTA Solem

PROTOENDODONTA Solem, 1977: 906. Type species: *Protoendodonta laddi* Solem, 1977, by original designation.

laddi. (Midway)
>   *Protoendodonta laddi* Solem, 1977: 907, pl. 2, figs. 2–6, pl. 3, figs. 1–4, pl. 4, figs. 1–6, text-figs. 1a–c, text-figs. 2a, b. Midway: "Sand Island hole (S-22) at 42.3-m depth in coarse foraminiferal sand."

Family PUNCTIDAE

The Punctidae, as treated by Solem (1983), are represented in the Hawaiian Islands by a single described species, although Solem (1976: 3) considered there to be material of 5–9 Hawaiian species in the collections of the Bishop Museum, the majority being undescribed. Only 1 other species of Punctidae is known from the Pacific Islands, *Punctum polynesicum* Solem, 1983 [not Hawaiian], which was referred to *Punctum* with some doubt by Solem (1983: 58–59).

Genus PUNCTUM Morse

PUNCTUM Morse, 1864a: 1 [1864b: 5, 27]. Type species: *Helix minutissima* Lea, 1841 (as "*Punctum minutissimum*") [= *Helix pygmaea* Draparnaud, 1801], by monotypy [see ICZN (1955: 67; Opinion 335)].

horneri. (H, O)
  Punctum horneri Ancey, 1904b: 66, pl. 5, figs. 11, 12. Hawaii: "Palihoukapapa, on the Hamakua slope of Mauna Kea, Kawaii [= Hawaii], at an elevation of 4,000 feet." (in the introduction to the paper).

Family SUCCINEIDAE

The Succineidae are found worldwide and constitute a major part of the land snail fauna of the Pacific. They have radiated dramatically in the Hawaiian Islands and occur in widely different habitats from high altitude rain forests, where they may be arboreal or ground-dwelling, to the low vegetation of the arid leeward shorelines (Christensen & Kirch, 1986; Zimmerman, 1948), contrasting with their generally semi-aquatic, stream or pond-side habitats in other parts of the world.

Subfamilies, genera and subgenera in this catalog follow Zilch (1959a), which differs from Vaught (1989) in tentatively placing *Laxisuccinea* in the Catinellinae rather than the Succineinae. However, these generic and subgeneric divisions remain uncertain without further anatomical studies. Likewise, placement of species in particular genera and subgenera in the Succineidae can rarely be done on the basis of shell characters alone and relies mainly on anatomical characters. Thus, unless species have been dissected they generally have remained in the genus *Succinea*. The work of Cooke (1921), Quick (1939), Odhner (1950) and Burch (1964) has allowed placement of some species in appropriate genera and these placements are followed here. Occurrences on different islands follow Caum (1928) and the original literature. Caum (1928) did not include *coccoglypta*, *henshawi* or *orophila* (all subg. *Succinea*). The Hawaiian fauna is in need of detailed study.

If *aperta* Lea, *pudorina* Gould and *oregonensis* Lea (see remarks under *oregonensis*), which may not be Hawaiian taxa, are excluded, the Hawaiian fauna consists of 42 species (including *inconspicua* Ancey, a junior homonym for which we do not provide a replacement name) and 7 infraspecific taxa. Solem (1990: 29) indicated 44 species but no infraspecific taxa.

Subfamily CATINELLINAE

Genus CATINELLA Pease

CATINELLA Pease, 1870a: 89. Type species: *Catinella rubida* Pease, 1870, by subsequent designation of Pease (1871b: 459).

baldwini. (M)
>*Succinea baldwini* Ancey, 1889a: 250. Maui: "Partie occidentale de l'île Maui [= west Maui]".

explanata. (K)
>*Succinea explanata* Gould, 1852: 13, pl. 2, figs. 31–31c. Kauai: [no additional details].

newcombi.
>*Succinea newcombi* Pfeiffer, 1855e: 297. Molokai: [no additional details].
>>*Remarks.* Synonym of *rotundata* Gould, *teste* Pfeiffer (1859d: 806) and Baldwin (1893: 24).

paropsis. (O)
>*Catinella paropsis* Cooke, 1921: 275, pl. 25, fig. 3. Oahu: "Kaipapau, near the summit of the Koolau Range".

patula.
>*Succinea patula* Mighels, 1845: 21. Oahu: [no additional details].
>>*Remarks*. Primary junior homonym of *Succinea patula* Blainville, 1827 [not Hawaiian]. Synonym of *rotundata* Gould, *teste* Johnson (1949: 227).

rotundata. (O, Mo, H)
>*Succinea rotundata* Gould, 1846d: 182. Oahu: "Mountains of Oahu".
>>*Remarks*. Possibly a synonym of *aperta* Lea [?not Hawaiian], *teste* Baldwin (1893: 24).

rubida. (K)
>*Catinella rubida* Pease, 1870a: 97. Kauai (in publication title): [no additional details].

tuberculata. (O)
>*Catinella tuberculata* Cooke, 1921: 275, pl. 25, fig. 2. Oahu: "Mount Kaala".

Genus LAXISUCCINEA Cooke

LAXISUCCINEA Cooke, 1921: 276. Type species: *Laxisuccinea libera* Cooke, 1921, by original designation.

haena. (K)
>*Laxisuccinea haena* Cooke, 1921: 277, pl. 25, fig. 5. Kauai: "in Pleistocene or Recent deposits in road cutting near the western extremity of the Haena Plain".

libera. (K)
>*Laxisuccinea libera* Cooke, 1921: 276, pl. 25, fig. 6. Kauai: "in Pleistocene or Recent deposits in road cutting near the southern extremity of the Hanamaulu flat".

Subfamily SUCCINEINAE

Genus SUCCINEA Draparnaud

SUCCINEA Draparnaud, 1801: 32, 55. Type species: *Helix putris* Linnaeus, 1758 [not Hawaiian], by subsequent designation of Fleming (1818: 312, implicitly; 1822: 574, explicitly).

ICZN (1926: 13; Opinion 94) and ICZN (1957a: 164, 185; Direction 72) indicated that the subsequent designation of *putris* as the type species was by Gray (1847: 171).

Subgenus SUCCINEA Draparnaud

SUCCINEA Draparnaud, 1801: 32, 55 (as genus). Type species: *Helix putris* Linnaeus, 1758 [not Hawaiian], by subsequent designation of Fleming (1818: 312, implicitly; 1822: 574, explicitly).

See comments under genus *Succinea*.

apicalis.
>*Succinea apicalis* Baldwin, 1893: 17. *Nom. nud.*

apicalis. (M)
> *Succinea apicalis* Ancey, 1904a: 118, pl. 7, fig. 3. Maui: "Makawao, E. Maui".

approximata. (Hawaiian Islands)
> *Succinea approximata* Sowerby *in* Reeve & Sowerby, 1872a: pl. 4, fig. 27a, b. Hawaiian Islands: [no additional details].
> *Remarks.* Caum (1928: 59) questioned the Hawaiian Islands as the locality.

aurulenta. (H)
> *Succinea aurulenta* Ancey, 1889a: 242. Hawaii: [no additional details].

bicolorata. (H)
> *Succinea bicolorata* Ancey, 1899: 271, pl. 12, fig. 2. Hawaii: "Waimea".

caduca. (O, Mo, L)
> *Succinea caduca* Mighels, 1845: 21. Oahu: [no additional details].
> *Remarks.* Verified as being in *Succinea* by Quick (1939: 298) on the basis of dissection.

canella. (Mo, M)
> *Succinea canella* Gould, 1846d: 184. Maui: [no additional details].

casta. (H)
> *Succinea casta* Ancey, 1899: 272, pl. 12, fig. 10. Hawaii: "Olaa".

cepulla. (O, Mo, H)
> *Succinea cepulla* Gould, 1846d: 182. Hawaii: [no additional details].

cinnamomea. (O)
> *Succinea cinnamomea* Ancey, 1889a: 247. Oahu: [no additional details].

+coccoglypta. (H)
> *Succinea tenerrima* var. *coccoglypta* Ancey, 1904a: 118. Hawaii: "Hilo".

+crassa. (M)
> *Succinea canella* var. *crassa* Ancey, 1889a: 246. Maui: "Partie orientale de l'île de Maui [= east Maui]".

delicata. (M)
> *Succinea delicata* Ancey, 1889a: 243. Maui: "Partie orientale de l'île de Maui [= east Maui]".

fragilis.
> *Succinea fragilis* Souleyet, 1852: 501, pl. 28, figs. 18–20. Hawaiian Islands: [no additional details].
> *Remarks.* Junior homonym of *Succinea fragilis* King & Broderip, 1831 [not Hawaiian]. Replaced by *souleyeti* Ancey. Synonym of *cepulla* Gould, *teste* Caum (1928: 60).

garrettiana. (H)
> *Succinea garrettiana* Ancey, 1899: 272, pl. 12, fig. 7. Hawaii: "Rainbow Falls, Hilo".

gibba. (H)
> *Succinea gibba* Henshaw, 1904: 62, pl. 5, figs. 7, 8. Hawaii: "Mana, Hamakua".

+henshawi. (H)
> *Succinea casta* var. *henshawi* Ancey, 1904a: 118. Hawaii: "Olaa, Hawaii, 2,425 ped. supra mare [= 2,425 feet above sea level]".

inconspicua. (H)
> *Succinea inconspicua* Ancey, 1899: 273, pl. 12, fig. 9. Hawaii: "Waimea".
> *Remarks.* Primary junior homonym of *Succinea inconspicua* Edwards, 1852 [not Hawaiian]. No new name proposed here, pending further study.

konaensis. (H)
> *Succinea konaensis* Sykes, 1897: 299. Hawaii: "Mt. Kona, Hawaii, at 4,000 feet".

kuhnsi. (H)
> *Succinea kuhnsi* Ancey, 1904a: 117, pl. 7, fig. 1. Hawaii: "Kaïwicki [= Kaiwiki], Hilo".

+lucida. (M)
: *Succinea canella* var. *lucida* Ancey, 1889a: 247. Maui: "Partie orientale de Maui [= east Maui]".

lumbalis. (K, H)
: *Succinea lumbalis* Gould, 1846d: 183. Kauai: [no additional details].

lutulenta. (M)
: *Succinea lutulenta* Ancey, 1889a: 244. Maui: [no additional details].

+mamillaris. (Mo)
: *Succinea canella* var. *mamillaris* Ancey, 1889a: 246. Molokai: [no additional details].

mauiensis. (M)
: *Succinea mauiensis* Ancey, 1889a: 248. Maui: "Partie orientale de Maui [= east Maui]".

maxima. (H)
: *Succinea maxima* Henshaw, 1904: 61, pl. 5, figs. 1, 2. Hawaii: "Mana, Hamakua".

mirabilis. (H)
: *Succinea mirabilis* Henshaw, 1904: 61, pl. 5, figs. 3, 4. Hawaii: "Palihoukapapa, Hamakua".

newcombiana. (H)
: *Succinea newcombianum* Garrett, 1857: 103. Hawaii: "District of Waimea".
: *Remarks.* "Anatomically a true *Succinea*" (Cooke, 1921: 273).

+obesula. (Mo)
: *Succinea canella* var. *obesula* Ancey, 1889a: 246. Molokai: [no additional details].

+orophila. (H)
: *Succinea casta* var. *orophila* Ancey, 1904a: 118. Hawaii: "Kaiwiki, Hawaii, 2,500 ped. s.m. [= 2,500 feet above sea level]".

pristina. (H)
: *Succinea pristina* Henshaw, 1904: 62, pl. 5, figs. 5, 6. Hawaii: "Mana, Hamakua".

protracta. (H)
: *Succinea protracta* Sykes, 1900: 388, pl. 11, fig. 25. Hawaii: "Kau".

punctata. (?H)
: *Succinea punctata* Pfeiffer, 1855e: 297. Hawaii: [no additional details].
: *Remarks.* Caum (1928: 60) gave the locality as "Hawaiian Islands" rather than specifically the island of Hawaii.

quadrata. (H)
: *Succinea quadrata* Ancey, 1904a: 119, pl. 7, fig. 5. Hawaii: "Olaa, Kaiwiki . . . 2,550 ped. supra mare [= 2,550 feet above sea level]".

souleyeti.
: *Succinea souleyeti* Ancey, 1889a: 255. Hawaiian Islands: [no additional details].
: *Remarks.* Proposed as a new name for *fragilis* Souleyet, 1852, preoccupied by *fragilis* King & Broderip, 1831 [not Hawaiian]. Synonym of *cepulla* Gould, *teste* Caum (1928: 60).

tahitensis. (Hawaiian Islands)
: *Succinea tahitensis* Pfeiffer, 1847a: 109. Tahiti.
: *Remarks.* Included by Caum (1928: 60, as "*tahietensis*") as a Hawaiian species.

tenerrima.
: *Succinea tenerrima* Baldwin, 1893: 18. *Nom. nud.*

tenerrima. (H)
: *Succinea tenerrima* Ancey, 1904a: 118, pl. 7, fig. 2. Hawaii: "Kawaiki, Hawaii, alt. 2500–2600 ped. s.m. [= 2500–2600 feet above sea level]".

tetragona. (M)
: *Succinea tetragona* Ancey, 1904a: 119, pl. 7, fig. 4. Maui: "Makawao, E. Maui".

thaanumi. (H)
    Succinea thaanumi Ancey, 1899: 272, pl. 12, fig. 3. Hawaii: "Olaa".
venusta. (H)
    Succinea venusta Gould, 1846e: 186. Hawaii: [no additional details].
vesicalis. (H)
    Succinea vesicalis Gould, 1846d: 183. Hawaii: "Mauna Kea".
waianaensis. (O)
    Succinea waianaensis Ancey, 1899: 273, pl. 12, fig. 12. Oahu: "Waianae Mountains".

Subgenus TRUELLA Pease

TRUELLA Pease, 1871b: 459. Type species: *Succinea elongata* Pease, 1870, by original designation.

elongata. (K)
    Succinea elongata Pease, 1870a: 96. Kauai: "à environ 4,000 pieds au-dessus du niveau de la mer [= at about 4,000 feet above sea level]".
rubella. (L)
    Succinea rubella Pease, 1871b: 460. Lanai: [no additional details].

Questionably Included SUCCINEIDAE in the Hawaiian Fauna

aperta. (?not Hawaiian)
    Succinea aperta Lea, 1838: 101, pl. 23, fig. 107. "Banks of Columbia River".
    Remarks. Possibly a synonym of *rotundata* Gould, *teste* Baldwin (1893: 24). Baldwin appears to be the only author to have included *aperta* in the Hawaiian fauna.
oregonensis. (?O)
    Succinea oregonensis Lea, 1841: 32. Oregon.
    Remarks. A number of specimens of *Succinea* in BPBM, all from Oahu, have been labelled with what is apparently a manuscript name of Ancey that was never published (*teste* unpublished notes of C.M. Cooke, Jr.). Cooke (unpublished notes) considered this material to be referable to *Succinea oregonensis* Lea, 1841, on the basis of material of *oregonensis* sent to him by H. Rehder with "Hawaii" as probable locality, and on the basis of correspondence with H. Rehder (H. Rehder *in litt.* to Cooke, 20 November 1947; 16 March 1948). Nothing definitive has been published indicating *oregonensis* as an indigenous Hawaiian species.
pudorina. (?not Hawaiian)
    Succinea pudorina Gould, 1846e: 186. "Mountains of Taheiti and Eimeo".
    Remarks. Reported from the Hawaiian Islands ["Sandwich Islands"] by Sowerby (*in* Reeve & Sowerby, 1872a: pl. 6, species 43), and mentioned as from Hawaii by Odhner (1922: 245), erroneously so according to Caum (1928: 60).

[153]

The "Zonitoid" Families HELICARIONIDAE and ZONITIDAE

Treatments by different authors (e.g., Baker, 1940, 1941; Boss, 1982; Riedel, 1980; Smith, 1992; Thiele, 1931; Tillier, 1989; Vaught, 1989; Zilch, 1959b) of what might loosely be called the "zonitoid" land snail families have differed widely. Baker (1940, 1941) was the last to revise the fauna of the Hawaiian Islands thoroughly and, in general, we follow his arrangement of taxa (cf. Solem, 1989: 543). Other authors have tended to treat the Helicarionidae and Zonitidae in narrower senses, raising certain subfamilies to family level (e.g. Vitrinidae, Euconulidae, etc.). Baker's treatment of subgenera and sections was not always clear. In some instances he treated a genus-group name as a subgenus and elsewhere in the text as a section of a subgenus. We treat all such names as subgenera. Correct spelling of Helicarionidae was clarified by ICZN (1992; Opinion 1678).

The "zonitoids" comprise one of the major land snail radiations on Pacific islands (Baker, 1938, 1940, 1941), although neither the Zonitidae nor the Helicarionidae is endemic to the region. The combined native "zonitoids" in the Hawaiian Islands comprise 70 species (and a single taxon, *exserta* Pfeiffer, that is of unclear familial placement and may not, in fact, be Hawaiian).

Family HELICARIONIDAE

The Hawaiian fauna consists of 60 species (6 in *Euconulus*, 13 in *Hiona*, 1 in *Kaala* and 40 in *Philonesia*) and ten infraspecific taxa (2 in *Euconulus*, 3 in *Hiona* and 5 in *Philonesia*).

Subfamily EUCONULINAE

Genus EUCONULUS Reinhardt

CONULUS Fitzinger, 1833 [not Hawaiian]. Type species: *Helix fulva* Müller, 1774 [not Hawaiian], by subsequent designation of Gray (1847: 173). [Preoccupied, Leske, 1778].

EUCONULUS Reinhardt, 1883: 86. Type species: *Helix fulva* Müller, 1774 [not Hawaiian], automatic [see ICZN (1955: 74; Opinion 335)]. [n.n. for *Conulus* Fitzinger, 1833].

Subgenus CHETOSYNA Baker

CHETOSYNA Baker, 1941: 213 (as *Euconulus* subg.), 217 (as *Euconulus* (*Nesoconulus*) sect.). Type species: *Euconulus thurstoni* Baker, 1941, by original designation.

thurstoni. (H)
Euconulus (*Chetosyna*) *thurstoni* Baker, 1941: 219, pl. 53, figs. 11, 12, pl. 62, fig. 12. Hawaii: "North Kona? . . . Puu Huluhulu (which?)".
Remarks. More than 1 location on the island of Hawaii bears the name "Puu Huluhulu"; Baker was uncertain which Puu Hulululu was the source of his specimens.

Subgenus NESOCONULUS Baker

NESOCONULUS Baker, 1941: 213 (list), 214 (key). Type species: *Helix subtilissima* Gould, 1846 (as "*E. subtilissimus*"), by original designation.

gaetanoi. (H)
Kaliella gaetanoi Pilsbry & Vanatta, 1908: 73, fig. 1. Hawaii: "Palihoukapapa".

+kaunakakai. (Mo, L)
Euconulus (*Nesoconulus*) *subtilissimus kaunakakai* Baker, 1941: 217, pl. 53, figs. 5, 6, pl. 62, fig. 3. Molokai: "Molokai (east) . . . landslide above cut, east side of Kaunakakai Valley" (from reference to type number in figure caption, p. 345).

konaensis. (H)
Kaliella konaensis Sykes, 1897: 299. Hawaii: "Mt. Kona, Hawaii, at 3,000 feet".

subtilissimus. (M)
Helix subtilissima Gould, 1846b: 177. Maui: [no additional details].

thaanumi. (H)
Kaliella thaanumi Ancey, 1904a: 119, pl. 7, fig. 6. Hawaii: "Olaa".

+vivens. (H)
Euconulus (*Nesoconulus*) *gaetanoi vivens* Baker, 1941: 218, pl. 43, fig. 5, pl. 53, figs. 9, 10, pl. 62, fig. 9. Hawaii: "South Hilo . . . large *kipuka* between Hilo and Kilauea trails from Humuula, alt. about 5,000 ft., 1855 flow" (from reference to type number in figure caption, p. 345).

Subgenus PELLUCIDOMUS Baker

PELLUCIDOMUS Baker, 1941: 213 (as *Euconulus* subg.), 217 (as *Euconulus* (*Nesoconulus*) sect.). Type species: *Kaliella lubricella* Ancey, 1904 (as "*E. lubricella*"), by original designation.

lubricellus. (H)
Kaliella lubricella Ancey, 1904a: 120, pl. 7, fig. 7. Hawaii: "Olaa".

Subfamily MICROCYSTINAE

Genus HIONA Baker

HIONA Baker, 1940: 106 (key), 163 (description). Type species: *Microcystis platyla* Ancey, 1889 (as "*H. platyla*"), by original designation.

HELICARIONIDAE [155]

Subgenus HIONA Baker

HIONA Baker, 1940: 106 (key), 163 (description) (as genus). Type species: *Microcystis platyla* Ancey, 1889 (as "*H. platyla*"), by original designation.

megodonta. (O)
    Hiona (*Hiona*) *megodonta* Baker, 1940: 183, pl. 21, fig. 10, pl. 41, fig. 3. Oahu: "Waianae Mountains, Waialua . . . east slopes of Kaala (Haleauau)" (from reference to type number in caption to pl. 41, fig. 3, p. 200).

platyla. (O)
    Microcystis platyla Ancey, 1889a: 196. Oahu: [no additional details].

waimanoi. (O)
    Hiona (*Hiona*?) *waimanoi* Baker, 1940: 184 (key), 186 (description), pl. 34, figs. 8, 9. Oahu: "Ewa, Koolau side . . . Waimano".

Subgenus HIONARION Baker

HIONARION Baker, 1940: 163 (list), 164 (key). Type species: *Hiona pilsbryi* Baker, 1940, by original designation.

pilsbryi. (K)
    Hiona (*Hionarion*) *pilsbryi* Baker, 1940: 178, pl. 26, figs. 13–15, pl. 38, fig. 4. Kauai: "Koloa . . . dam-swamp, Wahiawa" (from reference to type number in caption of pl. 38, fig. 4, p. 199).

wahiawae. (K)
    Hiona (*Hionarion*) *wahiawae* Baker, 1940: 178 (key), 179 (description), pl. 26, figs. 11, 12, pl. 34, figs. 6, 7. Kauai: "Koloa . . . back of Wahiawa" (from reference to type number in caption of pl. 34, figs. 6, 7, p. 198).

Subgenus HIONELLA

HIONELLA Baker, 1940: 163 (as *Hiona* subg.), 180 (as *Hiona* (*Hiona*) sect.). Type species: *Microcystis rufobrunnea* Ancey, 1904 (as "*H. rufobrunnea*"), by original designation.

+maunahoomae. (M)
    Hiona (*Hionella*) *perkinsi maunahoomae* Baker, 1940: 180 (key), 182 (description), pl. 35, figs. 1, 2. Maui: "West Maui (Lahaina) . . . high above camp, Maunahooma".

perkinsi. (L)
    Macrochlamys perkinsi Sykes, 1896: 126. Lanai: [no additional details].

+poholuae. (Mo)
    Hiona (*Hionella*) *perkinsi poholuae* Baker, 1940: 180 (key), 182 (description), pl. 35, figs. 3, 4. Molokai: "Molokai (eastern) . . . east of ridge, just below swamp, Poholua, Kalamaula".

rufobrunnea. (H)
    Microcystis rufobrunnea Ancey, 1904a: 119. Hawaii: "Olaa".

subdola. (Mo)
    Hiona (*Hionella*) *subdola* Baker, 1940: 180 (in key), 182 (description), pl. 26, figs. 19, 20, pl. 35, figs. 5, 6. Molokai: "Molokai (eastern) . . . pipeline above spring, Kalihi, Kaunakakai".

Subgenus NESOCYCLUS Baker

NESOCYCLUS Baker, 1940: 163 (as *Hiona* subg.), 184 (as *Hiona* (*Hiona*) sect.). Type species: *Helix exaequata* Gould, 1846, by original designation.

disculus.
    Helix disculus Pfeiffer, 1850b: 68. Hawaiian Islands: [no additional details].
        Remarks. Primary junior homonym of *Helix disculus* Deshayes, 1850 [not Hawaiian]. Replaced by *obtusangula* Pfeiffer, 1851. Synonym of *exaequata* Gould, *teste* Baker (1940: 187). We consider the use of "*discus* . . ." of Pfeiffer" by Tryon (1886b: 114) as an error for *disculus* Pfeiffer.

exaequata. (K)
    Helix exaequata Gould, 1846a: 171. Kauai: [no additional details].

kipui. (K)
    Hiona (*Nesocyclus*) *kipui* Baker, 1940: 187 (key), 189 (description), pl. 41, fig. 5. Kauai: "Lihue . . . above cliffs, front of Haupu" (from reference to type number in figure caption, p. 200).

meineckei. (K)
    Hiona (*Nesocyclus*) *meineckei* Baker, 1940: 187 (key), 189 (description), pl. 21, figs. 13, 14, pl. 42, figs. 10, 11. Kauai: "west Hanalei . . . northeast of big waterfall, Waiahuakua".

milolii. (K)
    Hiona (*Nesocyclus*) *milolii* Baker, 1940: 186, pl. 35, figs. 13, 14. Kauai: "northwest Waimea . . . Milolii Valley".

obtusangula.
    Helix obtusangula Pfeiffer, 1851: 153. Hawaiian Islands [Pfeiffer, 1850b: 68]: [no additional details].
        Remarks. Proposed as a new name for *disculus* Pfeiffer, 1850; preoccupied by *disculus* Deshayes, 1850 [not Hawaiian]. Synonym of *exaequata* Gould, *teste* Baker (1840: 187).

+waimeae. (K)
    Hiona (*Nesocyclus*) *exaequata waimeae* Baker, 1940: 187 (key), 188 (description), pl. 35, figs. 10, 11. Kauai: "north Waimea . . . west side of Halemanu Valley" (from reference to type number in figure caption, p. 199).

Subgenus NEUTRA Baker

NEUTRA Baker, 1940: 163 (list), 164 (key). Type species: *Microcystis lymanniana* Ancey, 1893 (as "*H. lymaniana*") [see *Code* Art. 67(d)], by original designation.

lymaniana.
    Hiona (*Neutra*) *lymaniana* Baker, 1940: 165, pl. 37, fig. 3. Unjustified emendation of *lymanniana* Ancey, 1893.
        Remarks. Objective synonym of *lymanniana* Ancey.

lymanniana. (O)
    Microcystis lymanniana Ancey, 1893: 329. Oahu: "Waialae".
        Remarks. Caum (1928: 63), referred to this species as "*lymaniana*", a spelling that was accepted by Baker (1940: 165) as an emendation, indicating that the collector's name was Lyman. However, Ancey's original publication spells the collector's name as "Lymann".

HELICARIONIDAE [157]

Genus KAALA Baker

KAALA Baker, 1940: 106 (key), 161. Type species: *Helix subrutila* Mighels, 1845 (as "*K. subrutila*"), by original designation.

subrutila. (O)
    Helix subrutila Mighels, 1845: 19. Oahu: [no additional details].

Genus PHILONESIA Sykes

PHILONESIA Sykes, 1900: 280. Type species: *Microcystis baldwini* Ancey, 1889, by original designation.

Subgenus AA Baker

AA Baker, 1940: 107 (list), 108 (key). Type species: *Philonesia waiheensis* Baker, 1940, by original designation.

abeillei. (Mo)
    Microcystis abeillei Ancey, 1889a: 199. Molokai: [no additional details].
gouveiana. (H)
    Philonesia (Aa) gouveiana Baker, 1940: 146 (key), 147 (description), pl. 24, figs. 11, 12, pl. 39, figs. 9, 10. Hawaii: "South Kona . . . alt. 6,000 ft., Honomalino" (from reference to type number in caption of pl. 39, figs. 9, 10, p. 200).
mapulehuae. (Mo)
    Philonesia (Aa) mapulehuae Baker, 1940: 146 (key), 147 (description), pl. 28, figs. 9, 10. Molokai: "east Molokai . . . Wailua ridge" (from reference to type number in figure caption, p. 196).
sericans. (H)
    Microcystis sericans Ancey, 1899: 268, pl. 12, fig. 5. Hawaii: "Olaa, Central Hawaii".
waiheensis. (M)
    Philonesia (Aa) waiheensis Baker, 1940: 146, pl. 24, fig. 13. Maui: "west Maui (Wailuku) . . . site of old camp below intake of upper ditch, Waihee" (from reference to type number by Baker [1941: caption of pl. 62, figs. 4, 5, p. 345]).

Subgenus HALEAKALA Baker

HALEAKALA Baker, 1940: 107 (as *Philonesia* subg.), 108, 138 (as *Philonesia (Aa)* sect.). Type species: *Philonesia diducta* Baker, 1940 (as subspecies of *turgida* Ancey), by original designation.

The type species is cited as *Philonesia diducta* Baker rather than *Philonesia turgida diducta* Baker, following *Code* Art. 61(d).

+diducta. (K, O, Mo, M, L, H)
    Philonesia (Haleakala) turgida diducta Baker, 1940: 140, pl. 24, figs. 2, 3, pl. 30, figs. 7, 8. Maui: "east Maui (Makawao) . . . above eucalyptus grove, Polipoli" (from reference to type number in caption of pl. 30, figs. 7, 8, p. 197).
        Remarks. We retain Baker's treatment of *diducta* as a subspecies of *turgida* Ancey.
guavarum. (M, Kah)
    Philonesia (Haleakala) guavarum Baker, 1940: 139 (key), 142 (description), pl. 30, figs. 9, 10. Maui: "west Maui (Wailuku) . . . site of old camp below intake of upper ditch, Waihee" (from reference to type number in figure caption, p. 197).

hahakeae. (M)
>*Philonesia (Haleakala) hahakeae* Baker, 1940: 138 (key), 141 (description), pl. 30, figs. 11, 12. Maui: "west Maui (Lahaina) . . . Hahakea" (from reference to type number in figure caption, p. 197).

indefinita. (M, Kah)
>*Microcystis indefinita* Ancey, 1889a: 203. Maui: "Partie orientale de Maui [= east Maui]".

interjecta. (L)
>*Philonesia (Haleakala) interjecta* Baker, 1940: 139 (key), 143 (description), pl. 28, figs. 3, 4. Lanai: "eastern end from base camp, Captain Sowles' house" (from reference to type number in figure caption, p. 196).

pusilla. (Mo)
>*Philonesia (Haleakala) pusilla* Baker, 1940: 139 (key), 143 (description), pl. 24, fig. 5, pl. 39, figs. 11, 12. Molokai: "Molokai (east) . . . east Myers Lake, Kalamaula" (from reference to type number in caption of pl. 39, figs. 11, 12, p. 200).

turgida. (M)
>*Microcystis turgida* Ancey, 1890: 339. Maui: [no additional details].

Subgenus HILOAA Baker

HILOAA Baker, 1940: 107 (as *Philonesia* subg.), 108 (as *Philonesia (Aa)* sect.). Type species: *Philonesia hiloi* Baker, 1940, by original designation.

hiloi. (H)
>*Philonesia (Hiloaa) hiloi* Baker, 1940: 143, pl. 24, figs. 7, 8, pl. 29, figs. 13, 14. Hawaii: "South Hilo . . . 4 miles out along Olaa road from Hilo" (from reference to type number in caption of pl. 29, figs. 13, 14, p. 197).

piihonuae. (H)
>*Philonesia (Hiloaa) piihonuae* Baker, 1940: 144, pl. 24, fig. 9, pl. 30, figs. 5, 6. Hawaii: "South Hilo . . . *kipuka* 4, Piihonua".

Subgenus KIPUA Baker

KIPUA Baker, 1940: 107. Type species: *Helix chamissoi* Pfeiffer, 1855 (as "*P. chamissoi*"), by original designation.

arenofunus. (K)
>*Philonesia (Kipua?) arenofunus* Baker, 1940: 114, pl. 27, figs. 1, 2. Kauai: "east Koloa . . . sand dunes close to and above west side of small bay, Aweoweonui, Mahaulepu".

chamissoi. (K)
>*Helix chamissoi* Pfeiffer, 1855f: 91. Hawaiian Islands: [no additional details].

Subgenus MAUKA Baker

MAUKA Baker, 1940: 107 (as *Philonesia* subg.), 108 (as *Philonesia (Philonesia)* sect.). Type species: *Philonesia welchi* Baker, 1940, by original designation.

polita. (O)
>*Philonesia (Mauka) polita* Baker, 1940: 134 (key), 135 (description), pl. 23, fig. 14, pl. 31, figs. 1, 2. Oahu: "Koolau Range, Koolauloa . . . first banana clump inland, Kaliuwaa, Kaluanui" (from reference to type number in caption of pl. 31, figs. 1, 2, p. 197).

HELICARIONIDAE   [159]

similaris. (O)
    Philonesia (*Mauka*) *similaris* Baker, 1940: 134 (key), 136 (description), pl. 31, figs. 3, 4. Oahu: "Waianae Mountains, Ewa-Waianae . . . first gap, Green Peak (Palikea)" (from reference to type number in figure caption, p. 197).

welchi. (O)
    Philonesia (*Mauka*) *welchi* Baker, 1940: 133, pl. 23, figs. 15, 16. Oahu: "Koolau Range, Waialua . . . division ridge between Kalaikoa and Waikakalaua Valleys, Waianaeuka" (from reference to type number by Baker [1941: caption of pl. 62, figs. 7, 8, p. 345]).

Subgenus PHILONESIA Sykes

PHILONESIA Sykes, 1900: 280 (as genus). Type species: *Microcystis baldwini* Ancey, 1889, by original designation.

ascendens. (O)
    Philonesia (*P.*) *ascendens* Baker, 1940: 128 (key), 130 (description), pl. 23, fig. 11, pl. 28, figs. 1, 2. Oahu: "Koolau Range, Honolulu . . . at waterfall, Waialae Iki" (from reference to type number in caption of pl. 28, figs. 1, 2, p. 196).

baldwini. (O)
    Microcystis baldwini Ancey, 1889a: 204. Oahu: [no additional details]; Maui: "partie occidentale de . . . Maui [= west Maui]" [?error = Oahu only (Baker, 1940: 131, 133)].

+boettgeriana. (H)
    Microcystis cicercula var. *boettgeriana* Ancey, 1889a: 206. Hawaii: "Kona".

cicercula. (H)
    Helix cicercula Gould, 1846a: 171. Hawaii: "Mountains of Hawaii".

cryptoportica. (O)
    Helix cryptoportica Gould, 1846a: 171. Type locality not given.

decepta. (M)
    Philonesia (*P.*) *decepta* Baker, 1940: 116 (key), 126 (description), pl. 23, figs. 6, 7, pl. 29, figs. 9, 10. Maui: "west Maui (Wailuku) . . . valley in front of needle, between base and bridge, Iao Valley".

+depressula. (O)
    Microcystis oahuensis var. *depressula* Ancey, 1889a: 203. Type locality not given.
        Remarks. Incorrectly spelled as *depressiuscula* by Sykes (1900: 284) and Cooke (1907: 13).

fallax. (O)
    Philonesia (*P.*) *fallax* Baker, 1940: 116 (key), 122 (description), pl. 22, figs. 13, 14, pl. 39, figs. 1, 2. Oahu: "Koolau Range, Honolulu . . . Waiomao Valley, Palolo (from reference to type number in caption of pl. 39, figs. 1, 2, p. 200).

glypha. (O)
    Philonesia (*P.*) *glypha* Baker, 1940: 128 (key), 133 (description), pl. 23, figs. 9, 10. Oahu: "Koolau Range, Ewa . . . at backbone (crest of range), overlooking Keahiakahoe, ridge between Moanalua and Halawa Valleys".

hartmanni. (O)
    Microcystis hartmanni Ancey, 1889a: 198. Oahu: [no additional details].

kauaiensis. (K)
    Philonesia (*P.*) *kauaiensis* Baker, 1940: 116 (key), 123 (description), pl. 23, figs. 1, 2, pl. 29, figs. 7, 8, pl. 39, fig. 15. Kauai: "Waimea . . . first valley below and slightly east of Puu Kapele" (from reference to type number in caption of pl. 39, fig. 15, p. 200).

konahuanui. (O)
> *Philonesia* (*P.*) *konahuanui* Baker, 1940: 128 (key), 130 (description), pl. 30, figs. 3, 4. Oahu: "Koolau Range, Honolulu . . . Konahuanui".

kualii. (O)
> *Philonesia* (*Philonesia*?) *kualii* Baker, 1940: 115, pl. 27, figs. 4, 5. Oahu: "Koolau Range, Honolulu . . . road cutting, 1 block below Kualii, Upper Manoa Road, Manoa".

maunalei. (L)
> *Philonesia* (*P.*) *maunalei* Baker, 1940: 116 (key), 126 (description), pl. 23, fig. 5, pl. 39, figs. 5, 6. Lanai: "northeast side . . . Maunalei Gulch" (from reference to type number in caption of pl. 39, figs. 5, 6, p. 200).

mokuleiae. (O)
> *Philonesia* (*P.*) *mokuleiae* Baker, 1940: 116 (key), 121 (description), pl. 36, fig. 2. Oahu: "Waianae Mountains, Waialua . . . Mokuleia Valley".

oahuensis. (O)
> *Microcystis oahuensis* Ancey, 1889a: 202. Oahu: [no additional details].

+palehuae. (O)
> *Philonesia* (*P.*) *hartmanni palehuae* Baker, 1940: 128 (key), 129 (description), pl. 39, figs. 7, 8. Oahu: "Waianae Mountains, Ewa . . . glen, southwest of house, Palehua" (from reference to type number in figure caption, p. 200).

perlucens. (M)
> *Microcystis perlucens* Ancey, 1889a: 207. Maui: "Partie orientale de l'île Maui [= east Maui]".

plicosa. (O)
> *Microcystis plicosa* Ancey, 1889a: 200. Oahu: [no additional details].

+popouwelae. (O)
> *Philonesia* (*P.*) *fallax popouwelae* Baker, 1940: 116 (key), 123 (description), pl. 29, figs. 1, 2. Oahu: "Waianae Mountains, Ewa . . . head of middle valley, Popouwela" (from reference to type number in figure caption, p. 196).

striata. (O)
> *Philonesia* (*P.*) *striata* Baker, 1940: 115 (key), 119 (description), pl. 27, figs. 14, 15. Oahu: "Koolau Range, Waialua . . . inland and east of flat, south Opaeula (Helemano), Paalaa" (from reference to type number in figure caption, p. 196).

waimanaloi. (O)
> *Philonesia* (*Philonesia*?) *waimanaloi* Baker, 1940: 116 (key), 123 (description), pl. 29, figs. 3, 4. Oahu: "Koolauloa . . . sand dunes along deep ditch, Waimanalo".

Subgenus PIENA Baker

PIENA Baker, 1940: 106 (as *Philonesia* subg.), 108 (as *Philonesia* (*Philonesia*) sect.). Type species: *Philonesia grandis* Baker, 1940, by original designation.

grandis. (O)
> *Philonesia* (*Piena*) *grandis* Baker, 1940: 134 (key), 136 (description), pl. 23, figs. 18, 19, pl. 31, figs. 5, 6, pl. 37, fig. 4. Oahu: "Koolau Range, Honolulu . . . top of Konahuanui" (from reference to type number in caption of pl. 37, fig. 4, p. 199).

palawai. (O)
> *Philonesia* (*Piena*) *palawai* Baker, 1940: 134 (key), 137 (description), pl. 24, fig. 1, pl. 31, figs. 7, 8. Oahu: "Waianae Mountains, Ewa . . . second valley east of Green Peak, south Palawai Valley".

parva. (O)
>*Philonesia (Piena) parva* Baker, 1940: 134 (key), 137 (description), pl. 23, fig. 17, pl. 37, fig. 5. Oahu: "Koolau Range, Honolulu . . . top of Konahuanui" (from reference to type number in caption of pl. 37, fig. 5, p. 199).

Subgenus WAIHOUA Baker

WAIHOUA Baker, 1940: 107. Type species: *Philonesia kaliella* Baker, 1940, by original designation.

kaliella. (H)
>*Philonesia*? (*Waihoua*) *kaliella* Baker, 1940: 114, pl. 36, fig. 5. Hawaii: "North Kona . . . inland of old branding pen along trail, Waihou, Puu Waawaa".

Family ZONITIDAE

The native Hawaiian fauna consists of 10 species, 3 each in *Striatura* (Gastrodontinae), *Godwinia* and *Nesovitrea* (both Zonitinae), and 1 in *Vitrina* (Vitrininae), and no infraspecific taxa. In addition, a single introduced taxon is included as it was described from Hawaiian material, although it is now considered a junior synonym of an extralimital taxon.

Subfamily GASTRODONTINAE

Genus STRIATURA Morse

STRIATURA Morse, 1864a: 1. Type species: *Helix milium* Whittemore, 1859 [not Hawaiian], by monotypy.

STRIATURA Morse, 1864b: 17. Type species: *Striatura ferrea* Morse, 1864 [not Hawaiian], by subsequent designation of Tryon (1865: 72).

Morse published *Striatura* as new twice; the later name (see Bibliography) is thus a junior homonym of the earlier name. Confusion has been generated by this duplicate proposal of *Striatura*. In the earlier publication (Morse, 1864a), 2 species were mentioned, *milium* Whittemore (as "Morse") and a new species, *ferrea* Morse, but the latter was a *nomen nudum*. Neither was explicitly designated as the type species in the original publication, so *milium* Whittemore, 1859 is the type species by monotypy. Authorship of *milium* is usually indicated as "Morse", but there is not adequate justification for this in the published work (Whittemore, 1859). In the second publication (Morse, 1864b), *Striatura* was again described as new, and *ferrea* was validly described; no type species was designated. Baker (1928: 33) correctly indicated *milium* Morse as the type species of the earlier *Striatura*, by monotypy. Riedel (1980: 24), however, cited the second description of *Striatura* only, but cited Baker (1928: 33) as having designated *milium* as its type species.

All 3 Hawaiian species were placed in the subgenus *Pseudohyalina* Morse, 1864, by Baker (1941). Riedel (1980: 25) regarded the assignment of *meniscus* and *discus* to that subgenus as uncertain.

Subgenus PSEUDOHYALINA Morse

PSEUDOHYALINA Morse, 1864a: 1 [1864b: 15] (as genus). Type species: *Helix exigua* Stimpson, 1850 (as "*Hyalina*") [not Hawaiian], by subsequent designation of Kobelt (1879b: 223).

discus. (O, Mo, M, L)

Striatura (Pseudohyalina) discus Baker, 1941: 326, pl. 60, figs. 7–9. Maui: "east Maui (Makawao) . . . Ainaho, half way down slope, top of Keanae Gap, Haleakala Crater" (from reference to type number in figure caption, p. 345).

meniscus. (H)
> *Pseudohyalinia meniscus* Ancey, 1904b: 65. Hawaii: "Palihoukapapa, on the Hamakua slope of Mauna Kea, Kawaii [= Hawaii], at an elevation of 4,000 feet" (in the introduction to the paper).

pugetensis. (K)
> *Patulastra*? (*Punctum*?) *pugetensis* Dall, 1895: 130. USA: "Seattle, Wash."
>> Remarks. *Striatura pugetensis* is 1 of probably only very few species not endemic to the Hawaiian Islands but nevertheless thought to be indigenous, the others being *Pupoidopsis hawaiensis* Pilsbry & Cooke, 1921 and perhaps *Vitrina tenella* Gould, 1846, *Succinea tahitensis* Pfeiffer, 1847, and a number of Ellobiidae. It was first reported as Hawaiian, from Kauai, as *Striatura* (*Pseudohyalina*) *pugetensis* (Dall, 1895), by Baker (1941: 325).

Subfamily VITREINAE

The single species, *Hawaiia minuscula* (Binney, 1840), is considered to have been introduced and is only included in this catalog because the genus *Hawaiia* Gude and its type species *kawaiensis* Reeve (now synonymized with *minuscula*) were described from Hawaiian material.

Genus HAWAIIA Gude

HAWAIIA Gude, 1911: 272. Type species: *Helix kawaiensis* Pfeiffer, 1855 [= *minuscula* Binney, 1840] [introduced], by original designation.

PSEUDOVITREA Baker, 1928: 25. Type species: *Helix minuscula* Binney, 1840 [introduced; not Hawaiian], by original designation.

In an erratum slip, probably published subsequent to the issue that included his establishment of *Hawaiia*, Gude stated that designation of "*Helix Kawaiiensis*, Pfr." as the type species of *Hawaiia* was a "*lapsus calami*" and that it should be "*Helix Hawaiiensis*, Ancey".

kawaiensis.
> *Helix kawaiensis* Reeve, 1854c, pl. 182, fig. 1256. Kauai (as "Kawai"): [no additional details].
>> Remarks. Synonym of *minuscula* Binney, 1840 [not Hawaiian], *teste* Baker (1941: 323).

kawaiensis.
> *Helix kawaiensis* Pfeiffer, 1855a: 52. Kauai (as "Kawai"): [no additional details].
>> Remarks. Primary junior homonym of *kawaiensis* Reeve. Synonym of *minuscula* Binney, 1840 [not Hawaiian], *teste* Baker (1941: 323).

minuscula. (Midway, K, O, Mo, M; introduced)
> *Helix minuscula* Binney, 1840 [not Hawaiian].
>> Remarks. *Hawaiia minuscula* is considered to have been introduced to the Hawaiian Islands by commerce (Baker, 1941: 323). It is included here only because the genus and its type species (now synonymized with *minuscula*) were described from Hawaiian material. *Hawaiia minuscula* was recorded from Midway by Conant *et al.* (1984: 83).

Subfamily VITRININAE

Genus VITRINA Draparnaud

VITRINA Draparnaud, 1801: 98. Type species: *Helix pellucida* Müller, 1774 [not Hawaiian], by monotypy [see ICZN (1931: 23; Opinion 119) and ICZN (1957a: 165, 175–76, 187; Direction 72)].

VITRINUS Montfort, 1810: 238. Unjustified emendation of *Vitrina* Draparnaud, 1801 [see ICZN (1957a: 170; Direction 72)].

tenella. (K, M, H)
    Vitrina tenella Gould, 1846c: 181. Kauai: [no additional details].
        Remarks. Baker (1941: 322) considered *V. tenella* as indigenous to the Hawaiian Islands but perhaps not endemic. (See also *Striatura pugetensis* (Dall), *Pupoidopsis hawaiensis* Pilsbry & Cooke, and *Succinea tahitensis* Pfeiffer.)

Subfamily ZONITINAE

Genus GODWINIA Sykes

GODWINIA Sykes, 1900: 277. Type species: *Vitrina caperata* Gould, 1846, by monotypy [*Vitrina tenella* Gould provisionally included].

Subgenus GODWINIA Sykes

GODWINIA Sykes, 1900: 277 (as genus). Type species: *Vitrina caperata* Gould, 1846, by monotypy [*Vitrina tenella* Gould provisionally included].

caperata. (K)
    Vitrina caperata Gould, 1846c: 181. Kauai: [no additional details].

haupuensis. (K)
    Godwinia haupuensis Cooke, 1921: 267, pl. 24, fig. 3, text fig. 3. Kauai: "Northern slope of Mount Haupu in the southeastern portion of the island".

Subgenus OMPHALOPS Baker

OMPHALOPS Baker, 1941: 332. Type species: *Helix newcombi* Reeve, 1854 (as "*G. newcombi*"), by original designation.

newcombi. (K, ?O)
    Helix newcombi Reeve, 1854d, pl. 189, fig. 1321. Oahu [?error = Kauai (Baker, 1941: 333)]: [no additional details].

newcombi.
    Helix newcombi Pfeiffer, 1855a: 51. Oahu: [no additional details].
        Remarks. Primary junior homonym of *newcombi* Reeve. Synonym of *newcombi* Reeve, *teste* Baker (1941: 333).

Genus NESOVITREA Cooke

NESOVITREA Cooke, 1921: 271. Type species: *Helix pauxillus* Gould, 1852 (as "*Vitrea*"), by original designation.

Baker (1941: 328) regarded *Nesovitrea* Cooke as a subgenus of *Retinella* Fischer, 1877. We follow Riedel (1980: 84) and Zilch (1959b: 245) who accorded it generic status.

baldwini.
>*Hyalinia baldwini* Ancey, 1889a: 192. West Maui.
>>*Remarks.* Synonym of *pauxilla* Gould, *teste* Baker (1941: 330).

hawaiiensis. (H)
>*Vitrea hawaiiensis* Ancey, 1904a: 120, pl. 7, figs. 8–8b. Hawaii: "Olaa".

lanaiensis.
>*Vitrea* (?) *lanaiensis* Sykes, 1897: 298. Lanai: "Mountains of Lanai, behind Koele".
>>*Remarks.* Synonym of *pauxilla* Gould, *teste* Baker (1941: 330).

molokaiensis. (Mo)
>*Vitrea* (?) *molokaiensis* Sykes, 1897: 298. Molokai: "Forest above Pelekunu".

pauxilla. (M, L)
>*Helix pauxillus* Gould, 1852: 40, pl. 3, figs. 46a–c. Maui [Gould, 1846a: 171]: [no additional details].
>>*Remarks.* Proposed as a new name for *pusillus* Gould, 1846, preoccupied by *pusilla* Lowe, 1831 [not Hawaiian].

pusillus.
>*Helix pusillus* Gould, 1846a: 171. Maui: [no additional details].
>>*Remarks.* Primary junior homonym of *Helix pusilla* Lowe, 1831 [not Hawaiian]. Replaced by *pauxilla* Gould, 1852.

Incertae Sedis in the "Zonitoid" Families

exserta. (?not Hawaiian)
>*Helix exserta* Pfeiffer, 1856e: 32. Hawaiian Islands: [no additional details].
>>*Remarks.* Listed in *Charopa* sect. *Thalassia* by Tryon (1886d: 215) in his treatment of the "Zonitidae", with "Sandwich Is." as the locality. Systematic position "doubtful" (Caum, 1928: 62). C.M. Cooke, Jr. (unpublished notes) examined the "type and only spm" and considered it a "small Zonitid", but he doubted that it belonged to the Hawaiian fauna.

Family MILACIDAE

There are no native slugs in the Hawaiian Islands but a number of taxa have been described from Hawaiian material of introduced species in both the Milacidae and Limacidae (below), and are therefore included here.

Genus MILAX Gray

MILAX Gray, 1855: 174. Type species: *Limax gagates* Draparnaud, 1801 [not Hawaiian], by subsequent designation of Kennard & Woodward (1926: 204) [also Hesse (1926: 31)] [see ICZN (1955: 57, 75; Opinion 335)].

AMALIA Moquin-Tandon, 1855a: 19 (as an infrageneric division of *Limax*). Type species: *Limax gagates* Draparnaud, 1801 [not Hawaiian], by subsequent designation of Wiktor (1987: 184).

Wiktor (*in litt.* to RHC, 3 October 1993) did not indicate an earlier type species designation for *Amalia*. Smith (1992: 261) stated that it is by original designation but this is incorrect; there is no type designation in the original work and there are 2 originally included species.

babori.
> *Amalia babori* Collinge, 1897: 294, figs. 4, 5. Maui: "Haleakala . . . 5,000 feet"; Hawaii: "Olaa to Kilauea, 2,000 to 4,000 feet".
>> *Remarks.* Synonym of *gagates* Draparnaud, *teste* Quick (1960: 149).

gagates. (M, H; introduced)
> *Limax gagates* Draparnaud, 1801 [not Hawaiian].
>> *Remarks.* Included here only because *babori* Collinge, described from the Hawaiian Islands, is now considered a junior synonym of *gagates* Draparnaud.

Family LIMACIDAE

See comments under Milacidae (above).

Genus DEROCERAS Rafinesque

DEROCERAS Rafinesque, 1820: 10 (as *Limax* subg.). Type species: *Limax gracilis* Rafinesque, 1820 [not Hawaiian], by monotypy.

AGRIOLIMAX Mörch, 1865: 378 (as *Limax* subg.). Type species: *Limax agrestis* Linnaeus, 1758 [not Hawaiian], by subsequent designation of Pilsbry (1922a: 80).

Although *Deroceras* was conditionally proposed by Rafinesque (1820: 10), it is available under *Code* Art. 11 (d)(i) and Art. 15.

Smith (1992: 252), in his genus heading indicated "Rafinesque, 1815" as the author and date of *Deroceras*, but cited Rafinesque's 1820 *Annals of Nature* . . . as the work in which *Deroceras* was first published. *Deroceras* does not appear in Rafinesque's 1815 *Analyse de la Nature* . . . . In addition, Smith (1992: 252) incorrectly indicated that the type species fixation was by original designation.

bevenoti.
>*Agriolimax bevenoti* Collinge, 1897: 295, fig. 9. Kauai: "4,000 feet".
>>*Remarks.* Synonym of *laeve* Müller, *teste* Quick (1960: 172).

globosum. (H; introduced)
>*Agriolimax globosus* Collinge, 1896: 47, fig. 1. Hawaii: "Mauna Loa".
>>*Remarks.* Quick (1960: 175) considered specimens labelled as *A. globosus* Collinge in the BMNH to belong to a form of *laeve* Müller, but he did not formally synonymize them.

laeve. (K, M; introduced)
>*Limax laevis* Müller, 1774 [not Hawaiian].
>>*Remarks.* Included here only because *bevenoti* Collinge, described from the Hawaiian Islands, is now considered a junior synonym of *laeve* Müller.

perkinsi. (L; introduced)
>*Agriolimax perkinsi* Collinge, 1896: 47, fig. 2. Lanai: "2000 feet".
>>*Remarks.* Quick (1960: 175) considered specimens labelled as *A. perkinsi* Collinge in the BMNH to belong to a form of *laeve* Müller, but he did not formally synonymize them.

Genus LIMAX Linnaeus

LIMAX Linnaeus, 1758: 652. Type species: *Limax maximus* Linnaeus, 1758 [not Hawaiian], by subsequent designation of Férussac (1819d: 67) [see ICZN (1926: 13; Opinion 94); ICZN (1957a: 164, 185; Direction 72)].

Férussac (1819d: 67) used the term "nobis" at the head of his treatment of *Limax*. This has led to some confusion as certain authors have taken this to indicate a new proposal of the name *Limax* rather than, as is clear from Férussac's text, a subsequent, more restricted usage of *Limax* Linnaeus. ICZN (1957a: 164, 185) accepted Férussac's type species designation as being for *Limax* Linnaeus (see also

Pilsbry, 1922b: 117; Kennard & Woodward, 1926: 189), yet in the same Direction (ICZN, 1957a: 169) stated that *Limax* Férussac is a junior homonym of *Limax* Linnaeus (i.e., a separate establishment of the genus), in which case Férussac's type species designation would not apply to *Limax* Linnaeus. Melville & Smith (1987: 117) confirmed Férussac's treatment of the genus as "a later usage of *Limax* Linnaeus, 1758". Additional confusion has been generated by the designation of *Limax rufus* as the type of *Limax* by Children (1823: 99), subsequently repeated by Gray (1847: 170). If this designation of *rufus* were valid it would lead to the unfortunate synonymizing of *Limax* Linnaeus with *Arion* Férussac, as acknowledged by Pilsbry (1922a: 78).

Smith (1992: 254) incorrectly indicated that the type of fixation of *Limax* Linnaeus was by original designation.

sandwichiensis. (Hawaiian Islands; introduced)
> *Limax sandwichiensis* Souleyet, 1852: 497, pl. 28, figs. 8–11. Hawaiian Islands: [no additional details].
>> *Remarks.* "A very doubtful species" (Collinge, 1896: 47).

Incertae Sedis in the Hawaiian Fauna

fornicata. (K)
> *Helix fornicata* Gould, 1846a: 172. Kauai: [no additional details].
>> *Remarks.* Caum (1928: 62) considered the systematic position of *fornicata* uncertain. Apparently, specimens exist (Johnson, 1964: 78) and it should be possible for future revisionary study to place this species correctly.

sandvicensis. (?Hawaiian Islands)
> *Helix sandvicensis* Pfeiffer, 1850a: 128. Hawaiian islands: [no additional details].
>> *Remarks.* Caum (1928: 62) listed this with Helicidae, but indicated that it had also been referred to *Charopa* (Punctoidea [= Endodontoidea]: Charopidae). He considered it not to occur in the Hawaiian Islands. Solem (1976, 1983) did not include it in his partial revision of the Hawaiian "endodontoids".

CHECKLIST

This checklist includes all the names listed in the main body of the catalog except misidentifications and incorrect spellings. Family-group and genus-group names appear in the same sequence as in the main catalog. Valid species are listed alphabetically within genera/subgenera. Within a species, valid infraspecific taxa are listed alphabetically, with no distinction of subspecies, varieties, color forms, etc., without implying any taxonomic judgment regarding their true status. Synonyms, homonyms and unavailable names (*nomina nuda*, etc.) are listed chronologically under the appropriate species-group name where justified. Otherwise, they are listed at the end of the appropriate subgenus, genus, family, etc. For full explanation of the treatments of the names in this list, refer to the main body of the catalog.

If a taxon was described as an infraspecific taxon of a species now synonymized with or considered a subspecies of another species, it is listed here as a subspecies of the latter.

Valid genus-group names are printed in boldface. Valid species-group names are printed in plain Roman type. Synonyms and unavailable names are in italics, indented. In 2 cases, *seminulum* Boettger (Pupillidae) and *inconspicua* Ancey (Succineidae), junior homonyms have not been replaced, so remain the names of valid taxa. As homonyms they appear in italic, but as the name of a valid taxon they appear flush left with other valid names. Names of introduced (or possibly introduced) taxa are not distinguished from native taxa in this checklist. The rationale for inclusion of certain introduced taxa is explained in the introduction to the catalog and names of these taxa are underlined in the main body of the catalog.

NERITIDAE
    Neritinae
        Neritina (**Neripteron**) Lesson, 1831
            vespertina Sowerby, 1849
                vespertina Jay, 1839
                solidissima Sowerby, 1849
        Neritina (**Neritona**) Martens, 1869
            granosa Sowerby, 1825
                papillosa Jay, 1839
                gigas Lesson, 1842
      Incertae Sedis within **Neritina**
            atra Jay, 1839
HYDROCENIDAE
        Georissa Blanford, 1864
            cookei Pilsbry, 1928
                neali Pilsbry, 1928
            kauaiensis Pilsbry, 1928
HELICINIDAE
        Orobophana Wagner, 1905
            baldwini baldwini Ancey, 1904
            baldwini lihueensis Neal, 1934

Orobophana Wagner, 1905 (continued)
    cookei Neal, 1934
    juddii Pilsbry & Cooke, 1908
    meineckei Neal, 1934
    stokesii praemagna Neal, 1934
    stokesii stokesii Neal, 1934
    uberta beta Pilsbry & Cooke, 1908
    uberta borealis Neal, 1934
    uberta bryani Neal, 1934
    uberta exanima Neal, 1934
    uberta hibrida Neal, 1934
    uberta lymaniana Pilsbry & Cooke, 1908
    uberta magdalenae Ancey, 1890
        rhodostoma Pfeiffer, 1850
    uberta makuaensis Neal, 1934
    uberta nuuanuensis Pilsbry & Cooke, 1908
    uberta percitrea Neal, 1934
    uberta subtenuis Neal, 1934
    uberta uberta Gould, 1847
    uberta wilderi Neal, 1934
Pleuropoma (**Aphanoconia**) Wagner, 1905
  Sphaeroconia Wagner, 1909
    kauaiensis kauaiensis Pilsbry & Cooke, 1908
    kauaiensis orientalis Neal, 1934
    niihauensis Neal, 1934
    rotelloidea knudseni Pilsbry & Cooke, 1908
    rotelloidea makalii Neal, 1934
    rotelloidea mauiensis Neal, 1934
    rotelloidea rotelloidea Mighels, 1845
        hawaiiensis Pilsbry & Cooke, 1908
    rotelloidea sola Neal, 1934
    sulculosa Ancey, 1904
Pleuropoma (**Pleuropoma**) Möllendorff, 1893
  Sturanya Wagner, 1905
  Sturanyella Pilsbry & Cooke, 1934
    laciniosa alpha Pilsbry & Cooke, 1908
    laciniosa bronniana Philippi, 1847
    laciniosa canyonensis Neal, 1934
    laciniosa delta Pilsbry & Cooke, 1908
    laciniosa ferruginea Neal, 1934
    laciniosa gamma Pilsbry & Cooke, 1908
    laciniosa globuloidea Neal, 1934
    laciniosa honokowaiensis Neal, 1934
    laciniosa kaaensis Neal, 1934
    laciniosa kahoolawensis Neal, 1934
    laciniosa kiekieensis Neal, 1934
    laciniosa konaensis Neal, 1934
    laciniosa kulaensis Neal, 1934
    laciniosa laciniosa Mighels, 1845
        berniceia Pilsbry & Cooke, 1908
    laciniosa laula Neal, 1934
    laciniosa matutina Neal, 1934
    laciniosa molokaiensis Neal, 1934
    laciniosa moomomiensis Neal, 1934

Pleuropoma (Pleuropoma) Möllendorff, 1893 (continued)
    laciniosa perparva Neal, 1934
    laciniosa piliformis Neal, 1934
    laciniosa praeparva Neal, 1934
    laciniosa pusilla Neal, 1934
    laciniosa sandwichiensis Souleyet, 1852
        dissotropis Ancey, 1904
    laciniosa signata Neal, 1934
    laciniosa spaldingi Neal, 1934
    nonouensis Neal, 1934
    oahuensis gemina Neal, 1934
    oahuensis oahuensis Pilsbry & Cooke, 1908
    subsculpta Neal, 1934

Questionably included HELICINIDAE in the Hawaiian fauna
    Helicina Lamarck, 1799
        antoni Pfeiffer, 1849
        constricta Pfeiffer, 1849
        crassilabris Philippi, 1847
        fulgora Gould, 1847
        pisum Philippi, 1847

HYDROBIIDAE
Incertae Sedis within HYDROBIIDAE
    porrecta Mighels, 1845

THIARIDAE
    Tarebia Adams & Adams, 1854
        granifera mauiensis Lea, 1856
            tahitensis Brot, 1877
    Thiara Röding, 1798
      Melania Lamarck, 1799
        baldwini Ancey, 1899
        indefinita Lea & Lea, 1851
            newcombii Lea, 1856
            contigua Pease, 1870
            oahuensis Brot, 1872
            paulla Brot, 1872
        kauaiensis Pease, 1870
        verrauiana Lea, 1856

ELLOBIIDAE
  Cassidulinae
    Allochroa Ancey, 1887
        bronnii Philippi, 1846
            sandwichiensis Souleyet, 1852
            brownii Adams & Adams, 1855
            conica Pease, 1863
    Auriculastra Martens, 1880
        elongata Küster, 1845
  Ellobiinae
    Ellobium Röding, 1798
      Auricula Lamarck, 1799
        fuscum Küster, 1844
  Melampodinae
    Melampus Montfort, 1810
      Pira Adams & Adams, 1855
        castaneus Megerle von Mühlfeld, 1816

        Melampus Montfort, 1810 (continued)
            lucidus Pease, 1869
            parvulus Pfeiffer, 1856
                parvulus Pfeiffer, 1854
            sculptus Pfeiffer, 1859
                semiplicatum Adams & Adams, 1854
                fricki Pfeiffer, 1859
                pseudocommodus Schmeltz, 1869
    Pedipedinae
        Laemodonta Philippi, 1846
            Lirator Beck, 1838
            Plecotrema Adams & Adams, 1854
            Enterodonta Sykes, 1894
            octanfracta Jonas, 1845
                multisulcatus Beck, 1838
                striata Philippi, 1846
                clausa Adams & Adams, 1854
                inaequalis Adams & Adams 1854
        Pedipes Férussac, 1821
            sandwicensis Pease, 1860
    Pythiinae
        Blauneria Shuttleworth, 1854
            gracilis Pease, 1860
LYMNAEIDAE
    Erinna Adams & Adams, 1855
        aulacospira Ancey, 1889
            hawaiensis Pilsbry, 1904
        newcombi Adams & Adams, 1855
    Lymnaea (**Pseudisidora**) Thiele, 1931
        producta Mighels, 1845
            reticulata Gould, 1847
            affinis Souleyet, 1852
            sinistrorsus Martens, 1866
            ambigua Pease, 1870
            compacta Pease, 1870
            turgidula Pease, 1870
            flavida Clessin, 1886
            hartmanni Clessin, 1886
            moreletiana Clessin, 1886
            naticoides Clessin, 1886
            peasei Clessin, 1886
            sandwichensis Clessin, 1886
            binominis Sykes, 1900
        rubella Lea, 1841
            sandwicensis Philippi, 1845
            volutata Gould, 1847
            oahouensis Souleyet, 1852
Incertae sedis within LYMNAEIDAE
        umbilicata Mighels, 1845
ANCYLIDAE
    Ferrissia (**Pettancylus**) Iredale
        sharpi Sykes, 1900

ACHATINELLIDAE
    Achatinellinae
        Achatinella (Achatinella) Swainson, 1828
            Apex Martens, 1860
                apexfulva alba Sykes, 1900
                    beata Pilsbry & Cooke, 1914
                apexfulva albipraetexta Welch, 1942
                apexfulva albofasciata Smith, 1873
                apexfulva aloha Pilsbry & Cooke, 1914
                apexfulva apexfulva Dixon, 1789
                    lugubris Chemnitz, 1795
                    lugubris Férussac, 1821
                    seminigra Lamarck, 1822
                    lugubris Férussac, 1825
                    pica Swainson, 1828
                apexfulva apicata Pfeiffer, 1856
                apexfulva aureola Welch, 1942
                apexfulva bakeri Welch, 1942
                apexfulva bruneola Welch, 1942
                apexfulva brunosa Welch, 1942
                apexfulva buena Welch, 1942
                apexfulva cervixnivea Pilsbry & Cooke, 1914
                apexfulva cestus Newcomb, 1854
                apexfulva chromatacme Pilsbry & Cooke, 1914
                apexfulva cinerea Sykes, 1900
                apexfulva coniformis Gulick, 1873
                apexfulva cookei Baldwin, 1895
                    cookii Baldwin, 1893
                apexfulva duplocincta Pilsbry & Cooke, 1914
                apexfulva ewaensis Welch, 1942
                apexfulva flavida Gulick, 1873
                apexfulva flavitincta Welch, 1942
                apexfulva forbesiana Pfeiffer, 1855
                apexfulva fumositincta Welch, 1942
                apexfulva fuscostriata Welch, 1942
                apexfulva glaucopicta Welch, 1942
                apexfulva globosa Pfeiffer, 1855
                apexfulva griseibasis Welch, 1942
                apexfulva gulickii Smith, 1873
                apexfulva hanleyana Pfeiffer, 1856
                apexfulva ihiihiensis Welch, 1942
                apexfulva innotabilis Smith, 1873
                apexfulva irwini Pilsbry & Cooke, 1914
                apexfulva kahukuensis Pilsbry & Cooke, 1914
                apexfulva kawaiiki Welch, 1942
                apexfulva laurani Welch, 1942
                apexfulva lemkei Welch, 1942
                apexfulva leucorrhaphe Gulick, 1873
                apexfulva leucozona Gulick, 1873
                    cinerosa Pfeiffer, 1855
                apexfulva lilacea Gulick, 1873
                apexfulva lineipicta Welch, 1942
                apexfulva meadowsi Welch, 1942
                apexfulva muricolor Welch, 1942

CHECKLIST [175]

Achatinella (Achatinella) Swainson, 1828 (continued)
    apexfulva napus Pfeiffer, 1855
        elongata Pfeiffer, 1868
    apexfulva nigripicta Welch, 1942
    apexfulva oioensis Welch, 1942
    apexfulva oliveri Welch, 1942
    apexfulva ovum Pfeiffer, 1857
    apexfulva paalaensis Welch, 1942
    apexfulva parvicolor Welch, 1942
    apexfulva paumaluensis Welch, 1942
    apexfulva perplexa Pilsbry & Cooke, 1914
    apexfulva pilsbryi Welch, 1942
    apexfulva poamohoensis Welch, 1942
    apexfulva polymorpha Gulick, 1873
        neglectus Smith, 1873
    apexfulva punicea Welch, 1942
    apexfulva roseata Welch, 1942
    apexfulva roseipicta Welch, 1942
    apexfulva rubidilinea Welch, 1942
    apexfulva rubidipicta Welch, 1942
    apexfulva simulacrum Pilsbry & Cooke, 1914
    apexfulva simulans Reeve, 1850
    apexfulva simulator Pilsbry & Cooke, 1914
    apexfulva steeli Welch, 1942
    apexfulva suturafusca Welch, 1942
    apexfulva suturalba Welch, 1942
    apexfulva tuberans Gulick, 1873
    apexfulva versicolor Gulick, 1873
    apexfulva vespertina Baldwin, 1895
        vespertina Baldwin, 1893
    apexfulva virgatifulva Welch, 1942
    apexfulva vittata Reeve, 1850
    apexfulva wahiawa Welch, 1942
    apexfulva waialaeensis Welch, 1942
    apexfulva wailelensis Welch, 1942
    apexfulva waimaluensis Welch, 1942
    concavospira concavospira Pfeiffer, 1859
    concavospira griseizona Pilsbry & Cooke, 1914
    concavospira turbiniformis Gulick, 1873
    decora Férussac, 1821
        sinistrorsus Chemnitz, 1795
        perversa Swainson, 1828
        tumefactus Gulick, 1873
        quernea Pilsbry & Cooke, 1914
    dolium Pfeiffer, 1855
    lorata lorata Férussac, 1825
        lorata Férussac, 1821
        alba Jay, 1839
        pallida Jay, 1839
        pallida Reeve, 1850
        ventrosa Pfeiffer, 1855
    lorata melanogama Pilsbry & Cooke, 1914
    lorata nobilis Pfeiffer, 1856
    lorata pulchella Pfeiffer, 1855

Achatinella (Achatinella) Swainson, 1828 (continued)
    mustelina altiformis Welch, 1938
    mustelina bicolor Pfeiffer, 1859
    mustelina brunibasis Welch, 1938
    mustelina brunicolor Welch, 1938
    mustelina christopherseni Welch, 1938
    mustelina collaris Welch, 1938
    mustelina dautzenbergi Welch, 1938
    mustelina decolor Welch, 1938
    mustelina diffusa Welch, 1938
    mustelina griseipicta Welch, 1938
    mustelina griseitincta Welch, 1938
    mustelina kaalaensis Welch, 1938
    mustelina kapuensis Welch, 1938
    mustelina lathropae Welch, 1938
    mustelina lymaniana Baldwin, 1895
        lymaniana Baldwin, 1893
    mustelina mailiensis Welch, 1938
    mustelina makahaensis Pilsbry & Cooke, 1914
    mustelina maxima Welch, 1938
    mustelina mixta Welch, 1938
    mustelina mustelina Mighels, 1845
        multilineata Newcomb, 1854
    mustelina nocturna Welch, 1938
    mustelina obesiformis Welch, 1938
    mustelina popouwelensis Welch, 1938
    mustelina russi Welch, 1938
    mustelina sordida Newcomb, 1854
    mustelina waianaeensis Welch, 1938
    turgida Newcomb, 1854
        swiftii Newcomb, 1854
        albospira Smith, 1873
    valida leucophaea Gulick, 1873
    valida valida Pfeiffer, 1855
    vestita Mighels, 1845

Achatinella (Achatinellastrum) Pfeiffer, 1854
  Helicteres Férussac, 1821
  Helicteres Beck, 1837
  Helicterella Gulick, 1873
    bellula Smith, 1873
        multizonata Baldwin, 1893
        multizonata Baldwin, 1895
    buddii Newcomb, 1854
        fuscozona Smith, 1873
    caesia caesia Gulick, 1858
        formosa Gulick, 1858
    caesia cervina Gulick, 1858
    caesia cognata Gulick, 1858
        scitula Gulick, 1858
    caesia concidens Gulick, 1858
    caesia littoralis Pilsbry & Cooke, 1914
    casta casta Newcomb, 1854
        cuneus Pfeiffer, 1856
        concolor Smith, 1873

Achatinella (Achatinellastrum) Pfeiffer, 1854 (continued)
    ligata Smith, 1873
    pygmaea Smith, 1873
  casta margaretae Pilsbry & Cooke, 1914
  curta Newcomb, 1854
    undulata Newcomb, 1855
    undulata Pfeiffer, 1856
    delta Gulick, 1858
    rhodorhaphe Smith, 1873
  dimorpha Gulick, 1858
    albescens Gulick, 1858
    contracta Gulick, 1858
    zonata Gulick, 1858
  fulgens ampla Newcomb, 1854
  fulgens fulgens Newcomb, 1854
    crassidentata Pfeiffer, 1855
    diversa Gulick, 1856
    plumata Gulick, 1856
    varia Gulick, 1856
    trilineata Gulick, 1856
    liliacea Pfeiffer, 1859
    augusta Smith, 1873
    angusta Paetel, 1873
  fulgens versipellis Gulick, 1856
    fuscolineata Smith, 1873
  juddii Baldwin, 1895
    juddii Baldwin, 1893
  juncea Gulick, 1856
  lehuiensis gulickiana Pilsbry & Cooke, 1914
  lehuiensis lehuiensis Smith, 1873
  lehuiensis meineckei Pilsbry & Cooke, 1921
  livida emmersonii Newcomb, 1854
  livida herbacea Gulick, 1858
  livida livida Swainson, 1828
    reevei Adams, 1851
  livida recta Newcomb, 1854
    glauca Gulick, 1858
  papyracea Gulick, 1856
  phaeozona Gulick, 1856
  solitaria Newcomb, 1854
  spaldingi Pilsbry & Cooke, 1914
  stewartii producta Reeve, 1850
    bilineata Reeve, 1850
    hybrida Newcomb, 1854
    venulata Newcomb, 1854
    dunkeri Pfeiffer, 1856
  stewartii stewartii Green, 1827
    aplustre Newcomb, 1854
    johnsoni Newcomb, 1854
  thaanumi Pilsbry & Cooke, 1914
  vulpina colorata Reeve, 1850
    ustulata Pfeiffer, 1859
    consanguinea Smith, 1873
  vulpina tricolor Smith, 1873

Achatinella (Achatinellastrum) Pfeiffer, 1854 (continued)
    vulpina vulpina Férussac, 1825
        vulpina Férussac, 1821
        adusta Reeve, 1850
        castanea Reeve, 1850
        olivacea Reeve, 1850
        prasinus Reeve, 1850
        analoga Gulick, 1856
        cucumis Gulick, 1856
        virens Gulick, 1858
        diluta Smith, 1873
        longispira Smith, 1873
        ernestina Baldwin, 1893
        ernestina Baldwin, 1895
        suturalis Pilsbry & Cooke, 1914
Incertae sedis within subg. **Achatinellastrum**
        olesonii Baldwin, 1893
        senistra Schmeltz, 1865
Achatinella (Bulimella) Pfeiffer, 1854
    abbreviata Reeve, 1850
        bacca Reeve, 1850
        nivosa Newcomb, 1854
        clementina Pfeiffer, 1856
    bulimoides albalabia Welch, 1954
    bulimoides arnemani Welch, 1958
    bulimoides bulimoides Swainson, 1828
    bulimoides caesiapicta Welch, 1954
    bulimoides circulospadix Welch, 1954
    bulimoides elegans Newcomb, 1854
        candida Pfeiffer, 1855
    bulimoides fricki Pfeiffer, 1855
        monacha Pfeiffer, 1855
    bulimoides fulvula Welch, 1954
    bulimoides glabra Newcomb, 1854
    bulimoides kaipapauensis Welch, 1958
    bulimoides kalekukei Welch, 1954
    bulimoides mistura Pilsbry & Cooke, 1913
    bulimoides obliqua Gulick, 1858
        oomorpha Gulick, 1858
    bulimoides oswaldi Welch, 1958
    bulimoides ovata Newcomb, 1853
    bulimoides papakokoensis Welch, 1958
    bulimoides rosea Swainson, 1828
    bulimoides rosealimbata Welch, 1954
    bulimoides rotunda Gulick, 1858
    bulimoides rufapicta Welch, 1958
    bulimoides spadicea Gulick, 1858
    bulimoides vidua Pfeiffer, 1855
        inelegans Pilsbry & Cooke, 1913
    bulimoides wheatleyana Pilsbry & Cooke, 1913
    byronii byronii Wood, 1828
        byronensis Beck, 1837
        limbata Gulick, 1858
    byronii nigricans Pilsbry & Cooke, 1913

Achatinella (Bulimella) Pfeiffer, 1854 (continued)
    byronii rugosa Newcomb, 1854
        capax Pilsbry & Cooke, 1913
        waimanoensis Pilsbry & Cooke, 1913
    decipiens decipiens Newcomb, 1854
        corrugata Gulick, 1858
        torrida Gulick, 1858
    decipiens kaliuwaaensis Pilsbry & Cooke, 1913
    decipiens planospira Pfeiffer, 1855
    decipiens swainsoni Pfeiffer, 1855
    faba Pfeiffer, 1859
    fuscobasis fuscobasis Smith, 1873
        luteostoma Baldwin, 1893
        luteostoma Baldwin, 1895
    fuscobasis lyonsiana Baldwin, 1895
        lyonsiana Baldwin, 1893
    fuscobasis wilderi Pilsbry, 1913
    lila Pilsbry, 1913
    pulcherrima nympha Gulick, 1858
    pulcherrima pulcherrima Swainson, 1828
        melanostoma Newcomb, 1854
        multicolor Pfeiffer, 1855
        mahogani Gulick, 1858
    pupukanioe Pilsbry & Cooke, 1913
    sowerbyana dextroversa Pilsbry & Cooke, 1914
    sowerbyana laiensis Pilsbry & Cooke, 1914
    sowerbyana roseoplica Pilsbry & Cooke, 1914
    sowerbyana sowerbyana Pfeiffer, 1855
        oviformis Newcomb, 1855
        oviformis Pfeiffer, 1856
    sowerbyana thurstoni Pilsbry & Cooke, 1914
    taeniolata Pfeiffer, 1846
        rubiginosa Newcomb, 1854
    viridans Mighels, 1845
        radiata Pfeiffer, 1846
        rutila Newcomb, 1854
        subvirens Newcomb, 1854
        macrostoma Pfeiffer, 1855
Incertae sedis within subgenus **Bulimella** Pfeiffer, 1854
        wheatleyi Newcomb, 1855
Incertae sedis within **Achatinella** s.l.
        agatha Schaufuss, 1869
        anacardiensis Paetel, 1883
        bensonia Schaufuss, 1869
        cingulata Paetel, 1873
        circulata Paetel, 1883
        compressa Paetel, 1873
        gravis Paetel, 1873
        havaiana Schaufuss, 1869
        hawaiana Paetel, 1883
        ignominiosus Paetel, 1873
        impressa Paetel, 1873
        magnifica Schaufuss, 1869
        scamnata Schaufuss, 1869

Incertae sedis within **Achatinella** *s.l.* (continued)
        semitecta Paetel, 1887
        sinistra Schaufuss, 1869
        subovata Schaufuss, 1869
        torquata Schaufuss, 1869
        turbinata Jay, 1839
Newcombia Pfeiffer, 1854
    canaliculata canaliculata Baldwin, 1895
        canaliculata Baldwin, 1893
    canaliculata wailauensis Pilsbry & Cooke, 1912
    cumingi Newcomb, 1853
    lirata gemma Pfeiffer, 1857
        costata Borcherding, 1901
    lirata lirata Pfeiffer, 1853
        plicata Pfeiffer, 1848
    perkinsi Sykes, 1896
    pfeifferi decorata Pilsbry & Cooke, 1912
    pfeifferi honomuniensis Pilsbry & Cooke, 1912
    pfeifferi pfeifferi Newcomb, 1853
        newcombianus Pfeiffer, 1853
        cinnamomea Pfeiffer, 1857
    pfeifferi ualapuensis Pilsbry & Cooke, 1912
    philippiana Pfeiffer, 1857
    sulcata Pfeiffer, 1857
Partulina (Baldwinia) Ancey, 1899
  Rugosella Coen, 1945
  Rugosella Mienis, 1987
    confusa Sykes, 1900
        phaeostoma Ancey, 1904
    dubia dubia Newcomb, 1853
        pexa Gulick, 1856
        platystyla Gulick, 1856
        morbida Pfeiffer, 1859
        fucosa Lyons, 1891
    dubia perantiqua Cooke & Kondo, 1952
    grisea Newcomb, 1854
    horneri candida Pilsbry & Cooke, 1914
    horneri fuscospira Pilsbry & Cooke, 1914
    horneri fuscozonata Pilsbry & Cooke, 1914
    horneri horneri Baldwin, 1895
        horneri Baldwin, 1893
    horneri kapuana Gouveia & Gouveia, 1920
    physa errans Pilsbry & Cooke, 1913
    physa konana Pilsbry & Cooke, 1914
    physa physa Newcomb, 1854
        hawaiiensis Baldwin, 1893
        hawaiiensis Baldwin, 1895
    thaanumiana Pilsbry, 1913
Incertae sedis within **Partulina (Baldwinia)**
        procera Ancey, 1904
Partulina (Eburnella) Pease, 1870
    anceyana Baldwin, 1895
        anceyana Baldwin, 1893
    germana Newcomb, 1854

Partulina (Eburnella) Pease, 1870 (continued)
    mighelsiana bella Reeve, 1850
    mighelsiana dixoni Borcherding, 1906
    mighelsiana hepatica Borcherding, 1906
    mighelsiana latizona Borcherding, 1906
    mighelsiana martensi Borcherding, 1906
    mighelsiana mighelsiana Pfeiffer, 1848
    mighelsiana polita Newcomb, 1853
    mutabilis Baldwin, 1908
    nattii Hartman, 1888
    porcellana flemingi Baldwin, 1906
    porcellana fulvicans Baldwin, 1906
    porcellana porcellana Newcomb, 1854
    porcellana wailuaensis Sykes, 1900
        cooperi Baldwin, 1906
    semicarinata hayseldeni Baldwin, 1896
    semicarinata semicarinata Newcomb, 1854
    variabilis lactea Gulick, 1856
        saccata Pfeiffer, 1859
    variabilis variabilis Newcomb, 1854
        fulva Pfeiffer, 1856
Partulina (Partulina) Pfeiffer, 1854
    Partulinella Hyatt, 1914
    aptycha Pfeiffer 1855
    arnemanni Cooke & Kondo, 1952
    carnicolor Baldwin, 1906
    crassa Newcomb, 1854
    crocea Gulick, 1856
    dolei Baldwin, 1895
    dwightii compta Pease, 1869
    dwightii concomitans Hyatt, 1912
    dwightii dwightii Newcomb, 1855
        dwighti Pfeiffer, 1856
    dwightii macrodon Borcherding, 1901
    dwightii mucida Baldwin, 1895
        mucida Baldwin, 1893
    dwightii occidentalis Pilsbry & Cooke, 1914
    fusoidea Newcomb, 1855
    induta Gulick, 1856
    kaaeana Baldwin, 1906
    lemmoni Baldwin, 1906
    marmorata Gould, 1847
        adamsi Newcomb, 1853
    montagui Pilsbry, 1913
    nivea kaupakaluana Pilsbry & Cooke, 1912
    nivea nivea Baldwin, 1895
        nivea Baldwin, 1893
    perdix perdix Reeve, 1850
        pyramidalis Gulick, 1856
    perdix undosa Gulick, 1856
    plumbea Gulick, 1856
    proxima multistrigata Pilsbry & Cooke, 1912
    proxima proxima Pease, 1862
    proxima schauinslandi Borcherding, 1901

Partulina (Partulina) Pfeiffer, 1854 (continued)
    radiata Gould, 1845
        gouldi Pfeiffer, 1846
        gouldi Pfeiffer, 1848
        densilineata Reeve, 1850
    redfieldi kamaloensis Pilsbry & Cooke, 1914
    redfieldi redfieldi Newcomb, 1853
    rufa idae Borcherding, 1901
    rufa rufa Newcomb, 1853
    splendida baileyana Gulick, 1856
    splendida splendida Newcomb, 1853
        solida Pfeiffer, 1859
    subpolita Pilsbry & Cooke, 1912
    talpina Gulick, 1856
        gouldi Newcomb, 1853
        myrrhea Pfeiffer, 1859
        perfecta Pilsbry, 1912
    tappaniana ampulla Gulick, 1856
    tappaniana dubiosa Adams, 1851
    tappaniana eburnea Gulick, 1856
    tappaniana fasciata Gulick, 1856
    tappaniana tappaniana Adams, 1851
        tuba Pfeiffer, 1859
        lutea Pfeiffer, 1876
    terebra attenuata Pfeiffer, 1855
    terebra corusca Pilsbry & Cooke, 1912
        corusca Pfeiffer, 1859
    terebra lignaria Gulick, 1856
        perforata Pfeiffer, 1859
    terebra longior Pilsbry & Cooke, 1912
    terebra terebra Newcomb, 1854
    tessellata meyeri Borcherding, 1901
    tessellata tessellata Newcomb, 1853
    theodorei Baldwin, 1895
        theodorii Baldwin, 1893
    ustulata Gulick, 1856
    virgulata halawaensis Borcherding, 1906
    virgulata kaluaahacola Pilsbry & Cooke, 1914
    virgulata virgulata Mighels, 1845
        rohri Pfeiffer, 1846
        insignis Reeve, 1850
    winniei Baldwin, 1908
Incertae sedis within **Partulina** *s. str.*
    dextra Schmeltz, 1865
Perdicella Pease, 1870
    carinella Baldwin, 1906
    fulgurans Sykes, 1900
    helena balteata Pilsbry & Cooke, 1912
    helena helena Newcomb, 1853
        minuscula Pfeiffer, 1857
    kuhnsi Pilsbry, 1912
    maniensis Pfeiffer, 1856
        mauiensis Pfeiffer, 1859
    ornata Newcomb, 1854

Perdicella Pease, 1870 (continued)
        thwingi Pilsbry & Cooke, 1914
        zebra Newcomb, 1855
        zebrina Pfeiffer, 1856
Incertae sedis within **Perdicella**
        manoensis Pease, 1870

Auriculellinae
Auriculella Pfeiffer, 1855
    Frickella Pfeiffer, 1855
        ambusta ambusta Pease, 1868
        ambusta obliqua Ancey, 1892
        amoena Pfeiffer, 1855
        armata Mighels, 1845
            westerlundiana Ancey, 1889
        auricula auricula Férussac, 1821
            owaihiensis Chamisso, 1829
            sinistrorsa Chamisso, 1829
            dumartroyii Souleyet, 1842
            triplicata Pease, 1868
            patula Smith, 1873
            perkinsi Sykes, 1900
        auricula pellucida Pilsbry & Cooke, 1915
            pellucida Gulick, 1905
        brunnea Smith, 1873
        canalifera Ancey, 1904
        castanea Pfeiffer, 1853
            lurida Pfeiffer, 1856
            solida Ancey, 1889
        cerea Pfeiffer, 1855
        chamissoi Pfeiffer, 1855
        crassula Smith, 1873
            solidissima Bland & Binney, 1873
            ponderosa Ancey, 1889
        diaphana cacuminis Pilsbry & Cooke, 1915
        diaphana diaphana Smith, 1873
        expansa expansa Pease, 1868
        expansa porcellana Ancey, 1889
        flavida Cooke, 1915
        kuesteri Pfeiffer, 1855
        lanaiensis Cooke, 1915
        malleata Ancey, 1904
        minuta Cooke & Pilsbry, 1915
        montana Cooke, 1915
        newcombi Pfeiffer, 1854
            obeliscus Pfeiffer, 1856
        olivacea Cooke, 1915
        perpusilla Smith, 1873
        perversa Cooke, 1915
        petitiana Pfeiffer, 1847
        pulchra Pease, 1868
        serrula Cooke, 1915
        straminea Cooke, 1915
        tantalus Pilsbry & Cooke, 1915
        tenella Ancey, 1889

Auriculella Pfeiffer, 1855 (continued)
 tenuis Smith, 1873
 turritella Cooke, 1915
 uniplicata jucunda Pilsbry & Cooke, 1915
  jucunda Bland & Binney, 1873
 uniplicata uniplicata Pease, 1868
Gulickia Cooke, 1915
 alexandri Cooke, 1915
Pacificellinae
 Lamellidea (Elamellidea) Cooke & Kondo, 1961
  tantalus Pilsbry & Cooke, 1915
 Lamellidea (Lamellidea) Pilsbry, 1910
  Lamellina Pease, 1861
  Lamellaria Liardet, 1876
  cylindrica cylindrica Sykes, 1900
  cylindrica kilohanana Pilsbry & Cooke, 1915
  gayi Cooke & Pilsbry, 1915
  gracilis Pease, 1871
   extincta Ancey, 1890
  lanceolata Cooke & Pilsbry, 1915
  oblonga Pease, 1865
   dentata Pease, 1871
  peponum Gould, 1847
  polygnampta kamaloensis Pilsbry & Cooke, 1915
  polygnampta polygnampta Pilsbry & Cooke, 1915
 Pacificella Odhner, 1922
  baldwini baldwini Ancey, 1889
  baldwini subrugosa Pilsbry & Cooke, 1915
  mcgregori Pilsbry & Cooke, 1915
Tornatellidinae
 Philopoa Cooke & Kondo, 1961
  singularis Cooke & Kondo, 1961
 Tornatellaria Pilsbry, 1910
  abbreviata abbreviata Ancey, 1903
  abbreviata hawaiiensis Pilsbry & Cooke, 1916
  adelinae Pilsbry & Cooke, 1915
  anceyana Cooke & Pilsbry, 1916
  baldwiniana Cooke & Pilsbry, 1916
  cincta Ancey, 1903
  convexior Pilsbry & Cooke, 1916
  henshawi Ancey, 1903
  lilae Cooke & Pilsbry, 1915
  newcombi Pfeiffer, 1857
  occidentalis Pilsbry & Cooke, 1916
  sharpi Pilsbry & Cooke, 1916
  smithi Cooke & Pilsbry, 1916
  sykesii illibata Cooke & Pilsbry, 1916
  sykesii sykesii Cooke & Pilsbry, 1916
  thaanumi Cooke & Pilsbry, 1915
  trochoides Sykes, 1900
  umbilicata Ancey, 1889
 Tornatellides (Aedituans) Cooke & Kondo, 1961
  neckeri Cooke & Kondo, 1961

Tornatellides (Tornatellides) Pilsbry, 1910
        attenuatus Cooke & Pilsbry, 1915
        bellus Cooke & Pilsbry, 1915
        brunneus Cooke & Pilsbry, 1915
        bryani Cooke & Pilsbry, 1915
        comes Pilsbry & Cooke, 1915
        compactus Sykes, 1900
        confusus Sykes, 1900
        cyphostyla Ancey, 1904
        diptyx Pilsbry & Cooke, 1915
        drepanophorus Cooke & Pilsbry, 1915
        euryomphala Ancey, 1889
        forbesi forbesi Cooke & Pilsbry, 1915
        forbesi nanus Cooke & Pilsbry, 1915
        frit Pilsbry & Cooke, 1915
        idae anisoplax Pilsbry & Cooke, 1915
        idae idae Cooke & Pilsbry, 1915
        inornatus Pilsbry & Cooke, 1915
        insignis Pilsbry & Cooke, 1915
        irregularis Cooke & Pilsbry, 1915
        kahoolavensis Cooke & Pilsbry, 1915
        kahukuensis Pilsbry & Cooke, 1915
        kamaloensis Pilsbry & Cooke, 1915
        kilauea Pilsbry & Cooke, 1915
        konaensis Cooke & Pilsbry, 1915
        leptospira Cooke & Pilsbry, 1915
        macromphala ada Pilsbry & Cooke, 1915
        macromphala macromphala Ancey, 1903
        macroptychia Ancey, 1903
        micromphala Pilsbry & Cooke, 1915
        moomomiensis Pilsbry & Cooke, 1915
        oahuensis Cooke & Pilsbry, 1915
        oncospira Cooke & Pilsbry, 1915
        oswaldi Cooke & Kondo, 1961
        perkinsi acicula Pilsbry & Cooke, 1915
        perkinsi perkinsi Sykes, 1900
        pilsbryi Cooke, 1914
        plagioptyx Pilsbry & Cooke, 1915
        popouelensis Pilsbry & Cooke, 1915
        prionoptychia Cooke & Pilsbry, 1915
        procerulus kailuanus Pilsbry & Cooke, 1915
        procerulus procerulus Ancey, 1903
        procerulus puukolekolensis Pilsbry & Cooke, 1915
        productus Ancey, 1903
        pyramidatus Ancey, 1903
        ronaldi Cooke & Pilsbry, 1915
        serrarius Pilsbry & Cooke, 1915
        spaldingi Cooke & Pilsbry, 1915
        stokesi Pilsbry & Cooke, 1916
        subangulatus Ancey, 1903
        terebra Ancey, 1903
        virgula Cooke & Pilsbry, 1915
            virgula Cooke, 1907
        vitreus Dohrn, 1863

Tornatellides (Tornatellides) Pilsbry, 1910 (continued)
  waianaensis Pilsbry & Cooke, 1915
Tornatellides (Waimea) Cooke & Pilsbry, 1915
  rudicostatus Ancey, 1904
Tornatellininae
Elasmias Pilsbry, 1910
  anceyanum Pilsbry & Cooke, 1915
  fuscum fuscum Ancey, 1903
  fuscum obtusum Pilsbry & Cooke, 1915
  luakahaense Pilsbry & Cooke, 1915
Incertae sedis within ACHATINELLIDAE
  clausinus Mighels, 1845
  kanaiensis Pfeiffer, 1857
    kauaiensis Pfeiffer, 1859
  sandwichensis Pfeiffer, 1857
Unplaced *nomina nuda* in ACHATINELLIDAE
  hawaiiensis Cooke, 1907
  leucozonalis Beck, 1837
  sulphuratus Beck, 1837
AMASTRIDAE
  Amastrinae
Amastra (Amastra) Adams & Adams, 1855
  affinis affinis Newcomb, 1854
    pupoidea Newcomb, 1854
    goniostoma Pfeiffer, 1856
    rustica Gulick, 1873
    abberans Hyatt & Pilsbry, 1911
    bigener Hyatt, 1911
    cinderella Hyatt, 1911
    subpulla Hyatt & Pilsbry, 1911
  affinis kaupakaluana Hyatt & Pilsbry, 1911
  albocincta Pilsbry & Cooke, 1914
  assimilis assimilis Newcomb, 1854
    deshaysii Morelet, 1857
  assimilis subassimilis Hyatt & Pilsbry, 1911
  aurostoma Baldwin, 1896
  baldwiniana baldwiniana Hyatt & Pilsbry, 1911
  baldwiniana kahakuloensis Pilsbry & Cooke, 1914
  biplicata Newcomb, 1854
  borcherdingi Hyatt & Pilsbry, 1911
  conifera Smith, 1873
  durandi Ancey, 1897
  elegantula Hyatt & Pilsbry, 1911
  goniops Pilsbry & Cooke, 1914
  grayana Pfeiffer, 1856
  hitchcocki Cooke, 1917
  humilis humilis Newcomb, 1855
    humilis Pfeiffer, 1856
  humilis moomomiensis Pilsbry & Cooke, 1914
  humilis sepulta Pilsbry & Cooke, 1914
  inopinata Cooke, 1933
  johnsoni Hyatt & Pilsbry, 1911
  kalamaulensis Pilsbry & Cooke, 1914
  kaunakakaiensis Pilsbry & Cooke, 1914

Amastra (Amastra) Adams & Adams, 1855 (continued)
    lahainana Pilsbry & Cooke, 1914
    lineolata Newcomb, 1853
    magna balteata Hyatt & Pilsbry, 1911
    magna magna Adams, 1851
        baldwinii Newcomb, 1854
        gigantea Newcomb, 1854
    makawaoensis Hyatt & Pilsbry, 1911
    malleata Smith, 1873
    mastersi Newcomb, 1854
    mirabilis Cooke, 1917
    modesta dimissa Hyatt & Pilsbry, 1911
    modesta modesta Adams, 1851
        pumila Pfeiffer & Clessin, 1879
    moesta longa Sykes, 1896
    moesta moesta Newcomb, 1854
    moesta obscura Newcomb, 1854
    montana Baldwin, 1906
    mucronata atroflava Hyatt & Pilsbry, 1911
    mucronata citrea Sykes, 1896
    mucronata mucronata Newcomb, 1853
        fusiformis Pfeiffer, 1855
        simularis Hartman, 1888
        maura Ancey & Sykes, 1899
    mucronata roseotincta Sykes, 1896
    mucronata semicarnea Ancey & Sykes, 1899
    nana Baldwin, 1895
        nana Baldwin, 1893
    neglecta Pilsbry & Cooke, 1914
    nigra Newcomb, 1855
        nigra Pfeiffer, 1856
        globosa Hyatt & Pilsbry, 1911
    nubifera dissimiliceps Hyatt & Pilsbry, 1911
    nubifera nubifera Hyatt & Pilsbry, 1911
    nubilosa georgii Pilsbry & Cooke, 1914
    nubilosa macerata Hyatt & Pilsbry, 1911
    nubilosa nubilosa Mighels, 1845
    nucula Smith, 1873
    pullata pullata Baldwin, 1895
        pullata Baldwin, 1893
    pullata subnigra Hyatt & Pilsbry, 1911
    pullata umbrosa Baldwin, 1895
        umbrosa Baldwin, 1893
    pusilla Newcomb, 1855
        pulla Pfeiffer, 1856
    rubristoma Baldwin, 1906
    seminuda Baldwin, 1906
        seminuda Baldwin, 1893
    subcrassilabris Hyatt & Pilsbry, 1911
    subobscura puella Pilsbry & Cooke, 1914
    subobscura subobscura Hyatt & Pilsbry, 1911
    sykesi Hyatt & Pilsbry, 1911
    tricincta Hyatt & Pilsbry, 1911
    uniplicata uniplicata Hartman, 1888

Amastra (Amastra) Adams & Adams, 1855 (continued)
    uniplicata vetuscula Cooke, 1917
    violacea violacea Newcomb, 1853
    violacea wailauensis Hyatt & Pilsbry, 1911
Amastra (Amastrella) Sykes, 1900
    abavus Hyatt & Pilsbry, 1911
    anthonii anthonii Newcomb, 1861
    anthonii meineckei Cooke, 1933
    anthonii remota Cooke, 1917
    anthonii subglobosa Cooke, 1933
    conica conica Baldwin, 1906
    conica gentilis Cooke, 1917
    conica gyrans Hyatt, 1911
    conica kohalensis Hyatt & Pilsbry, 1911
    decorticata Gulick, 1873
        ?*solida* Pease, 1869[1]
    elliptica Gulick, 1873
    flavescens emortua Cooke, 1917
    flavescens flavescens Newcomb, 1854
    flavescens henshawi Baldwin, 1903
    flavescens saxicola Baldwin, 1903
    fossilis Baldwin, 1903
    fragosa Cooke, 1917
    hawaiiensis Hyatt & Pilsbry, 1911
    inflata Pfeiffer, 1856
    luctuosa luctuosa Pfeiffer 1856
    luctuosa sulphurea Ancey, 1904
    melanosis kauensis Pilsbry & Cooke, 1915
    melanosis melanosis Newcomb, 1854
    nucleola Gould, 1845
        brevis Pfeiffer, 1846
    ovatula Cooke, 1933
    pagodula Cooke, 1917
    petricola Newcomb, 1855
        petricola Pfeiffer, 1856
    porcus Hyatt & Pilsbry, 1911
    rubens castanea Hyatt & Pilsbry, 1911
    rubens corneiformis Hyatt & Pilsbry, 1911
    rubens infelix Hyatt & Pilsbry, 1911
    rubens kahana Hyatt & Pilsbry, 1911
    rubens rubens Gould, 1845
    rubens rubinia Hyatt & Pilsbry, 1911
    rubida Gulick, 1873
    rugulosa annosa Cooke, 1917
    rugulosa fastigata Cooke, 1917
    rugulosa janeae Cooke, 1933
    rugulosa normalis Hyatt & Pilsbry, 1911
    rugulosa rugulosa Pease, 1870
    seminigra Hyatt & Pilsbry, 1911
    senilis Baldwin, 1903
    spicula Cooke, 1917
    tenuilabris rubicunda Baldwin, 1895
        rubicunda Baldwin, 1893

---

[1] According to Pilsbry & Cooke (1914b: 31), *solida* Pease is a synonym of either *subrostrata* Pfeiffer (subg. *Metamastra*) or *decorticata* Gulick (subg. *Amastrella*).

CHECKLIST

Amastra (**Amastrella**) Sykes, 1900 (continued)
    tenuilabris tenuilabris Gulick, 1873
    tristis Férussac, 1825
        tristis Férussac, 1821
        fuliginosa Gould, 1845
    viriosa Cooke, 1917
    whitei Cooke, 1917
Amastra (**Armiella**) Hyatt & Pilsbry, 1911
    knudsenii Baldwin, 1895
    ricei armillata Cooke, 1917
    ricei ricei Cooke, 1917
Amastra (**Cyclamastra**) Pilsbry & Vanatta, 1905
    agglutinans Newcomb, 1854
        carinata Gulick, 1873
    antiqua antiqua Baldwin, 1895
        antiqua Baldwin, 1893
    antiqua kawaihapaiensis Pilsbry & Cooke, 1914
    cyclostoma cyclostoma Baldwin, 1895
        cyclostoma Baldwin, 1893
    cyclostoma gregoryi Cooke, 1933
    delicata Cooke, 1933
    elephantina Cooke, 1917
    extincta Pfeiffer, 1856
        hartmani Newcomb, 1888
    fragilis Pilsbry & Cooke, 1914
    globosa Cooke, 1933
    gouveii Cooke, 1917
    juddii Cooke, 1917
    metamorpha debilis Pilsbry & Cooke, 1914
    metamorpha metamorpha Pilsbry & Cooke, 1914
    modicella Cooke, 1917
    morticina Hyatt & Pilsbry, 1911
    obesa aurora Pilsbry & Cooke, 1914
    obesa obesa Newcomb, 1853
    problematica Cooke, 1933
    similaris Pease, 1870
    sola Hyatt & Pilsbry, 1911
    sphaerica Pease, 1870
    thurstoni bembicodes Cooke, 1933
    thurstoni thurstoni Cooke, 1917
    ultima Pilsbry & Cooke, 1914
    umbilicata arenarum Pilsbry & Cooke, 1914
    umbilicata pluscula Cooke, 1917
    umbilicata umbilicata Pfeiffer, 1856
Amastra (**Heteramastra**) Pilsbry, 1911
    dwightii Cooke, 1933
    elongata Newcomb, 1853
        acuta Newcomb, 1854
        sororem Pfeiffer, 1868
    flemingi Cooke, 1917
    fraterna Sykes, 1896
    hutchinsonii Pease, 1862
        villosa Sykes, 1896
    implicata Cooke, 1933

Amastra (Heteramastra) Pilsbry, 1911 (continued)
 laeva Baldwin, 1906
 nannodes Cooke, 1933
 nubigena Pilsbry & Cooke, 1914
 perversa Hyatt & Pilsbry, 1911
 pilsbryi Cooke, 1913
 sinistrorsa Baldwin, 1906
 soror interjecta Hyatt & Pilsbry, 1911
 soror laticeps Hyatt & Pilsbry, 1911
 soror soror Newcomb, 1854
 subsoror auwahiensis Pilsbry & Cooke, 1914
 subsoror subsoror Hyatt & Pilsbry, 1911
Amastra (Kauaia) Sykes, 1900
 Carinella Pfeiffer, 1875
 kauaiensis Newcomb 1860
Amastra (Metamastra) Hyatt & Pilsbry, 1911
 aemulator Hyatt & Pilsbry, 1911
 albolabris Newcomb, 1854
 badia Baldwin, 1895
 breviata Baldwin, 1895
 caputadamantis Hyatt & Pilsbry, 1911
 cornea Newcomb, 1854
 crassilabrum Newcomb, 1854
 davisiana Cooke, 1908
 eos Pilsbry & Cooke, 1914
 forbesi Cooke, 1917
 gulickiana dichroma Cooke, 1933
 gulickiana gulickiana Hyatt & Pilsbry, 1911
 irwiniana Cooke, 1908
 montagui Pilsbry, 1913
 montivaga Cooke, 1917
 oswaldi Cooke, 1933
 paulula Cooke, 1917
 pellucida Baldwin, 1895
  pellucida Baldwin, 1893
 praeopima Cooke, 1917
 reticulata conspersa Pfeiffer, 1855
 reticulata dispersa Hyatt & Pilsbry, 1911
 reticulata errans Hyatt & Pilsbry, 1911
 reticulata orientalis Hyatt & Pilsbry, 1911
 reticulata reticulata Newcomb, 1854
 reticulata vespertina Pilsbry & Cooke, 1914
 sericea anaglypta Cooke, 1917
 sericea sericea Pfeiffer, 1859
 spaldingi Cooke, 1908
 subcornea Hyatt & Pilsbry, 1911
 subrostrata acuminata Cooke, 1933
 subrostrata subrostrata Pfeiffer, 1859
  ?*solida* Pease, 1869[1]
 textilis kaipaupauensis Hyatt & Pilsbry, 1911
 textilis media Hyatt & Pilsbry, 1911

---

[1] According to Pilsbry & Cooke (1914b: 31), *solida* Pease is a synonym of either *subrostrata* Pfeiffer (subg. *Metamastra*) or *decorticata* Gulick (subg. *Amastrella*).

Amastra (Metamastra) Hyatt & Pilsbry, 1911 (continued)
    textilis textilis Férussac, 1825
        textilis Férussac, 1821
        microstoma Gould, 1845
        ellipsoidea Gould, 1847
        cookei Hyatt & Pilsbry, 1911
    thaanumi Hyatt & Pilsbry, 1911
    transversalis bryani Pilsbry & Cooke, 1914
    transversalis transversalis Pfeiffer, 1856
    undata Baldwin, 1895
        undata Baldwin, 1893
    vetusta Baldwin, 1895
        vetusta Baldwin, 1893
Amastra (Paramastra) Hyatt & Pilsbry, 1911
    cylindrica Newcomb, 1854
        intermedia Newcomb, 1854
        porphyrea Newcomb, 1854
        grossa Pfeiffer, 1856
        conicospira Smith, 1873
    micans Pfeiffer, 1859
        erecta Pease, 1869
        frosti Ancey, 1892
    porphyrostoma Pease, 1869
    spirizona chlorotica Pfeiffer, 1856
        albida Pfeiffer, 1856
    spirizona nigrolabris Smith, 1873
    spirizona rudis Pfeiffer, 1855
    spirizona spirizona Férussac, 1825
        spirizona Férussac, 1821
        acuta Swainson, 1828
        baetica Jay, 1850
    tenuispira Baldwin, 1895
        tenuispira Baldwin, 1893
    turritella aiea Hyatt & Pilsbry, 1911
    turritella turritella Férussac, 1821
        oahuensis Green, 1827
        inornata Mighels, 1845
    turritella waiawa Hyatt & Pilsbry, 1911
        unicolor Ancey, 1899
    variegata Pfeiffer, 1849
        decepta Adams, 1851
Incertae sedis within **Amastra** s.l.
    amicta Smith, 1873
    farcimen Pfeiffer, 1857
    luteola Férussac, 1825
    peasei Smith, 1873
Unplaced *nomina nuda* in **Amastra** s.l.
        breviana Baldwin, 1893
        ferruginea Baldwin, 1893
        testudinea Baldwin, 1893
Carelia Adams & Adams, 1855
    anceophila Cooke, 1931
    bicolor angulata Pease, 1871

Carelia Adams & Adams, 1855 (continued)
    bicolor bicolor Jay, 1839
        adusta Gould, 1845
        fuliginea Pfeiffer, 1854
        suturalis Ancey, 1904
        minor Borcherding, 1910
        zonata Borcherding, 1910[2]
        zonata Borcherding, 1910[2]
        hyperleuca Hyatt & Pilsbry, 1911
    cochlea Reeve, 1849
        rigida Hyatt, 1911
    cumingiana cumingiana Pfeiffer, 1855
    cumingiana meineckei Cooke, 1931
    dolei dolei Ancey, 1893
        kobelti Borcherding, 1910
    dolei isenbergi Cooke, 1931
    evelynae Cooke & Kondo, 1952
    glossema Cooke, 1931
    hyattiana Pilsbry, 1911
    kalalauensis Cooke, 1931
    knudseni Cooke, 1931
    lirata Cooke, 1931
    lymani Cooke, 1931
    mirabilis Cooke, 1931
    necra necra Cooke, 1931
    necra spaldingi Cooke, 1931
    olivacea baldwini Cooke, 1931
    olivacea infrequens Cooke, 1931
    olivacea moloaaensis Cooke & Kondo, 1952
    olivacea olivacea Pease, 1866
        variabilis Pease, 1871
        viridans Pease, 1871
        viridis Pease, 1871
    olivacea priggei Cooke, 1931
    olivacea propinquella Cooke, 1931
    paradoxa magnapustulata Cooke & Kondo, 1952
    paradoxa paradoxa Pfeiffer, 1854
    paradoxa thaanumi Cooke, 1931
    paradoxa waipouliensis Cooke & Kondo, 1952
    periscelis Cooke, 1931
    pilsbryi pilsbryi Sykes, 1909
    pilsbryi tsunami Cooke & Kondo, 1952
    sinclairi Ancey, 1892
        extincta Hyatt & Pilsbry, 1911
    tenebrosa Cooke, 1931
    turricula Mighels, 1845
        obeliscus Reeve, 1850
        newcombi Pfeiffer, 1853
        azona Ancey, 1904
Laminella Pfeiffer, 1854
    alexandri alexandri Newcomb, 1865
    alexandri duoplicata Baldwin, 1908
    aspera Baldwin, 1908

---

[2] Borcherding (1910: 244) used the name *zonata* to refer to two different entities (see main catalog).

Laminella Pfeiffer, 1854 (continued)
    bulbosa Gulick, 1858
    citrina citrina Pfeiffer, 1848
        semivenulata Borcherding, 1906
    citrina helvina Baldwin, 1895
        helvina Baldwin, 1893
    concinna circumcincta Hyatt & Pilsbry, 1911
    concinna concinna Newcomb, 1854
    depicta depicta Baldwin, 1895
        depicta Baldwin, 1893
    depicta kamaloensis Pilsbry & Cooke, 1915
    gravida aurantium Pilsbry & Cooke, 1915
    gravida dimondi Adams, 1851
        lata Adams, 1851
    gravida gravida Férussac, 1825
        gracilis Férussac, 1825
    gravida kalihiensis Pilsbry & Cooke, 1915
    gravida suffusa Reeve, 1850
        concolor Martens, 1860
    gravida waianaensis Pilsbry & Cooke, 1915
    kuhnsi Cooke, 1908
    picta Mighels, 1845
        picta Pfeiffer, 1846
    remyi Newcomb, 1855
        remyi Pfeiffer, 1856
    sanguinea leucoderma Pilsbry & Cooke, 1915
    sanguinea sanguinea Newcomb, 1854
        ferussaci Pfeiffer, 1856
    straminea Reeve, 1850
    tetrao gracilior Hyatt & Pilsbry, 1911
    tetrao tetrao Newcomb, 1855
        tetrao Pfeiffer, 1856
    venusta muscaria Hyatt & Pilsbry, 1911
    venusta orientalis Hyatt & Pilsbry, 1911
    venusta semivestita Hyatt & Pilsbry, 1911
    venusta venusta Mighels, 1845
Planamastra Pilsbry, 1911
    digonophora Ancey, 1889
    peaseana Pilsbry, 1911
    spaldingi koolauensis Cooke, 1933
    spaldingi spaldingi Cooke, 1933
Tropidoptera Ancey, 1889
  Pterodiscus Pilsbry, 1893
  Helicamastra Pilsbry & Vanatta, 1905
    alata alata Pfeiffer, 1856
    alata lita Pilsbry, 1911
    discus Pilsbry & Vanatta, 1905
    heliciformis Ancey, 1890
    rex Sykes, 1904
        cookei Hyatt & Pilsbry, 1911
        thaanumi Pilsbry, 1911
    wesleyi ewaensis Pilsbry, 1911
    wesleyi wesleyi Sykes, 1896

Leptachatininae
- **Armsia** Pilsbry, 1911
  - petasus Ancey, 1899
- **Leptachatina (Angulidens)** Pilsbry & Cooke, 1914
  - anceyana Cooke, 1910
  - cookei Pilsbry, 1914
  - fossilis Cooke, 1910
  - hyperodon Pilsbry & Cooke, 1914
  - microdon Pilsbry & Cooke, 1914
  - subcylindracea Cooke, 1910
- **Leptachatina (Ilikala)** Cooke, 1911
  - fraterna Cooke, 1911
  - fusca fusca Newcomb, 1853
  - fusca striatella Gulick, 1856
  - irregularis Pfeiffer, 1856
    - *irregularis* Pfeiffer, 1856
  - nematoglypta Pilsbry & Cooke, 1914
  - petila Gulick, 1856
- **Leptachatina (Labiella)** Pfeiffer, 1854
  - callosa Pfeiffer, 1857
  - labiata Newcomb, 1853
    - *dentata* Pfeiffer, 1855
  - lagena Gulick, 1856
  - lenta Cooke, 1911
- **Leptachatina (Leptachatina)** Gould, 1847
  - accineta Mighels, 1845
    - *accincta* Gould, 1852
    - *margarita* Pfeiffer, 1856
    - *granifera* Gulick, 1856
  - acuminata Gould, 1847
  - antiqua Pease, 1870
  - approximans Ancey, 1897
  - arborea Sykes, 1900
  - attenuata Cooke, 1911
  - baldwini Cooke, 1910
  - balteata Pease, 1870
  - brevicula brevicula Pease, 1869
  - brevicula micra Cooke, 1910
  - captiosa Cooke, 1910
  - cerealis Gould, 1847
  - cingula Mighels, 1845
    - *fumosa* Newcomb, 1854
    - *vitrea* Newcomb, 1854
  - compacta Pease, 1869
  - concolor Cooke, 1910
  - conicoides Sykes, 1900
  - conspicienda Cooke, 1910
  - convexiuscula Sykes, 1900
  - corneola Pfeiffer, 1846
  - coruscans coruscans Hartman, 1888
  - coruscans dissimilis Cooke, 1910
  - costulata Gulick, 1856
  - costulosa Pease, 1870
  - crystallina Gulick, 1856

Leptachatina (Leptachatina) Gould, 1847 (continued)
    cuneata Cooke, 1910
    cylindrata Pease, 1869
    deceptor Cockerell, 1927
    defuncta Cooke, 1910
    dimidiata Pfeiffer, 1856
    dormitor Pilsbry & Cooke, 1914
    emerita Sykes, 1900
    exilis Gulick, 1856
    exoptabilis Cooke, 1910
    extensa Pease, 1870
    fulgida Cooke, 1910
    fumida Gulick, 1856
    gayi Cooke, 1911
    glutinosa Pfeiffer, 1856
        lacrima Gulick, 1856
    gracilis Pfeiffer, 1855
        elevata Pfeiffer, 1856
    grana Newcomb, 1853
    gummea Gulick, 1856
        fragilis Gulick, 1856
    guttula Gould, 1847
    haenensis Cockerell, 1927
    illimis Cooke, 1910
    imitatrix Sykes, 1900
    impressa Sykes, 1896
    isthmica Ancey & Sykes, 1899
    knudseni Cooke, 1910
    konaensis konaensis Sykes, 1900
    konaensis olaaensis Cooke, 1910
    kuhnsi Cooke, 1910
    laevigata Cooke, 1910
    laevis Pease, 1870
    lanaiensis Cooke, 1911
    lanceolata Cooke, 1911
    leiahiensis Cooke, 1910
    lepida Cooke, 1910
    leucochila Gulick, 1856
    longiuscula Cooke, 1910
    lucida Pease, 1870
    maniensis Pfeiffer, 1855
    marginata Gulick, 1856
    mcgregori Pilsbry & Cooke, 1914
    molokaiensis Cooke, 1910
    nitida nitida Newcomb, 1853
    nitida occidentalis Cooke, 1910
    obsoleta Pfeiffer, 1857
    obtusa Pfeiffer, 1856
    octogyrata Gulick, 1856
    opipara manana Pilsbry & Cooke, 1914
    opipara opipara Cooke, 1910
    oryza avus Pilsbry & Cooke, 1914
    oryza hesperia Pilsbry & Cooke, 1914
    oryza oryza Pfeiffer, 1856

Leptachatina (Leptachatina) Gould, 1847 (continued)
    ovata Cooke, 1910
    pachystoma brevis Cooke, 1910
    pachystoma cylindrella Cooke, 1910
    pachystoma pachystoma Pease, 1869
    pachystoma turgidula Pease, 1870
    perkinsi Sykes, 1896
    persubtilis Cooke, 1910
    pilsbryi Cooke, 1910
    popouwelensis Pilsbry & Cooke, 1914
    praestabilis Cooke, 1910
    pulchra Cooke, 1910
    pumicata Mighels, 1845
    pupoidea Cooke, 1911
    pyramis Pfeiffer, 1846
    resinula Gulick, 1856
    saccula Hartman, 1888
    sagittata Pilsbry & Cooke, 1914
    sandwicensis Pfeiffer, 1846
        obclavata Pfeiffer, 1855
    saxatilis Gulick, 1856
    sculpta Pfeiffer, 1856
    scutilus Mighels, 1845
    semipicta Sykes, 1896
    simplex Pease, 1869
    smithi Sykes, 1896
    somniator Pilsbry & Cooke, 1914
    stiria Gulick, 1856
    striata Newcomb, 1861
    striatula Gould, 1845
        clara Pfeiffer, 1846
    subovata Cooke, 1910
    subula Gulick, 1856
    succincta Newcomb, 1855
        succincta Pfeiffer, 1856
    supracostata Sykes, 1900
    tenebrosa Pease, 1870
    tenuicostata Pease, 1869
    terebralis Gulick, 1856
    teres Pfeiffer, 1856
    triticea Gulick, 1856
    turrita Gulick, 1856
    vana Sykes, 1900
    varia Cooke, 1910
    ventulus Férussac, 1825
        ventulus Férussac, 1821
        melampoides Pfeiffer, 1853
        manoaensis Pfeiffer, 1859
    vitreola parvula Gulick, 1856
    vitreola vitreola Gulick, 1856
Incertae sedis within **Leptachatina** *s. str.*
        octavula Paetel, 1873

Leptachatina (Thaanumia) Ancey, 1899
        fuscula Gulick, 1856
        dulcis Cooke, 1911
        henshawi Sykes, 1903
        morbida Cooke, 1911
        omphalodes Ancey, 1899
        optabilis Cooke, 1911
        perforata Cooke, 1911
        thaanumi Cooke, 1911
Pauahia Cooke, 1911
        artata Cooke, 1911
        chrysallis Pfeiffer, 1855
            columna Ancey, 1889
        tantilla Cooke, 1911
Questionably included AMASTRIDAE in the Hawaiian fauna
        semicostata Pfeiffer, 1856
PUPILLIDAE
  Gastrocoptinae
    Gastrocopta (Gastrocopta) Wollaston, 1878
        servilis kailuana Pilsbry, 1917
        servilis servilis Gould, 1843
            lyonsiana Ancey, 1892
    Gastrocopta (Sinalbinula) Pilsbry, 1916
        pediculus nacca Gould, 1862
  Nesopupinae
    Lyropupa (Lyropupa) Pilsbry, 1900
        clathratula Ancey, 1904
        lyrata fossilis Cooke & Pilsbry, 1920
        lyrata gouldi Pilsbry & Cooke, 1920
        lyrata lyrata Gould, 1843
            magdalenae Ancey, 1892
            carbonaria Ancey, 1904
        lyrata uncifera Cooke & Pilsbry, 1920
        microthauma Ancey, 1904
        prisca Ancey, 1904
        rhabdota baldwiniana Cooke, 1920
        rhabdota lanaiensis Cooke, 1920
        rhabdota pluris Pilsbry & Cooke, 1920
        rhabdota rhabdota Cooke & Pilsbry, 1920
        striatula Pease, 1871
        thaanumi Cooke & Pilsbry, 1920
        truncata Cooke, 1908
    Lyropupa (Lyropupilla) Pilsbry & Cooke, 1920
        anceyana Cooke & Pilsbry, 1920
        antiqua Cooke & Pilsbry, 1920
        hawaiiensis Ancey, 1904
        mirabilis Ancey, 1890
        scabra Pilsbry and Cooke, 1920
        spaldingi Pilsbry & Cooke, 1920
        sparna sinulifera Pilsbry & Cooke, 1920
        sparna sparna Cooke & Pilsbry, 1920

Lyropupa (Mirapupa) Cooke & Pilsbry, 1920
    costata costata Pease, 1871
        kahoolavensis Pilsbry & Cooke, 1920
    costata puukolekolensis Pilsbry & Cooke, 1920
    cubana Dall, 1890
    cyrta Cooke & Pilsbry, 1920
    micra maunaloae Pilsbry & Cooke, 1920
    micra micra Cooke & Pilsbry, 1920
    micra percostata Pilsbry & Cooke, 1920
    ovatula kona Pilsbry & Cooke, 1920
    ovatula moomomiensis Pilsbry & Cooke, 1926
    ovatula ovatula Cooke & Pilsbry, 1920
    perlonga cylindrata Pilsbry & Cooke, 1920
    perlonga filicostata Cooke & Pilsbry, 1920
    perlonga interrupta Pilsbry & Cooke, 1920
    perlonga perlonga Pease, 1871
    plagioptyx Pilsbry & Cooke, 1920
    thaumasia Cooke & Pilsbry, 1920
Nesopupa (Infranesopupa) Cooke & Pilsbry, 1920
    anceyana Cooke & Pilsbry, 1920
    bishopi Cooke & Pilsbry, 1920
    dubitabilis dubitabilis Cooke & Pilsbry, 1920
    dubitabilis kaalaensis Cooke & Pilsbry, 1920
    forbesi Cooke & Pilsbry, 1920
    infrequens Cooke & Pilsbry, 1920
    limatula Cooke & Pilsbry, 1920
    subcentralis Cooke & Pilsbry, 1920
Nesopupa (Limbatipupa) Cooke & Pilsbry, 1920
    alloia Cooke & Pilsbry, 1920
    kauaiensis Ancey, 1904
    newcombi angusta Cooke & Pilsbry, 1920
    newcombi disjuncta Pilsbry & Cooke, 1920
    newcombi gnampta Cooke & Pilsbry, 1920
    newcombi interrupta Cooke & Pilsbry, 1920
    newcombi multidentata Cooke & Pilsbry, 1920
    newcombi newcombi Pfeiffer, 1853
        costulosa Pease, 1871
    seminulum Boettger, 1881
    oahuensis Cooke & Pilsbry, 1920
    singularis Cooke & Pilsbry, 1920
Nesopupa (Nesodagys) Cooke & Pilsbry, 1920
    thaanumi Ancey, 1904
    wesleyana gouveiae Cooke & Pilsbry, 1920
    wesleyana rhadina Cooke & Pilsbry, 1920
        kamaloensis Cooke & Pilsbry, 1920
    wesleyana tryphera Cooke & Pilsbry, 1920
    wesleyana wesleyana Ancey, 1904
Nesopupa (Nesopupilla) Pilsbry & Cooke, 1920
    bacca Pease, 1871
    baldwini baldwini Ancey, 1904
    baldwini centralis Ancey, 1904
    baldwini lanaiensis Pilsbry & Cooke, 1920
    baldwini subcostata Pilsbry & Cooke, 1920
    dispersa Cooke & Pilsbry, 1920

Nesopupa (Nesopupilla) Pilsbry & Cooke, 1920 (continued)
    litoralis Cooke & Pilsbry, 1920
    plicifera Ancey, 1904
    waianaensis Cooke & Pilsbry, 1920
Pronesopupa (Edentulopupa) Cooke & Pilsbry, 1920
    admodesta Mighels, 1845
Pronesopupa (Pronesopupa) Iredale, 1913
    acanthinula Ancey, 1892
    boettgeri boettgeri Cooke & Pilsbry, 1920
    boettgeri spinigera Cooke & Pilsbry, 1920
    hystricella Cooke & Pilsbry, 1920
Pronesopupa (Sericipupa) Cooke & Pilsbry, 1920
    frondicola corticicola Cooke & Pilsbry, 1920
    frondicola frondicola Cooke & Pilsbry, 1920
    incerta Cooke & Pilsbry, 1920
    lymaniana Cooke & Pilsbry, 1920
    molokaiensis Cooke & Pilsbry, 1920
    orycta Cooke & Pilsbry, 1920
    sericata Cooke & Pilsbry, 1920
Pupillinae
  Pupoidopsis Pilsbry & Cooke, 1921
    hawaiensis Pilsbry & Cooke, 1921
Vertigininae
  Columella Westerlund, 1878
    alexanderi Cooke & Pilsbry, 1906
    olaaensis Pilsbry, 1926
    sharpi Pilsbry & Cooke, 1906

FERUSSACIIDAE
  Cecilioides Férussac, 1814
    Caecilianella Bourguignat, 1856
    baldwini Ancey, 1892

SUBULINIDAE
  Subulininae
    Allopeas Baker, 1935
      gracile Hutton, 1834
        junceus Gould, 1846
        pyrgiscus Pfeiffer, 1861
      clavulinum hawaiiense Sykes, 1904
    Opeas Albers, 1850
      opella Pilsbry & Vanatta, 1906
    Paropeas Pilsbry, 1906
      achatinaceum Pfeiffer, 1846
        henshawi Sykes, 1904

ENDODONTIDAE
  Cookeconcha Solem, 1976
    antiqua Solem, 1977
    contorta Férussac, 1825
      intercarinata Mighels, 1845
    cookei Cockerell, 1933
    decussatula Pease, 1866
    elisae Ancey, 1889
    henshawi Ancey, 1904
    hystricella Pfeiffer, 1859

Cookeconcha Solem, 1976 (continued)
    hystrix Pfeiffer, 1846
        setigera Gould, 1844
    jugosa Mighels, 1845
        rubiginosa Gould, 1846
    lanaiensis Sykes, 1896
    luctifera Pilsbry & Vanatta, 1905
    nuda Ancey, 1899
    paucicostata Pease, 1871
        filicostata Pease, 1871
    paucilamellata Ancey, 1904
    ringens Sykes, 1896
    stellula Gould, 1844
    thaanumi Pilsbry & Vanatta, 1905
    thwingi Ancey, 1904
Endodonta Albers, 1850
    apiculata Ancey, 1889
    binaria Pfeiffer, 1856
    concentrata Pilsbry & Vanatta, 1906
    ekahanuiensis Solem, 1976
    fricki Pfeiffer, 1858
    kalaeloana Christensen, 1982
    kamehameha Pilsbry & Vanatta, 1906
    lamellosa Férussac, 1825d
        lamellosa Férussac, 1821
    laminata Pease, 1866
    marsupialis Pilsbry & Vanatta, 1906
    rugata Pease, 1866
Nesophila Pilsbry, 1893
    baldwini albina Ancey, 1889
    baldwini baldwini Ancey, 1889
    capillata Pease, 1866
    distans Pease, 1866
    tiara Mighels, 1845
Protoendodonta Solem, 1977
    laddi Solem, 1977
PUNCTIDAE
  Punctum Morse, 1864
    horneri Ancey, 1904
SUCCINEIDAE
  Catinellinae
  Catinella Pease, 1870
    baldwini Ancey, 1889
    explanata Gould, 1852
    paropsis Cooke, 1921
    rotundata Gould, 1846
        patula Mighels, 1845
        newcombi Pfeiffer, 1855
    rubida Pease, 1870
    tuberculata Cooke, 1921
  Laxisuccinea Cooke, 1921
    haena Cooke, 1921
    libera Cooke, 1921

Succineinae
    Succinea (Succinea) Draparnaud, 1801
        apicalis Ancey, 1904
            apicalis Baldwin, 1893
        approximata Sowerby, 1872
        aurulenta Ancey, 1889
        bicolorata Ancey, 1899
        caduca Mighels, 1845
        canella canella Gould, 1846
        canella crassa Ancey, 1889
        canella lucida Ancey, 1889
        canella mamillaris Ancey, 1889
        canella obesula Ancey, 1889
        casta casta Ancey, 1899
        casta henshawi Ancey, 1904
        casta orophila Ancey, 1904
        cepulla Gould, 1846
            fragilis Souleyet, 1852
            souleyeti Ancey, 1889
        cinnamomea Ancey, 1889
        delicata Ancey, 1889
        garrettiana Ancey, 1899
        gibba Henshaw, 1904
        inconspicua Ancey, 1899
        konaensis Sykes, 1897
        kuhnsi Ancey, 1904
        lumbalis Gould, 1846
        lutulenta Ancey, 1889
        mauiensis Ancey, 1889
        maxima Henshaw, 1904
        mirabilis Henshaw, 1904
        newcombiana Garrett, 1857
        pristina Henshaw, 1904
        protracta Sykes, 1900
        punctata Pfeiffer, 1855
        quadrata Ancey, 1904
        tahitensis Pfeiffer, 1847
        tenerrima coccoglypta Ancey, 1904
        tenerrima tenerrima Ancey, 1904
            tenerrima Baldwin, 1893
        tetragona Ancey, 1904
        thaanumi Ancey, 1899
        venusta Gould, 1846
        vesicalis Gould, 1846
        waianaensis Ancey, 1899
    Succinea (Truella) Pease, 1871
        elongata Pease, 1870
        rubella Pease, 1871
Questionably included SUCCINEIDAE in the Hawaiian fauna
        aperta Lea, 1838
        oregonensis Lea, 1841
        pudorina Gould, 1846

HELICARIONIDAE
  Euconulinae
    Euconulus (Chetosyna) Baker, 1941
      thurstoni Baker, 1941
    Euconulus (Nesoconulus) Baker, 1941
      gaetanoi gaetanoi Pilsbry & Vanatta, 1908
      gaetanoi vivens Baker, 1941
      konaensis Sykes, 1897
      subtilissimus kaunakakai Baker, 1941
      subtilissimus subtilissimus Gould, 1846
      thaanumi Ancey, 1904
    Euconulus (Pellucidomus) Baker, 1941
      lubricellus Ancey, 1904
  Microcystinae
    Hiona (Hiona) Baker, 1940
      megodonta Baker, 1940
      platyla Ancey, 1889
      waimanoi Baker, 1940
    Hiona (Hionarion) Baker, 1940
      pilsbryi Baker, 1940
      wahiawae Baker, 1940
    Hiona (Hionella) Baker, 1940
      perkinsi maunahoomae Baker, 1940
      perkinsi perkinsi Sykes, 1896
      perkinsi poholuae Baker, 1940
      rufobrunnea Ancey, 1904
      subdola Baker, 1940
    Hiona (Nesocyclus) Baker, 1940
      exaequata exaequata Gould, 1846
        disculus Pfeiffer, 1850
        obtusangula Pfeiffer, 1851
      exaequata waimeae Baker, 1940
      kipui Baker, 1940
      meineckei Baker, 1940
      milolii Baker, 1940
    Hiona (Neutra) Baker, 1940
      lymanniana Ancey, 1893
        lymaniana Baker, 1940
    Kaala Baker, 1940
      subrutila Mighels, 1845
    Philonesia (Aa) Baker, 1940
      abeillei Ancey, 1889
      gouveiana Baker, 1940
      mapulehuae Baker, 1940
      sericans Ancey, 1899
      waiheensis Baker, 1940
    Philonesia (Haleakala) Baker, 1940
      guavarum Baker, 1940
      hahakeae Baker, 1940
      indefinita Ancey, 1889
      interjecta Baker, 1940
      pusilla Baker, 1940
      turgida diducta Baker, 1940
      turgida turgida Ancey, 1890

Philonesia (Hiloaa) Baker, 1940
    hiloi Baker, 1940
    piihonuae Baker, 1940
Philonesia (Kipua) Baker, 1940
    arenofunus Baker, 1940
    chamissoi Pfeiffer, 1855
Philonesia (Mauka) Baker, 1940
    polita Baker, 1940
    similaris Baker, 1940
    welchi Baker, 1940
Philonesia (Philonesia) Sykes, 1900
    ascendens Baker, 1940
    baldwini Ancey, 1889
    cicercula boettgeriana Ancey, 1889
    cicercula cicercula Gould, 1846
    cryptoportica Gould, 1846
    decepta Baker, 1940
    fallax fallax Baker, 1940
    fallax popouwelae Baker, 1940
    glypha Baker, 1940
    hartmanni hartmanni Ancey, 1889
    hartmanni palehuae Baker, 1940
    kauaiensis Baker, 1940
    konahuanui Baker, 1940
    kualii Baker, 1940
    maunalei Baker, 1940
    mokuleiae Baker, 1940
    oahuensis depressula Ancey, 1889
    oahuensis oahuensis Ancey, 1889
    perlucens Ancey, 1889
    plicosa Ancey, 1889
    striata Baker, 1940
    waimanaloi Baker, 1940
Philonesia (Piena) Baker, 1940
    grandis Baker, 1940
    palawai Baker, 1940
    parva Baker, 1940
Philonesia (Waihoua) Baker, 1940
    kaliella Baker, 1940

ZONITIDAE
  Gastrodontinae
    Striatura (Pseudohyalina) Morse, 1864
        discus Baker, 1941
        meniscus Ancey, 1904
        pugetensis Dall, 1895
  Vitreinae
    Hawaiia Gude, 1911
      Pseudovitrea Baker, 1928
        minuscula Binney, 1840
            kawaiensis Reeve, 1854
            kawaiensis Pfeiffer, 1855

Vitrininae
    Vitrina Draparnaud, 1801
        Vitrinus Montfort, 1810
            tenella Gould, 1846
Zonitinae
    Godwinia (Godwinia) Sykes, 1900
        caperata Gould, 1846
        haupuensis Cooke, 1921
    Godwinia (Omphalops) Baker, 1941
        newcombi Reeve, 1854
            newcombi Pfeiffer, 1855
    Nesovitrea Cooke, 1921
        hawaiiensis Ancey, 1904
        molokaiensis Sykes, 1897
        pauxilla Gould, 1852
            pusillus Gould, 1846
            baldwini Ancey, 1889
            lanaiensis Sykes, 1897
Incertae sedis within HELICARIONIDAE and ZONITIDAE
        exserta Pfeiffer, 1856
MILACIDAE
    Milax Gray, 1855
        Amalia Moquin-Tandon, 1855
        gagates Draparnaud, 1801
            babori Collinge, 1897
LIMACIDAE
    Deroceras Rafinesque, 1820
        Agriolimax Mörch, 1865
        globosum Collinge, 1896
        laeve Müller, 1774
            bevenoti Collinge, 1897
        perkinsi Collinge, 1896
    Limax Linnaeus, 1758
        sandwichiensis Souleyet, 1852
Incertae sedis within the Hawaiian fauna
        fornicata Gould, 1846
        sandvicensis Pfeiffer, 1850

BIBLIOGRAPHY

by Neal L. Evenhuis & Robert H. Cowie

A concerted effort has been made to see all the references listed in this bibliography, by examination of the original or photocopy, to ensure accurate citation of author, date, title, and pagination. In addition, precise date of publication was researched for all the literature examined in this study. Citation is given verbatim. If a paper represents a note presented at a meeting, the title is placed in square brackets and, in some cases, paraphrased for citation purposes. The date of publication, as accurately as could be ascertained, is listed in square brackets at the end of each citation. When dating could not be found in the publication itself, outside resources were consulted. These included various publisher's advertising journals, reviews in scientific journals, and library receipts. The dates recorded here are the earliest found for each citation. If the actual year of publication was found to be different from the printed date in the publication itself, the actual year of publication is placed in square brackets after the author. Where no date other than the year could be found, the publication date should be treated as 31 December until such time as earlier publication evidence is discovered. Sources for the dates listed here are deposited in the Bishop Museum Department of Natural Sciences. When an author published more than 1 paper in a particular year, each paper is listed chronologically and the year given a letter suffix, which corresponds to the citation in the catalog. In cases where tabular collation is given for a publication issued in parts, we have given the date letter to each part in the "Date of publication" column. Authorship of works is listed as on the title page. In the case of anonymous works, the actual author is listed in square brackets if evidence could be found to identify the responsible person. Opinions of the International Commission on Zoological Nomenclature are cited with authorship as "ICZN" in the catalog text, but spelled out in full in the Bibliography.

Abbott, R.T. 1948. Handbook of medically important mollusks of the Orient and the western Pacific. *Bulletin of the Museum of Comparative Zoology at Harvard College* **100**(3): 245–328, pls. 1–5. [April]

Abbott, R.T. 1952. A study of an intermediate snail host (*Thiara granifera*) of the oriental lung fluke (*Paragonimus*). *Proceedings of the United States National Museum* **102**[= No. 3292]: 71–116, pls. 8, 9. [26 February]

Adams, A. 1855. Descriptions of two new genera and several new species of Mollusca, from the collection of Hugh Cuming, Esq. *Proceedings of the Zoological Society of London* **23**: 119–24. [Publication split: p. 119–20 (13 August); p. 121–24 (1 December)]

Adams, C.B. 1851a. Descriptions of new species of *Partula* and *Achatinella*. *Contributions to Conchology* **8**: 125–28. [March]

Adams, C.B. 1851b. Descriptions of new species of *Partula* and *Achatinella*. *Annals of the Lyceum of Natural History of New York* **5**: 41–44. [May]

Adams, H. & Adams, A. 1853–1858. *The genera of recent Mollusca; arranged according to their organization.* 2 vols. J. Van Voorst, London.

Published in parts as follows. Dating is from p. 661 of Adams & Adams, vol. 2:

Volume I:

Part	Plates	Pages	Date of publication
1	1–4	1–32	January 1853a
2	5–8	33–64	February 1853b
3	9–12	65–96	June 1853c
4	13–16	97–128	August 1853d
5	17–20	129–160	September 1853e
6	21–24	161–192	October 1853f
7	25–28	193–224	November 1853g

Part	Plates	Pages	Date of publication
8	29–32	225–256	December 1853h
9	33–36	257–288	January 1854a
10	37–40	289–320	February 1854b
11	41–44	321–352	March 1854c
12	45–48	353–384	April 1854d
13	49–52	385–416	May 1854e
14	53–56	417–448	June 1854f
15	57–60	449–484	July 1854g

Volume II:

Part	Plates	Pages	Date of publication
16	61–64	1–28	September 1854h
17	65–68	29–60	October 1854i
18	69–72	61–92	November 1854k
19	73–76	93–124	January 1855b
20	77–80	125–156	February 1855c
21	81–84	157–188	April 1855d
22	85–88	189–220	June 1855e
23	89–92	221–252	September 1855f
24	93–96	253–284	November 1855g
25	97–100	285–316	March 1856a
26	101–104	317–348	June 1856b
27	105–108	349–380	August 1856c
28	109–112	381–412	November 1856d
29	113–116	413–444	March 1857a
30	117–120	445–476	April 1857b
31	121–124	477–508	September 1857c
32	125–128	509–540	December 1857d
33	129–132	541–572	January 1858a
34	133–136	573–604	May 1858b
35–36	137–138	605–661	November 1858c

Adams, H. & Adams, A. 1854j. Monograph of *Plecotrema*, a new genus of gasteropodous mollusks, belonging to the family Auriculidae, from specimens in the collection of Hugh Cuming, Esq. *Proceedings of the Zoological Society of London* **21**: 120–22. [14 November]

Adams, H. & Adams, A. 1854l. Monographs of *Ellobium* and *Melampus*, two genera of pulmoniferous Mollusca. *Proceedings of the Zoological Society of London* **22**: 7–13. [30 December]

Adams, H. & Adams, A. [1855]a. Contributions towards the natural history of the Auriculidae, a family of pulmoniferous Mollusca; with descriptions of many new species from the Cumingian collection. *Proceedings of the Zoological Society of London* **22**[1854]: 30–37. [10 January]

Albers, J.C. 1850. *Die Heliceen, nach natürlicher Verwandtschaft.* T.C.F. Enslin, Berlin. 262 p. [before 7 November]

Ancey, C.F. 1887. Nouvelles contributions malacologiques. III. Considérations sur le genre *Opisthostoma* et les diplommatinacées. IV. Auriculacées d'Aden (Arabie). V. Descriptions de clausilies exotiques nouvelles. VI. Études sur la faune malacologique des îles Galapagos. *Bulletins de la Société Malacologique de France* **4**: 273–99.

Ancey, C.F. 1889a. Étude sur la faune malacologique des Iles Sandwich. *Bulletins de la Société Malacologique de France* **6**: 171–258. [June]

Ancey, C.F. 1889b. Diagnoses de mollusques nouveaux. *Le Naturaliste* **1889**: 266. [15 November]

Ancey, C.F. 1889c. Descriptions de mollusques nouveaux. *Le Naturaliste* **1889**: 290–291. [15 December]

Ancey, C.F. 1890. Mollusques nouveaux de l'archipel d'Hawai, de Madagascar et de l'Afrique équatoriale. *Bulletins de la Société Malacologique de France* **7**: 339–47. [June]

Ancey, C.F. 1892. Études sur la faune malacologique des Iles Sandwich. *Mémoires de la Société Zoologique de France* **5**: 708–22.

Ancey, C.F. 1893. Études sur la faune malacologique des Iles Sandwich. *Mémoires de la Société Zoologique de France* **6**(4): 321–30.
Ancey, C.F. 1897a. Description de deux nouvelles espèces de mollusques (Achatinellidae). *Le Naturaliste* **1897**: 178. [1 August]
Ancey, C.F. 1897b. Description d'un mollusque nouveau. *Le Naturaliste* **1897**: 222. [1 October]
Ancey, C.F. 1899. Some notes on the non-marine molluscan fauna of the Hawaiian Islands, with diagnoses of new species. *Proceedings of the Malacological Society of London* **3**(5): 268–74, pls. 12–13. [15 July]
> Plate 13 is by Sykes.

Ancey, C.F. 1903. Études sur la faune malacologique des Iles Sandwich. *Journal de Conchyliologie* **51**(4): 295-307, pl. 12. [25 March]
Ancey, C.F. 1904a. On some non-marine Hawaiian Mollusca. *Proceedings of the Malacological Society of London* **6**(2): 117–28, pl. 7. [23 June]
Ancey, C.F. 1904b. Report on semi-fossil land shells found in the Hamakua District, Hawaii. *Journal of Malacology* **11**(3): 65–71, pl. 5.
Athens, J.S. Erkelens, C., Ward, J.V., Cowie, R.H., & Pietrusewsky, M. 1994. The archaeological investigation of inadvertantly discovered human remains at the Piikoi and Kapiolani intersection, Kewalo, Waikiki, Oaho, Hawaii. *International Archaeological Research Institute Report*, vi + 56 + [13] p. [June]
Athens, J.S. & Ward, J.V. 1993. Paleoenvironmental investigations at Hamakua Marsh, Kailua, Oʻahu, Hawaiʻi. *International Archaeological Research Institute Report*, viii + 50 p. [September]
Baker, H.B. 1922. Notes on the radula of the Helicinidae. *Proceedings of the Academy of Natural Sciences of Philadelphia* **74**: 29–67, pls. 3–7. [8 August]
Baker, H.B. 1923. Notes on the radula of the Neritidae. *Proceedings of the Academy of Natural Sciences of Philadelphia* **75**: 117–78, pls. 9–15. [15 May]
Baker, H.B. 1928. Minute American Zonitidae. *Proceedings of the Academy of Natural ciences of Philadelphia* **80**: 1–44, pls. 1–8. [16 May]
Baker, H.B. 1935. Jamaican land snails, 3. *The Nautilus* **48**(3): 83–88, pl. 3. [19 January]
Baker, H.B. 1938. Zonitid snails from Pacific islands part 1. 1. Southern genera of Microcystinae. *Bernice P. Bishop Museum Bulletin* **158**: 1–102, pls. 1–20. [10 October]
Baker, H.B. 1940. Zonitid snails from Pacific islands part 2. 2. Hawaiian genera of Microcystinae. *Bernice P. Bishop Museum Bulletin* **165**: 103–201, pls. 21–42. [20 January]
Baker, H.B. 1941. Zonitid snails from Pacific islands parts 3 and 4. 3. Genera other than Microcystinae. 4. Distribution and indexes. *Bernice P. Bishop Museum Bulletin* **166**: 203–370, pls. 43–65. [5 February]
Baker, H.B. 1945. Some American Achatinidae. *The Nautilus* **58**(3): 84–92. [19 February]
Baldwin, D.D. 1893. *Catalogue. Land and fresh water shells of the Hawaiian Islands*. Press Publishing Co., Honolulu. 25 p. [1 May]
Baldwin, D.D. 1895. Descriptions of new species of Achatinellidae from the Hawaiian Islands. *Proceedings of the Academy of Natural Sciences of Philadelphia* **1895**: 214–36, pls. 10–11. [2 July]
Baldwin, D.D. 1896. Description of two new species of Achatinellidae from the Hawaiian Islands. *The Nautilus* **10**(3): 31–32. [2 July]
Baldwin, D.D. 1903. Descriptions of new species of Achatinellidae from the Hawaiian Islands. *The Nautilus* **17**(3): 34–36. [3 July]
Baldwin, D.D. 1906a. Description of new species of Achatinellidae from the Hawaiian Islands. *The Nautilus* **19**(10): 111–13. [8 February]

Baldwin, D.D. 1906b. Description of new species of Achatinellidae from the Hawaiian Islands. *The Nautilus* **19**(12): 135–38. [5 April]

Baldwin, D.D. 1908. Descriptions of new species of Achatinellidae, from the Hawaiian Islands. *The Nautilus* **22**(7): 67–69. [14 November]

Beck, H. 1837–1838. *Index Molluscorum praesentis aevi musei principis augustissimi Christiani Frederici. Fasciculus primus et secundus. Mollusca gastraepoda pulmonata.* [Published by the author], Hafniae [= Copenhagen]. 124 p. [Publication split: p. 1–100 [= fasciculus primus] (1837); p. 101–24 [= fasciculus secundus] (1838)]

Bieler, R. 1992. Gastropod phylogeny and systematics. *Annual Review of Ecology and Systematics* **23**: 311–38. [November]

Bland, T. & Binney, W.G. 1873. On the lingual dentition and anatomy of *Achatinella* and other Pulmonata. *Annals of the Lyceum of Natural History of New York* **10**: 331–51, pls. 15–16. [November]

Blanford, W.T. 1864. On the classification of the Cyclostomacea of eastern Asia. *Annals and Magazine of Natural History* (3) **13**: 441–65. [June]

Boettger, O. 1881. Die *Pupa*-Arten Oceaniens. *Conchologische Mittheilungen* **1**(4): 45–72, pls. 10–12.

Borcherding, F. 1901. Diagnosen neuer Achatinellen-Formen von der Sandwich-Insel Molokai. *Nachrichtsblatt der Deutschen Malakologischen Gesellschaft* **33**(3–4): 52–58. [30 March]

Borcherding, F. 1906. Achatinellen-Fauna der Sandwich-Insel Molokai, nebst einem Verzeichnis der übrigen daselbst vorkommenden Land- und Süßwassermollusken. *Zoologica* **48**, vii + 195 p., 10 pls.

Borcherding, F. 1910. Monographie der auf der Sandwichinsel Kauai lebenden Molluskengattung *Carelia* H. und A. Adams. *Abhandlungen der Senckenbergischen Naturforschenden Gesellschaft* **32**: 227–51, pls. 19–20. [20 February]

Boss, K.J. 1982. Mollusca, p. 945–1166. *In:* Parker, S.P., ed., *Synopsis and classification of living organisms.* Volume 1. McGraw-Hill, New York.

Boss, K.J. & Bieler, R. 1991. Johannes Thiele and his contributions to Zoology. Part 2. Genus-group names (Mollusca). *Nemouria. Occasional Papers of the Delaware Museum of Natural History* **39**: 1–77. [30 September]

Bourguignat, J.R. 1856. Aménités malacologiques. *Revue et Magasin de Zoologie* (2) **8**: 378–86. [3 September]

[Bronn, H.G.] 1847. *Preis-Verzeichniss auslandischer Konchylien, welche einzeln verkauft werden bei dem "Zoologischen Museum der Universität Heidelberg".* No. IV. 8 p. [April]

Brot, A.L. 1872. *Matériaux pour servir a l'étude de la famille des Mélaniens. III. Notice sur les Mélanies de Lamarck conservées dans le Musée Delessert et sur quelques espéces nouvelles ou peu connues.* Georg, Libraire-éditeur, Genève. 55 p., 4 pls.

Brot, A.L. 1874–[1879]. Die Melaniaceen (Melanidae) in Abbildungen nach der Natur mit Beschreibungen. *In:* Küster, H.C., *Systematisches Conchylien-Cabinet von Martini und Chemnitz. Neu herausgegeben und vervollständigt.* Band I. Abtheilung XXIV. Bauer & Raspe, Nürnberg [= Nuremberg]. 488 p., 49 pls.

Published in Lieferungen as follows:

Lieferung	Pages	Plates	Date of publication
229	1–32	1–6	1874
235	33–80	7–12	1875a
244	81–128	13–18	1875b
249	129–192	19–24	1876
259	193–272	25–30	1877a
264	273–352	31–36	1877b
271	353–400	37–42	1878
280	401–456	43–48	1879a
283	457–488	49	1879b

BIBLIOGRAPHY [209]

Brown, R.W. 1956. *Composition of scientific words. A manual of methods and a lexicon of materials for the practice of logotechnics.* Smithsonian Institution, Washington, D.C. 882 p.

Burch, J.B. 1964. Chromosomes of the succineid snail *Catinella rotundata*. *Occasional Papers of the Museum of Zoology, University of Michigan* **638**: 1–8. [17 June]

Carson, H.L. 1987. The process whereby species originate. *BioScience* **37**(10): 715–720. [November]

Caum, E.L. 1928. Check list of Hawaiian land and fresh water Mollusca. *Bernice P. Bishop Museum Bulletin* **56**: 1–79.

> Probably published at the end of December 1928. Received at the Bishop Museum Library on 3 January 1929.

Chamisso, A. de. 1829. Species novas conchyliorum terrestrium ex insulis, Sandwich dictus. *Nova Acta Physico-Medica Academiae Caesareae Leopoldino Carolinae Naturae Curiosorum* **14**(2): 639–40, pl. 36.

Chemnitz, J.H. 1795. *Neues Systematisches Conchylien-Cabinet.* Vol. XI. G.N. Raspe, Nürnberg [= Nuremberg]. [xx] + 310 + [ii] p., pls. 174–213.

Children, J.G. 1823. Lamarck's genera of shells. *Quarterly Journal of Science, Literature, and the Arts* **15**: 216–58, pls. 7–8. [July]

Christensen, C.C. 1982. A new species of *Endodonta* (Pulmonata, Endodontidae) from Oahu, Hawaii. *Malacological Review* **15**: 135–36. [4 May]

Christensen, C.C. 1985. Endangered species information system species workbook. Part 1 - Species Distribution. Part II - Species Biology. Unpublished report prepared for the U.S. Department of Interior, Fish and Wildlife Service, Office of Endangered Species. [iii] + 37 + [6] + [iv] + 42 p.

Christensen, C.C. & Kirch, P.V. 1981. Nonmarine mollusks from archaeological sites on Tikopia, southeastern Solomon Islands. *Pacific Science* **35**(1): 75–88. [30 October]

Christensen, C.C. & Kirch, P.V. 1986. Nonmarine mollusks and ecological change at Barbers Point, Oʻahu, Hawaiʻi. *Bishop Museum Occasional Papers* **26**: 52–80. [May]

Clarke, A.H. 1958. Status of Newcomb's achatinellid names. *The Nautilus* **71**(4): 148–51. [24 April]

Clench, W.J. & Turner, R.D. 1950. The western Atlantic marine mollusks described by C.B. Adams. *Occasional Papers on Mollusks* **1**(15): 233–403. [26 June]

Climo, F.M. [1974]. The systematics, biology and zoogeography of the land snail fauna of Great Island, Three Kings Group, New Zealand. *Journal of the Royal Society of New Zealand* **3**(4)[1973]: 565–627. [January]

Cockerell, T.D.A. 1927. Two fossil species of *Leptachatina* from the island of Kauai. *Journal of Conchology* **18**(4): 117. [January]

Cockerell, T.D.A. 1933. A new *Endodonta* from the Hawaiian Islands. *The Nautilus* **47**(2): 58. [1 November]

Coen, G.S. 1945. Catalogo dei gasteropodi polmonati della collezione Coen. *Pontificiae Academiae Scientiarum Scripta Varia* **3**: 1–99.

Collinge, W.E. 1896. On a collection of slugs from the Sandwich Islands. *Proceedings of the Malacological Society of London* **2**(1): 46–51. [19 April]

Collinge, W.E. 1897. On a further collection of slugs from the Hawaiian (or Sandwich) Islands. *Proceedings of the Malacological Society of London* **2**(6): 293–97. [27 November]

Conant, S., Christensen, C.C., Conant, P., Gagné, W.C. & Goff, M.L. 1984. The unique terrestrial biota of the Northwestern Hawaiian Islands, p. 77–94. *In*: Grigg, R.W. & Tanoue, K.Y., eds., *Proceedings of the second symposium on resource investigations in the Northwestern Hawaiian Islands.* Volume 1. University of Hawaii Sea Grant College Program, Honolulu. [April]

Cooke, C.M., Jr. 1907. Dr. Cooke's report. *Occasional Papers of the Bernice P. Bishop Museum* **2**(5): 11–16. [3 May]
Cooke, C.M., Jr. 1908a. A new species of *Lyropupa* from Hawaii. *Occasional Papers of the Bernice P. Bishop Museum* **3**(2): 211–12. [24 July]
Cooke, C.M., Jr. 1908b. Three new species of *Amastra* from Oahu. *Occasional Papers of the Bernice P. Bishop Museum* **3**(2): 213–16. [24 July]
Cooke, C.M., Jr. 1908c. *Amastra* (*Laminella*) *kuhnsi*. *Occasional Papers of the Bernice P. Bishop Museum* **3**(2): 217–18. [24 July]
Cooke, C.M., Jr. 1913. A new sinistral *Amastra*. *The Nautilus* **27**(6): 68–69. [9 October]
Cooke, C.M., Jr. 1914. Description of a new species of *Tornatellides*. *The Nautilus* **28**(7): 79–80. [20 November]
Cooke, C.M., Jr. 1917. Some new species of *Amastra*. *Occasional Papers of the Bernice P. Bishop Museum* **3**(3): 221–50, pls. 5–7 [18 January]
Cooke, C.M., Jr. 1921. Notes on Hawaiian Zonitidae and Succineidae. *Occasional Papers of the Bernice P. Bishop Museum* **7**(12): 263–77, pls. 24–25.
Cooke, C.M., Jr. 1928. Three *Endodonta* from Oahu. *Bernice P. Bishop Museum Bulletin* **47**: 13–27. [8 March]
Cooke, C.M., Jr. 1931. The land snail genus *Carelia*. *Bernice P. Bishop Museum Bulletin* **85**: 1–97, 18 pls. [August]
Cooke, C.M., Jr. 1932. The genus *Armsia*. *The Nautilus* **45**(4): 125–26. [9 April]
Cooke, C.M., Jr. 1933. New species of Amastridae. *Occasional Papers of the Bernice P. Bishop Museum* **10**(6): 1–27, 2 pls. [24 March]
Cooke, C.M., Jr. & Kondo, Y. 1952. New fossil forms of *Carelia* and *Partulina* (Pulmonata) from Hawaiian Islands. *Occasional Papers of the Bernice P. Bishop Museum* **20**(20): 329–46. [9 June]
Cooke, C.M., Jr. & Kondo, Y. [1961]. Revision of Tornatellinidae and Achatinellidae (Gastropoda, Pulmonata). *Bernice P. Bishop Museum Bulletin* **221** [1960]: 1–303. [15 February]

> Received by the Bishop Museum Library on this date. As there has been no earlier receipt date found, it is most probable that there was a delay in publication after the printed date of "1960" and that this volume was in fact published in 1961.

Cooke, C.M., Jr. & Neal, M.C. 1928. Distribution and anatomy of *Pupoidopsis hawaiiensis*. *Bernice P. Bishop Museum Bulletin* **47**: 28–33. [8 March]
Cowie, R.H. 1992. Evolution and extinction of Partulidae, endemic Pacific Island land snails. *Philosophical Transactions of the Royal Society, London* (B) **335**: 167–91. [29 February]
Cowie, R.H., Christensen, C.C. & Evenhuis, N.L. 1994. *Nesopupa* Pilsbry, 1900 (Mollusca; Gastropoda): proposed conservation. *Bulletin of Zoological Nomenclature* **51**(3): in press.
Dall, W.H. 1890. Description of a new species of land shell from Cuba—*Vertigo cubana*. *Proceedings of the United States National Museum* **13**: 1–2. [1 July]
Dall, W.H. 1895. New species of land shells from Puget Sound. *The Nautilus* **8**(11): 129–30. [4 March]
Deshayes, G.P. 1830. *In*: *Encyclopédie méthodique, ou par ordre de matières. Histoire naturelle des vers*. Tome second [part 1]. V. Agasse, Paris. vii + p. 1–144. [6 February]
Deshayes, G.P. 1851. *In*: Férussac, A.E.J.P.J.F.d'A. de & Deshayes, G.P., *Histoire naturelle générale et particulière des mollusques terrestres et fluviatiles tant des espèces que l'ou trouve aujourd'hui vivantes, que des dépouilles fossiles de celles qui n'existent plus; classés d'après les caractères essentiels que présentent ces animaux et leurs coquilles. Accompagnée d'un atlas de 247 planches gravées.* "1820–1850". Tome deuxième. Part 2. J.-B. Baillière, Paris. 260 p. [30 July]

Dixon, G. 1789. *A voyage round the world; but more particularly to the north-west coast of America: performed in 1785, 1786, 1787, and 1788, in the* King George *and* Queen Charlotte, *Captains Portlock and Dixon.* G. Goulding, London. xxix + [2] + 360 + 47 p., 17 pls. [after 7 February]

Dohrn, H. 1863. Ueber *Tornatellina. Malakozoologische Blätter* **10**: 156–62. [August]

Dohrn, H. & Pfeiffer, L. 1857. Neue Landschnecken. *Malakozoologische Blätter* **4**: 85–89. [May]

Draparnaud, J. [1801]. *Tableau des mollusques terrestres et fluviatiles de la France.* "An IX." Renaud, Montpellier; Bossange, Masson & Besson, Paris. 116 p. [14 July]

Férussac, A.E.J.P.J.F.d'A. de. 1814. *Mémoires géologiques sur les terrains formés sous l'eau douce par les débris fossiles des mollusques vivant sur la terne, ou dans l'eau non salée.* J.-B. Baillière, Paris. 76 p. [13 August]

Férussac, A.E.J.P.J.F.d'A. de. [1819–1820]. *Histoire naturelle générale et particulière des mollusques terrestres et fluviatiles tant des espèces que l'on trouve aujourd'hui vivantes, que des dépouilles fossiles de celles qui n'existent plus; classés d'après les caractères essentiels que présentent ces animaux et leurs coquilles. Accompagnée d'un atlas de 247 planches gravées.* "1820–1850". Tome deuxième. Part 1. J.-B. Baillière, Paris. xvi + 128 p.

> Issued in livraisons, of which the Baron Férussac, the son, edited 1–28 (1819–1832). Deshayes completed the work, editing the remaining livraisons, 29–42 (1838–1851). The younger Férussac and Deshayes are listed as joint authors on the title page of the completed work. The collation and dating for volume 2, part 1 is as follows:

> | Livraison | Pages | Date of publication |
> | --- | --- | --- |
> | 1 | i–xvi | 6 March 1819a |
> | 2 | 1–16 | 5 June 1819b |
> | 3 | 17–56 | 10 July 1819c |
> | 4 | 57–72 | 18 September 1819d |
> | 5 | 73–96 | 4 December 1819e |
> | 6 | 97–128 | 6 March 1820 |

Férussac, A.E.J.P.J.F.d'A. de. 1821. *Tableaux systématiques des animaux mollusques classés en familles naturelles . . . .* Deuxième partie. (Première section.). *Tableaux particuliers des mollusques terrestres et fluviatiles, présentant pour chaque famille les genres et espèces qui la composent.* J.-B. Baillière, Paris.

> This paper forms a part of the author's *Histoire naturelle générale et particulière des mollusques terrestres et fluviatiles* . . . and was issued in the livraisons of that publication. Two versions of this paper are known. One in folio, the other in quarto. The dates of issue of both editions are given below. Pages cited in the catalog text refer to the folio version.

> | Livr. | Folio Pages | Quarto Pages | Date of publication |
> | --- | --- | --- | --- |
> | 9 | 1–32 | 1–24 | 6 April 1821a |
> | 10 | 33–56 | 25–48 | 26 May 1821b |
> | 11 | 57–76 | 49–72 | 13 July 1821c |
> | 12 | 77–92 | 73–88 | 21 September 1821d |
> | 13 | 93–114 | 89–111 | 10 November 1821e |

Fleming, J. 1818. Conchology. *In: Encyclopaedia Britannica. Supplement to the fourth, fifth and sixth editions.* Vol. 3. [Part 1.] A. Constable, Edinburgh. 79 + 316 p., pls. 54–65. [February]

Fleming, J. 1822. Mollusca, p. 567–84. *In: Encyclopaedia Britannica. Supplement to the fourth, fifth and sixth editions.* Vol. 5. [Part 2.] A. Constable, Edinburgh. P. 163–586 + [2], pls. 93–100. [May]

Garrett, A. 1857. Mr. Garrett's paper on new species of marine shells of the Sandwich Islands. *Proceedings of the California Academy of Natural Sciences* **1**: 102–03.

Gould, A.A. 1843. [The conclusion of the "Monograph of the Pupadae [sic] of the United States"]. *Proceedings of the Boston Society of Natural History* **1**: 138–39. [October]

Gould, A.A. 1844. [Two species of *Helix* from the Sandwich Islands]. *Proceedings of the Boston Society of Natural History* **1**: 174. [May]

Gould, A.A. 1845. [Descriptions of species of land shells, from the Sandwich Islands, supposed to be hitherto undescribed]. *Proceedings of the Boston Society of Natural History* **2**: 26–28. [January]

Gould, A.A. 1846a. [Descriptions of new shells, collected by the United States Exploring Expedition, and belonging to the genus *Helix*]. *Proceedings of the Boston Society of Natural History* **2**: 170–73. [September]

Gould, A.A. 1846b. [Descriptions of the shells of the Exploring Expedition]. *Proceedings of the Boston Society of Natural History* **2**: 177–79. [after 16 December]

Gould, A.A. 1846c. [Descriptions of species of *Vitrina*, from the collection of the Exploring Expedition]. *Proceedings of the Boston Society of Natural History* **2**: 180–81. [after 16 December]

Gould, A.A. 1846d. [Descriptions of shells collected by the U.S. Exploring Expedition]. *Proceedings of the Boston Society of Natural History* **2**: 182–84. [after 16 December]

Gould, A.A. 1846e. [Descriptions of *Succinea*]. *Proceedings of the Boston Society of Natural History* **2**: 185-87. [after 16 December]

Gould, A.A. 1846f. [Description of shells from the Exploring Expedition.] *Proceedings of the Boston Society of Natural History* **2**: 190–92. [after 16 December]

Gould, A.A. 1847a. [Descriptions of species of *Partula*, *Pupa*, and *Balea*, collected by the Exploring Expedition]. *Proceedings of the Boston Society of Natural History* **2**: 196–98. [March]

Gould, A.A. 1847b. [Descriptions of the Expedition Shells of the genera *Achatinella* and *Helicina*]. *Proceedings of the Boston Society of Natural History* **2**: 200–03. [March]

Gould, A.A. 1847c. [Descriptions of species of Limniadae, from the collection of the Exploring Expedition]. *Proceedings of the Boston Society of Natural History* **2**: 210–12. [June]

Gould, A.A. 1847d. [Descriptions of species of *Physa*, from the collection of the Exploring Expedition]. *Proceedings of the Boston Society of Natural History* **2**: 214–15. [June]

Gould, A.A. 1852. *United States Exploring Expedition.* Vol. 12. Mollusca & Shells. Gould & Lincoln, Boston. xvi + 510 p. [December]

Gould, A.A. 1862. Descriptions of new genera and species of shells. *Proceedings of the Boston Society of Natural History* **8**: 280–84. [February]

Gouveia, J.J. & Gouveia, A. 1920. A new variety of *Partulina horneri*. *Occasional Papers of the Bernice P. Bishop Museum* **7**(6): 53, pl. 15.

Gray, J.E. 1847. A list of the genera of Recent Mollusca, their synonyma and types. *Proceedings of the Zoological Society of London* **15**: 129–219. [November]

Gray, J.E. 1855. *Catalogue of Pulmonata or air-breathing Mollusca in the collection of the British Museum.* Part 1. British Museum, London. 192 p. [18 April]

Green, J. 1827. Description of two new species of *Achatina*, from the Sandwich Islands—with some remarks on the *Ti*, the plant on which these shells are commonly found. *Contributions of the Maclurian Lyceum to the Arts and Sciences* **1**(2): 47–50, pl. 4. [July]

Gude, G.K. 1911. Note on some preoccupied molluscan generic names and proposed new genera of the family Zonitidae. *Proceedings of the Malacological Society of London* **9**(4): 269–73. [30 March]

Gude, G.K. 1921. Mollusca.—III. Land operculates (Cyclophoridae, Truncatellidae, Assimineidae, Helicinidae). *In*: Shipley, A.E. & Marshall, G.A.K., eds., *The Fauna of British India, including Ceylon and Burma.* Taylor & Francis, London. xiv + 386 p. [February]

Gulick, J.T. 1856. Descriptions of new species of *Achatinella*, from the Hawaiian Islands [part]. *Annals of the Lyceum of Natural History of New York* **6**: 173–230. [December]

Reprinted in 1857, Craighead Publ., New York, with different pagination.

Gulick, J.T. 1858. Descriptions of new species of *Achatinella*, from the Hawaiian Islands [concl.]. *Annals of the Lyceum of Natural History of New York* **6**: 231–55. [February]

Gulick, J.T. 1873a. On the classification of the Achatinellinae. *Proceedings of the Zoological Society of London* **1873**: 89–91. [June]

Gulick, J.T. 1873b. On diversity of evolution under one set of external conditions. *Proceedings of the Linnean Society of London* (Zoology) **11**: 496–505. [18 July]

Gulick, J.T. 1905. Evolution, racial and habitudinal. *Carnegie Institution of Washington Publication* **25**, xii + 269 p., 5 pls. [9 September]

Gulick, J.T. & Smith, E.A. 1873. Descriptions of new species of Achatinellinae. *Proceedings of the Zoological Society of London* **1873**: 73–89, pls. 9–10 [June]

Hadfield, M.G. 1986. Extinction in Hawaiian achatinelline snails. *Malacologia* **27**(1): 67–81. [7 March]

Hadfield, M.G. & Miller, S.E. 1989. Demographic studies on Hawaii's endangered tree snails: *Partulina proxima*. *Pacific Science* **43**(1): 1–16. [March]

Hadfield, M.G., Miller, S.E. & Carwile, A.H. 1993. The decimation of endemic Hawai'ian tree snails by alien predators. *American Zoologist* **33**: 610–22.

Hadfield, M.G. & Mountain, B.S. [1981]. A field study of a vanishing species, *Achatinella mustelina* (Gastropoda, Pulmonata), in the Waianae Mountains of Oahu. *Pacific Science* **34**(4)[1980]: 345–58. [7 July]

Hartman, W.D. 1888a. A bibliographic and synonymic catalogue of the genus *Auriculella* Pfeiffer. *Proceedings of the Academy of Natural Sciences of Philadelphia* **1888**: 14–15. [10 April]

Hartman, W.D. 1888b. A bibliographic and synonymic catalogue of the genus *Achatinella*. *Proceedings of the Academy of Natural Sciences of Philadelphia* **1888**: 16–56, 1 pl. [Publication split: p. 16–40 (10 April); p. 41–56 (24 April)]

Hartman, W.D. 1888c. New species of shells from the New Hebrides and Sandwich Islands. *Proceedings of the Academy of Natural Sciences of Philadelphia* **1888**: 250–52. [23 October]

Haynes, A. 1988. Notes on the stream neritids (Gastropoda; Prosobranchia) of Oceania. *Micronesica* **21**: 93-102. [December]

Henshaw, H.W. 1904. On certain deposits of semi-fossil shells in Hamakua District, Hawaii, with descriptions of new species. *Journal of Malacology* **11**(3): 56–64, pl. 5. [29 September]

Herrmannsen, A.N. 1846–1849. *Indicis generum malacozoorum primordia. Nomina subgenerum, generum, familiarum, tribum, ordinem, classium, adjectis auctoribus, temporibus, locis systematicis atque literariis, etymis, synonymis. Praetermittuntur Cirripedia, Tunicata et Rhizopoda.* 2 vols. T. Fischer, Casselis [= Cassel].

> Published in 10 Lieferungen, 1846–1849. Dates other than year that could be found are as follows:
>
> | Volume | Lieferung | Pages | Date of publication |
> | --- | --- | --- | --- |
> | I "1846" | 1 | i–xxvii, 1–104 | 1846 |
> | | 2 | 105–232 | March 1847a |
> | | 3–5 | 233–637 | October 1847b |
> | II "1847– | 6–7 | xxix–xlii, 1–352 | October 1847c |
> | 1849" | 8–9 | 352–492 | 1848 |
> | | 10 | 493–717 | 10 January 1849 |

Hesse, P. 1926. Die Nacktschnecken der Palaearktischen Region. *Abhandlungen des Archiv für Molluskenkunde* **2**(1): 1–152. [after March]

Hubendick, B. 1952. Hawaiian Lymnaeidae. *Occasional Papers of the Bernice P. Bishop Museum* **20**(19): 307–28. [9 June]

Hubendick, B. 1967. Studies on Ancylidae. The Australian, Pacific and Neotropical form-groups. *Acta Regiae Societatis Scientiarum et Litterarum Gothoburgensis. Zoologica* **1**: 1–52. [after 30 January]

Hyatt, A., & Pilsbry, H.A. [1910]–1911. *Manual of Conchology. Structural and systematic. With illustrations of the species. Second series: Pulmonata.* Vol. XXI. Achatinellidae (Amastrinae). Academy of Natural Sciences, Philadelphia. xxii + 387 p., 56 pls.

> The series was begun by Tryon. Hyatt & Pilsbry are the authors of the material contained in this volume except as noted below. Published in parts as follows:
>
> | Part | Pages | Plates | Date of publication |
> |---|---|---|---|
> | 81 | 1–64 | 1–9 | 30 July 1910 |
> | 82 | 65–128 | 10–23 | 14 March 1911a |
> | 83 | 129–240 | 24–36 | 23 August 1911b |
> | 84 | 241–387, i–xxii | 37–56 | December 1911c |
>
> The article "Genus *Leptachatina* Gould, 1847", p. 1–92, is by C. Montague Cooke., Jr.

International Commission on Zoological Nomenclature. 1926. Opinion 94. Twenty-two mollusk and tunicate names placed in the Official List of Generic Names, p. 12–13. *In*: Opinions rendered by the International Commission on Zoological Nomenclature. Opinions 91 to 97. *Smithsonian Miscellaneous Collections* **73**(4), 30 p. [8 October]

International Commission on Zoological Nomenclature. 1931. Opinion 119. Six molluscan generic names placed in the Official List of Generic Names, p. 23–28. *In*: Opinions rendered by the International Commission on Zoological Nomenclature. Opinions 115 to 123. *Smithsonian Miscellaneous Collections* **73**(7), 36 p. [10 January]

International Commission on Zoological Nomenclature. 1955. Opinion 335. Addition to the Official List of Generic Names in Zoology of the names of thirty-four non-marine genera of the Phylum Mollusca. *Opinions and Declarations Rendered by the International Commission on Zoological Nomenclature* **10**(2): 45–76. [17 March]

International Commission on Zoological Nomenclature. 1957a. Direction 72. Completion and in certain cases correction of entries relating to the names of genera of the Phyla Mollusca, Brachiopoda, Echinodermata and Chordata made on the Official List of Generic Names in Zoology by rulings given in *Opinions* rendered in the period up to the end of 1936. *Opinions and Declarations Rendered by the International Commission on Zoological Nomenclature* **1E**(11): 161–92. [20 September]

International Commission on Zoological Nomenclature. 1957b. Opinion 495. Designation under the Plenary Powers of a type species in harmony with accustomed usage for the nominal genus *Unio* Philipsson, 1788 (Class Pelecypoda) and validation under the same Powers of the family-group name Margaritiferidae Haas, 1940. *Opinions and Declarations Rendered by the International Commission on Zoological Nomenclature* **17**(17): 287–322. [10 December]

International Commission on Zoological Nomenclature. 1985. *International code of zoological nomenclature. Third edition adopted by the XX General Assembly of the International Union of Biological Sciences.* International Trust for Zoological Nomenclature, London; University of California Press, Berkeley & Los Angeles. xx + 338 p. [February]

International Commission on Zoological Nomenclature. 1992. Opinion 1678. *Helicarion* Férussac, 1821 (Mollusca, Gastropoda): conserved, and *Helicarion cuvieri* Férussac, 1821 designated as the type species. *Bulletin of Zoological Nomenclature* **49**(2): 160–61. [25 June]

Iredale, T. 1913. The land Mollusca of the Kermadec Islands. *Proceedings of the Malacological Society of London* **10**(6): 364–88, pl. 18. [22 September]

Iredale, T. 1943. A basic list of the fresh water Mollusca of Australia. *The Australian Zoologist* **10**(2): 188–230. [30 April]

Jacobson, M.K. & Boss, K.J. 1973. The Jamaican land shells described by C.B. Adams. *Occasional Papers on Mollusks* **3**(47): 305–519. [29 November]

Jay, J.C. 1839. *A catalogue of the shells, arranged according to the Lamarckian system; together with descriptions of new or rare species, contained in the collection of John C. Jay, M.D.* Third Edition. Wiley & Putnam, New York. 125 + [1] p., 10 pls. [after April]

Jay, J.C. 1850. *A catalogue of the shells, arranged according to the Lamarckian system, with their authorities, synonymes, and references to works where figured or described, contained in the collection of John C. Jay, M.D.* Fourth Ed. Craighead, New York. 459 p.

Johnson, R.I. 1949. Jesse Wedgewood Mighels with a bibliography and a catalogue of his species. *Occasional Papers on Mollusks* **1**(14): 213–31. [30 March]

Johnson, R.I. 1964. The recent Mollusca of Augustus Addison Gould. *United States National Museum Bulletin* **239**, 182 p., 45 pls. [28 July]

Johnson, R.I. & Boss, K.J. 1972. The fresh-water, brackish, and non-Jamaican land mollusks described by C.B. Adams. *Occasional Papers on Mollusks* **3**(43): 193–233. [2 June]

Jonas, J.H. 1845. Neue Conchylien. *Zeitschrift für Malakozoologie* **1845**: 168–73. [December]

Jutting, W.S.S. van Benthem. 1952. Systematic studies on the non-marine Mollusca of the Indo-Australian archipelago. III. Critical revision of the Javanese pulmonate land-snails of the families Ellobiidae to Limacidae, with an appendix on Helicarionidae. *Treubia* **21**(2): 291–435. [1 August]

Kay, E.A. 1965a. The Reverend John Lightfoot, Daniel Solander, and the Portland Catalogue. *The Nautilus* **79**(1): 10–19. [9 July]

Kay, E.A. 1965b. Marine molluscs in the Cuming collection, British Museum (Natural History) described by William Harper Pease. *Bulletin of the British Museum (Natural History) Zoology Supplement* **1**, 96 p., 14 pls. [December]

Kay, E.A. 1979. *Hawaiian Marine Shells*. Bishop Museum Press, Honolulu. xviii+ 653 p. [29 November]

Kennard, A.S. & Woodward, B.B. 1926. *Synonymy of the British non-marine Mollusca (Recent and post-Tertiary)*. British Museum (Natural History), London. xxiv + 447 p. [27 March]

Kinzie, R.A., III. 1992. Predation by the introduced carnivorous snail *Euglandina rosea* (Férussac) on endemic aquatic lymnaeid snails in Hawaii. *Biological Conservation* **60**: 149–55.

Knight, J.B., Cox, L.R., Keen, A.M., Batten, R.L., Yochelson, E.L. & Robertson, R. 1960. Systematic descriptions, p. 169–310. *In*: Moore, R.C. & Pitrat, C.W., eds., *Treatise on Invertebrate Paleontology. Part I. Mollusca 1*. University of Kansas Press, Lawrence, Kansas; Geological Society of America, New York.

Kobelt, W. 1879. *Illustrirtes Conchylienbuch*. Parts 6–8. G.N. Raspe, Nürnberg [= Nuremberg]. P. 145–264, pls. 51–80.

> The entire series covers the years 1876–1881, issued in parts. The parts above were issued in 1879 as follows:
>
> | Part | Pages | Plates | Date of publication |
> |---|---|---|---|
> | 6 | 145–176 | 51–60 | June–July 1879a |
> | 7–8 | 177–264 | 61–80 | after July 1879b |

Kobelt, W. 1907. Mollusca für 1905. Geographische Verbreitung, Systematik und Biologie. *Archiv für Naturgeschichte* **67**(2): 197–256. [June]

Küster, H.C. [1841–1845]. Die Ohrschnecken (Auriculacea.). In Abbildungen nach der Natur mit Beschreibungen. *In*: Küster, H.C., *Systematisches Conchylien-Cabinet von Martini und Chemnitz. Neu herausgegeben und vervollständigt.* Band I. Abtheilung XVI. Theil 1. G.N. Raspe, Nürnberg [= Nuremburg]. vi + 76 p., 10 pls.

Published in Lieferungen. The collation per year for this section is as follows:

Lieferung	Pages	Plates	Date of publication
25	1–22, 24	2	1841a
30		3	1841b
41	23, 25–30	—	1843a
42		4–6	1843b
49	31–46	1, 7–9	1844
53	v–vi, 47–76	—	1845

Küster, H.C. 1852–[1853]. Die Gattungen *Paludina*, *Hydrocaena* und *Valvata*. In Abbildungen nach der Natur mit Beschreibungen. *In*: Küster, H.C., ed., *Systematisches Conchylien-Cabinet von Martini und Chemnitz. Neu herausgegeben und vervollständigt.* Band I. Abtheilung XXI. G.N. Raspe, Nürnberg [= Nuremburg]. 96 p., 14 pl.

Published in Lieferungen. The collation per year for this section is as follows:

Lieferung	Pages	Plates	Date of publication
113	1–24	1–2	1852a
115	25–56	3–8	1852b
119	57–96	9–14	1853

Küster, H.C., Dunker, W. & Clessin, S. [1841]–1886. Die Familie der Limnaeiden enthaltend die Genera *Planorbis*, *Limnaeus*, *Physa* und *Amphipeplea*. In Abbildungen nach der Natur mit Beschreibungen. *In*: Küster, H.C., *Systematisches Conchylien-Cabinet von Martini und Chemnitz. Neu herausgegeben und vervollständigt.* Band I. Abtheilung XVII. Theil 1. G.N. Raspe, Nürnberg [= Nuremburg]. 430 p., 55 pls.

This section was begun by Küster & Dunker (i.e., 1841–1850). After Küster's death, it was completed by Clessin (i.e., p. 63–430). In 1878, Clessin replaced the original 1850 pages 29–34 with an updated set of pages 29–36a (i.e., 2 additional pages indicated below by an asterisk). Published in Lieferungen. The collation per year for this section is as follows:

Lieferung	Pages	Plates	Date of publication
32	1–8	—	1841
42	—	1	1843
47	9–20	2–3, 16	1844a
49	—	4	1844b
90	21–62	5–10	1850
270	29–36a*	—	1878
319	63–94	11–15, 17	1882
320	95–110	18–22	1883
328	111–150	23–27	1884a
331	151–182	28–33	1884b
332	183–222	34–39	1884c
334	223–278	40–44	1885a
336	279–310	45–50	1885b
338	311–358	51–55	1886a
339	359–430	—	1886b

Lamarck, J.B.P.A. de M. de. 1799. Prodrome d'une nouvelle classification des coquilles, comprenant une rédaction appropriée des caractères génériques, et l'établissement d'un grand nombre de genres nouveaux. *Mémoires de la Société d'Histoire Naturelle de Paris* **1**: 63–91. [May]

Lamarck, J.B.P.A. de M. de. 1801. *Système des animaux sans vertèbres; ou, tableau général des classes, des ordres, et des genres de ces animaux... precédée du discours d'ouverture du cours de zoologie, donné dans le Museum National d'Histoire Naturelle l'an 8 de la République.* Déterville, Paris. viii + 432 p. [21 January]

Lamarck, J.B.P.A. de M. de. 1816. *Encyclopédie méthodique. Tableau Encyclopédique et méthodique des trois règnes de la nature. Vingt-troisième partie. Liste des objets representés dans les planches de cette livraison.* V. Agasse, Paris. 16 p., pls. 391–488. [14 December]

> This is the 84th livraison, which contains plates and 16 pages of explanations to the plates in the "Liste des objets".

Lamarck, J.B.P.A. de M. de. 1822. *Histoire naturelle des animaux sans vertèbres, présentant les caractères généraux et particuliers de ces animaux, leur distribution, leurs classes, leurs familles, leurs genres, et la citation des principales espèces qui s'y rapportent; précédée d'une introduction offrant la détermination des caratères essentiels de l'animal, sa distinction du végétal et des autres corps naturels; enfin, l'exposition des principes fondamentaux de la zoologie.* Tome septième. Verdière, Paris. 711 p. [August]

Lea, I. 1838. Description of new freshwater and land shells [part]. *Transactions of the American Philosophical Society* (n.s.) **6**: 1–153, 23 pls. [15 June]

> This article was issued in parts from 1838–1853, published in the Society's *Transactions*. It was reprinted with separate pagination in his "Observations on the genus *Unio*, together with descriptions of new genera and species". The articles in the *Transactions* take priority over publication in his "Observations . . ."

Lea, I. 1841. [Fifty-seven new species; nearly the whole of them from this country]. *Proceedings of the American Philosophical Society* **2**(17): 30–35. [21 May]

Lea, I. 1844. Description of new freshwater and land shells [part]. *Transactions of the American Philosophical Society* (n.s.) **9**: 1–32.

> See also note under Lea, 1838.

Lea, I. 1856. Description of fifteen new species of exotic Melaniana. *Proceedings of the Academy of Natural Sciences of Philadelphia* **8**: 144–45. [3 October]

Lea, I. & Lea, H.C. 1851. Description of a new genus of the family Melaniana, and of many new species of the genus *Melania*, chiefly collected by Hugh Cuming, Esq., during his zoological voyage in the East, and now first described. *Proceedings of the Zoological Society of London* **18**: 179–97. [28 February]

Lesson, R.P. 1830–1831. *Voyage autour du monde, exécuté par ordre du Roi, sur la corvette de sa Majesté, La Coquille, pendant les années 1822, 1823, 1824 et 1825, sous le ministère et conformément aux instructions de S.E.M. Le Marquis de Clermont-Tonnerre, Ministre de la Marine; et publié sous les auspices de son Excellence Mgr. Le Cte De Chabrol, Ministre de la Marine et des Colonies.* Histoire naturelle. Zoologie. Vol. 2, pt. 1. A. Bertrand, Paris.

> Published in parts as follows:
>
> | Livraison | Pages | Date of publication |
> |---|---|---|
> | 16 | 1–24 | 12 June 1830 |
> | 25 | 25–240 | 12 November 1831a |
> | 26 | 241–471 | 10 December 1831b |

Lesson, R.P. 1842. Description d'une espèce nouvelle de Nériptère. *Revue Zoologique, par la Société Cuvierienne* **5**: 187–88. [3 July]

Liardet, E.A. 1876. On the land-shells of Taviuni, Fiji Islands, with descriptions of new species. *Proceedings of the Zoological Society of London* **1876**(1): 99–101, pl. 5. [June]

Linnaeus, C. 1758. *Systema naturae per regna tria naturae, secundum classes, ordines, genera, species, cum caracteribus, differentiis, synonymis, locis. Tomus I. Editio decima, reformata.* L. Salvii, Holmiae [= Stockholm]. [iv] + 824 p. [1 January]

Lyons, A.B. 1891. A few Hawaiian land shells, p. 103–109, pls. 1–2. *In*: Thrum, T.G, *Hawaiian Almanac and Annual for 1892.* Press Publishing Co., Honolulu. [October]

Maciolek, J.A. 1978. Shell character and habitat of nonmarine Hawaiian neritid snails. *Micronesica* **14**(2): 209–14. [December]
Martens, E. von. 1860. *Die Heliceen nach natürlicher Verwandtschaft systematisch geordnet von Joh. Christ. Albers.* Zweite Ausgabe. W. Engelmann, Leipzig. xviii + 359 p.
Martens, E. von. 1866. Conchological gleanings. *Annals and Magazine of Natural History* (3) **17**: 202–13.
Martens, E. von. 1869. Ueber die Deckel der Schneckengattungen *Neritina, Nerita* und *Navicella*. *Sitzungberichte der Gesellschaft Naturforschender Freunde zu Berlin* **1869**: 21–23.
Megerle von Mühlfeld, J.C. 1816. Beschreibung einiger neuen Conchylien. *Magazin. Gesellschaft Naturforschender Freunde zu Berlin* **8**(1): 3–11, pls. 1, 2.
Melville, R.V. & Smith, J.D.D. 1987. *Official lists and indexes of names and works in zoology.* International Trust for Zoological Nomenclature, London. [iv] + 366 p.
Menke, C.T. 1830. *Synopsis methodica molluscorum generum omnium et specierum earum, quae in Museo Menkeano adservatur; cum synonymia critica et novarum specierum diagnosibus.* Editio altera, auctior et emendatior. G. Uslar, Pyrmont. xvi + 168 + [1] p. [after 12 April]
Mienis, H.K. 1987. On the identity and synonymy of *Rugosella, Porrectus* and *Impervia*, three (sub)generic names introduced by Coen in 1945. *Levantina* **71**: 729–31. [December]
Mighels, J.W. 1845. Descriptions of shells from the Sandwich Islands, and other localities. *Proceedings of the Boston Society of Natural History* **2**: 18–25. [January]
Möbius, K., Richters, F. & von Martens, E. 1880. *Beiträge zur Meeresfauna der Insel Mauritius und der Seychellen.* Gutmann, Berlin. vi + 352 p., 22 pls.

    Von Martens is the author of the Mollusca article in this work.

Möllendorff, O.F. von. 1893. Materialien zur Fauna der Philippinen. XI. Die Insel Leyte. *Bericht über die Senckenbergische Naturforschende Gesellschaft in Frankfurt am Main* **24**: 51–154, pls. 3–5.
Mörch, O.A.L. 1865. Quelques mots sur un arrangement des mollusques pulmonés terrestres (Géophiles, Fér.) basé sur le système naturel (suite). *Journal de Conchyliologie* **13**: 376–96. [5 October]
Montfort, D. de. 1810. *Conchyliologie systématique, et classification méthodique des coquilles; offrant leurs figures, leur arrangement générique, leurs descriptions caractéristiques, leurs noms; ainsi que leur synonymie en plusieurs langues. Ouvrage destiné à faciliter l'étude des coquilles, ainsi que leur disposition dans les cabinets d'histoire naturelle. Coquilles univalves, non cloisonnées.* Tome second. F. Schoell, Paris. 676 p.
Moquin-Tandon, A. 1855–1856. *Histoire naturelle des mollusques terrestres et fluviatiles de France.* Vol. II. J.-B. Ballière, Paris. 646 p.

    Published in parts as follows:

Volume	Pages	Plate	Date of publication
I	i–viii, 1–256	—	14 July 1855a
	257–416	9	18 August 1855b
II	1–368	—	19 January 1856a
	369–646	—	26 April 1856b

Morelet, A. 1857. Testacea nova Australiae. *Bulletin de la Société d'Histoire Naturelle du Département de la Moselle* **8**: 26–33.
Morrison, J.P.E. 1952. World relations of the melanians. *American Malacological Bulletin and Report* **1951**: 6–8. [early 1952]
Morrison, J.P.E. 1954. The relationships of old and new world melanians. *Proceedings of the United States National Museum* **103**[= No. 3325]: 357–94, pl. 11. [20 April]

Morrison, J.P.E. [1969]. Notes on Hawaiian Lymnaeidae. *Malacological Review* **1** [1968]: 31–33. [7 June]

Morse, E.S. 1864a. *Synopsis of the fluviatile and terrestrial Mollusca of the State of Maine.* Published by the author, Portland, Maine. 3 p. [before 17 March]

> Treated here as published prior to Morse (1864b) because its title and pagination were cited in detail by Morse (1864b). This work was intended to be a preliminary list, and mentions (p. 3) that descriptions of the species indicated as new (placed in italics) are in preparation—they are described in Morse, 1864b. However, some generic names are validated by being proposed in association with described species.

Morse, E.S. 1864b. Observations on the terrestrial Pulmonifera of Maine, including a catalogue of all the species of terrestrial and fluviatile Mollusca known to inhabit the State. *Journal of the Portland Society of Natural History* **1**: 1–63, 10 pls. [17 March]

Motteler, L.S. 1986. Pacific Island Names. *Bishop Museum Miscellaneous Publication* **34**, [iv] + 91 p.

Naggs, F. 1992. Case 2833. *Tortaxis* Pilsbry, 1906 and *Allopeas* Baker, 1935 (Mollusca, Gastropoda): proposed conservation by the designation of a neotype for *Achatina erecta* Benson, 1842. *Bulletin of Zoological Nomenclature* **49**(4): 258–60. [17 December]

Naggs, F. 1994. The reproductive anatomy of *Paropeas achatinaceum* and a new concept of *Paropeas* (Pulmonata: Achatinoidea: Subulinidae). *Journal of Molluscan Studies* **60**(2): 175–91. [May]

Neal, M.C. 1928. Anatomical studies of Achatinellidae. *Bernice P. Bishop Museum Bulletin* **47**: 34–49. [8 March]

Neal, M.C. 1934. Hawaiian Helicinidae. *Bernice P. Bishop Museum Bulletin* **125**: 1–102. [26 October]

Newcomb, W. 1853. Descriptions of new species of *Achatinella* from Sandwich Islands. *Annals of the Lyceum of Natural History of New York* **6**: 18–30. [May]

Newcomb, W. 1854a. Descriptions of seventy-nine new species of *Achatinella*, (Swains.) a genus of pulmoniferous mollusks, in the collection of Hugh Cuming, Esq. Zoological Society of London. P. 3–31, 3 pls. [before June]

> We follow Clarke (1958) in treating this separately printed pamphlet of Newcomb (1854b) as published before Pfeiffer (1854b) (which was issued in June). This article is not strictly a preprint or offprint of Newcomb (1854b), as there are differences in spelling when compared with the journal article, including a slightly different title.

Newcomb, W. [1854]b. Descriptions of seventy-nine new species of *Achatinella*, Swains., a genus of pulmoniferous mollusks, in the collection of Hugh Cuming, Esq. *Proceedings of the Zoological Society of London* **21**[1853]: 128–57, pls. 22–24 [14 November]

Newcomb, W. [1855]a. Abstract of descriptions of some animals of *Achatinella*, and other remarks. *Proceedings of the Zoological Society of London* **22**[1854]: 310–311. [8 May]

Newcomb, W. 1855b. [Five species of *Achatinella*]. *Proceedings of the Boston Society of Natural History* **5**: 218–20. [September]

Newcomb, W. 1855c. Descriptions of new species of *Achatinella*. *Annals of the Lyceum of Natural History of New York* **6**: 142–47. [October]

Newcomb, W. 1858. Synopsis of the genus *Achatinella*. *Annals of the Lyceum of Natural History of New York* **6**: 303–36. [September]

Newcomb, W. 1860. Descriptions of new species of the genera *Achatinella*, and *Pupa*. *Annals of the Lyceum of Natural History of New York* **7**: 145–47. [April]

Newcomb, W. 1861. [Descriptions of new shells]. *Proceedings of the California Academy of Natural Sciences* **2**: 91–94. [after 4 February]

Newcomb, W. 1865. Description of new species of land shells. *Proceedings of the California Academy of Natural Sciences* **3**: 179–82. [January]

Newcomb, W. 1866. Descriptions of Achatinellae. *American Journal of Conchology* **2**(3): 209–17. [1 July]

Odhner, N.H. 1922. Mollusca from Juan Fernandez and Easter Island, p. 219–54, pls. 8, 9. *In:* Skottsberg, C., ed., *The Natural History of Juan Fernandez and Easter Island.* Vol. III. Zoology. Almqvist & Wiksells, Uppsala. 688 p.

Odhner, N.H. 1950. Succineid studies: genera and species of subfamily Catinellinae nov. *Proceedings of the Malacological Society of London* **28**(4–5): 200–10. [15 December]

d'Orbigny, A. 1835–1846. *Voyage dans l'Amérique méridionale (le Brésil, la République orientale de l'Uruguay, la République Argentine, la Patagonie, la République du Chili, la République de Bolivia, la République du Pérou), exécuté pendant les années 1826, 1827, 1828, 1829, 1830, 1831, 1832 et 1833, par Alcide d'Orbigny.* Vol. V. Part 3. Mollusques. P. Bertrand, Paris & V. Levrault, Strasbourg. 758 p.

> The entire work consists of 7 volumes of text and two volumes of plates, all published in livraisons, each livraison of which contained portions of some of the volumes (i.e., Géographie, Insectes, Botanie, Planches, Mollusques, etc.). The printed dates on the wrappers are often incorrect, there being delays in publication, or wrapper dates were not changed from one livraison to another. The Académie des Sciences de Paris received this work in a regular fashion until the end of 1844, with virtually all livraisons issued in chronological order (the receipt dates at the Académie sometimes differing by a year or more from the printed wrapper date). Receipt of livraisons by the Académie des Sciences de Paris, when known, are taken here as correct dates of publication.

Livr.	Pages	Plates	Wrapper date	Acad. Sci. Paris date*
1	—	1, 2	1834	15 May 1835a (*Bibliogr. Fr.*)
3	—	4	1835	15 May 1835a (*Bibliogr. Fr.*)
4	—	3	1835	Before 31 August 1835b
5	—	5, 6, 7	1835	31 August 1835c
6	1–48	10, 12	1834	14 September 1835d
7	49–72	—	1835	23 November 1835e
8	73–104	—	1834	7 December 1835f
9	105–128	9, 11, 13	1834	4 January 1836g
11	129–152	17, 21	1835	18 April 1836h
12	153–176	8	1835	30 May 1836a
13	—	18, 19, 22	1835	—
14	—	20, 25	1835	11 July 1836c
15	—	23	1835	1 August 1836d
16	—	15, 16	1834	26 September 1836e
17	177–184	27, 28	1836	3 October 1836f
18	—	14, 26	1836	7 November 1836g
21	—	31	1836	—
22	—	24, 35	1836	27 February 1837a
23	—	30, 32, 34	1836	3 April 1837b
24	—	35, 37	1836	5 June 1837c
25	—	38, 41	1837	19 June 1837d
26	—	38, 39	1837	7 August 1837e
27	—	40, 45	1837	18 September 1837f
28	—	29, 46	1837	—
29	—	40, 42, 43	1837	6 November 1837h
31	185–232	44	1837	5 March 1838a
32	233–280	47	1837	23 April 1838b
33	281–328	48, 52	1837	6 May 1838c
34	329–376	—	1837	11 June 1838d
35	—	49, 50, 51	1837	15 October 1838e
36	—	55	1835	12 November 1838f
37	—	56	1834	8 April 1839a
38	—	57	1837	29 April 1839b
39	—	58	1836	24 June 1839c
42	—	59	1839	11 November 1839d

* except where noted

Livr.	Pages	Plates	Wrapper date	Acad. Sci. Paris date
43	—	64, 65	1839	21 November 1839e
44	—	54, 60–63	1839	6 September 1841a
46	—	66	1839	8 November 1841b
47	—	68, 69	1839	8 November 1841c
48	—	70	1840	8 November 1841d
49	377–408	—	1840	15 November 1841e
50	—	53, 67, 71	1840	15 November 1841f
51	409–424	72	1841	15 November 1841g
52	425–472	73, 74, 79	1841	15 November 1841h
53	473–488	75, 76, 80	1841	14 February 1842
82	489–528	—	1846	—
83	529–600	—	1845	—
84	601–656	—	1846	—
85	657–704	—	1846	—
86	705–728	—	1846	—
—	729–758	—	—	(no livraison known)
88	—	83, 85	1842	—
89	—	78, 81	1847	—
90	—	79, 82	1847	—

Pace, G.L. 1973. The freshwater snails of Taiwan (Formosa). *Malacological Review Supplement* **1**, 118 p. [25 October]

Paetel, F. 1873. *Catalog der Conchylien-Sammlung von Fr. Paetel.* Gebrüder Paetel, Berlin. [i] + 172 p.

Paetel, F. 1883. *Catalog der Conchylien-Sammlung von Fr. Paetel.* Gebrüder Paetel, Berlin. [i] + 271 p.

Paetel, F. 1889–1890. *Catalog der Conchylien-Sammlung von Fr. Paetel.* Vierte Neubearbeitung. Zweite Abtheilung: die Land- und Süsswasser-Gastropoden. Gebrüder Paetel, Berlin.

Published in Lieferungen. Collation per year is as follows:

Pages	Date of publication
1–160	1889
161–505, i–xii	1890

Pease, W.H. 1860. Descriptions of new species of Mollusca from the Sandwich Islands. *Proceedings of the Zoological Society of London* **28**: 141–48. [February–May]

Pease, W.H. [1861]. Descriptions of six new species of land shells, from the islands of Ebon, Marshall's group, in the collection of H. Cuming. *Proceedings of the Zoological Society of London* **28**[1860]: 439-40. [31 March]

Pease, W.H. 1862. Descriptions of two new species of *Helicter* (*Achatinella*, Swains.), from the Sandwich Islands, with a history of the genus. *Proceedings of the Zoological Society of London* **1862**(1): 3–7. [30 June]

Pease, W.H. [1863]. Descriptions of new species of marine shells from the Pacific Inlands [sic]. *Proceedings of the Zoological Society of London* **1862**(3): 240–43. [April]

Pease, W.H. [1865]. Descriptions of new species of land shells from the islands of the central Pacific. *Proceedings of the Zoological Society of London* **1864**(3): 668–76. [31 May]

Pease, W.H. 1866. Descriptions of new species of land shells, inhabiting Polynesia. *American Journal of Conchology* **2**(4): 289–93. [1 October]

Pease, W.H. 1868. Descriptions d'espèces nouvelles d'*Auriculella*, provenant des îles Hawaï. *Journal de Conchyliologie* **16**: 342–47, pl. 14. [16 October]

Pease, W.H. 1869a. Description d'espèces nouvelles du genre *Helicter*, habitant les îles Hawaii. *Journal de Conchyliologie* **17**: 167–76. [16 July]

Pease, W.H. 1869b. Descriptions of new species of marine Gasteropodae inhabiting Polynesia. *American Journal of Conchology* **5**(2): 64–79. [7 October]

Pease, W.H. 1870a. Observations sur les espèces de coquilles terrestres qui habitent l'île de Kauai (îles Hawaii), accompagnées de descriptions d'espèces nouvelles. *Journal de Conchyliologie* **18**: 87–97. [10 January]

Pease, W. H. [1870]b. On the classification of the Helicterinae. *Proceedings of the Zoological Society of London* **1869**(3): 644–52. [30 April]

Pease, W.H. 1870c. Remarks on the species of *Melania* and *Limnaea* inhabiting the Hawaiian Islands, with descriptions of new species. *American Journal of Conchology* **6**(1): 4–7. [7 July]

Pease, W.H. 1871a. Remarques sur certaines espèces de coquilles terrestres, habitant la Polynésie, et description d'espèces nouvelles. *Journal de Conchyliologie* **18**: 393–403. [22 January]

Pease, W.H. 1871b. Catalogue of the land-shells inhabiting Polynesia, with remarks on their synonymy, distribution, and variation, and descriptions of new genera and species. *Proceedings of the Zoological Society of London* **1871**: 449–77. [31 August]

Pfeiffer, L. [1846]a. Remarks on the genus *Achatinella*, Swainson, and descriptions of six new species from Mr. Cuming's collection. *Proceedings of the Zoological Society of London* **13**[1845]: 89–90. [January]

Pfeiffer, L. 1846b. Descriptions of thirty new species of Helicea, belonging to the collection of H. Cuming, Esq. *Proceedings of the Zoological Society of London* **14**: 28–34. [May]

Pfeiffer, L. 1846c. Descriptions of twenty new species of Helicea, in the collection of H. Cuming, Esq. *Proceedings of the Zoological Society of London* **14**: 37–41. [July]

Pfeiffer, L. 1846d. Ueber neue Landschnecken von Jamaika und den Sandwichsinseln. *Zeitschrift für Malakozoologie*. **1846**: 113–20. [August]

Pfeiffer, L. 1846e. *Symbolae ad historiam heliceorum. Sectio tertia.* T. Fischer, Cassellis [= Cassel]. 100 p. [October]

This is the last of three parts, the entire work being published 1841–1846.

Pfeiffer, L. [1847]a. [Descriptions of 38 new species of Land-shells, in the collection of Hugh Cuming, Esq]. *Proceedings of the Zoological Society of London* **14** [1846]: 109–16. [26 January]

Pfeiffer, L. 1847b. Diagnosen neuer Landschnecken. *Zeitschrift für Malakozoologie* **4**: 145–51. [October]

Pfeiffer, L. [1848]a. Descriptions of nineteen new species of Helicea, from the collection of H. Cuming, Esq. *Proceedings of the Zoological Society of London* **15** [1847]: 228–32. [29 March]

Pfeiffer, L. 1848. *Monographia heliceorum viventium. Sistens descriptiones systematicas et criticas omnium huius familiae generum et specierum hodie cognitarum. Volumen secundum.* F.A. Brockhaus, Lipsiae [= Leipzig]. xxxii + 484 p.

Published in parts. All parts of volume 2 ssued in 1848 as follows:

Lieferung	Pages	Date of publication
4	1–160	Before September 1848b
5–7	161–594	Before December 1848c

Pfeiffer, L. 1849a. Methodische Anordnung aller bekannten Helicinaceen. *Zeitschrift für Malakozoologie* **5**[1848]: 81–89. [after 10 January]

This is the June 1848 issue, which has an advertisement on the last page dated 10 January 1849.

Pfeiffer, L. [1849]b. Description of twenty-nine new species of *Helicina*, from the collection of H. Cuming, Esq. *Proceedings of the Zoological Society of London* **16** [1848]: 119–25. [25 April]

Pfeiffer, L. 1849c. Nachträge zu L. Pfeiffer Monographia Heliceorum [part]. *Zeitschrift für Malakozoologie* **6**(6): 81–95. [November]

BIBLIOGRAPHY [223]

Pfeiffer, L. [1850]a. Descriptions of twenty-four new species of Helicea, from the collection of H. Cuming, Esq. *Proceedings of the Zoological Society of London* **17**[1849]: 126–31. [30 June]

Pfeiffer, L. 1850b. Beschreibungen neuer Landschnecken. *Zeitschrift für Malakozoologie* **7**(5): 65–80. [July]

Pfeiffer, L. [1850–1853]. Die gedeckelten Lungenschnecken. (Helicinacea et Cyclostomacea.). In Abbildungen nach der Natur mit Beschreibungen. *In*: Küster, H.C., ed., *Systematisches Conchylien-Cabinet von Martini und Chemnitz*. Band I. Abtheilung XVIII. Theil I. Bauer & Raspe, Nürnberg [= Nuremberg]. 78 p., 10 pls.

> The title page to some bound volumes of this work has "1846". This has led to some confusion in dating because this date pertains merely to the beginning of the issuance of the Lieferungen for Theil II in the same volume [= Abtheilung XVIII] and not to the Lieferungen in Theil I, which appeared *after* Theil II. The collation per year for Theil I is as follows:

> | Lieferung | Pages | Plates | Date of publication |
> |---|---|---|---|
> | 94 | 1–24 | 1–6 | 1850c |
> | 100 | 25–64 | 7–9 | 1850d |
> | 114 | — | 10 | 1852 |
> | 122 | 65–78 | — | 1853b |

Pfeiffer, L. 1851. Einige Bemerkungen über Deshayes's Bearbeitung des Ferussacschen Werkes. *Zeitschrift für Malakozoologie* **7**[1850]: 145–60. [March]

Pfeiffer, L. [1853]a. Description of fifty-four new species of Helicea, from the collection of Hugh Cuming, Esq. [part]. *Proceedings of the Zoological Society of London* **19**[1851]: 252–63. [Publication split: p. 252–56 (26 July); p. 257–63 (7 December).]

Pfeiffer, L. 1853c. *Monographia heliceorum viventium. Sistens descriptiones systematicas et criticas omnium huius familiae generum et specierum hodie cognitarum. Volumen tertium. Supplementum. Sistens enumerationem auctam omnium huius familiae generum et specierum hodie cognitarum, accedentibus descriptionibus novarum specierum et enumeratione fossilium*. F.A. Brockhaus, Lipsiae [= Leipzig]. viii + 711 p. [after May]

Pfeiffer, L. [1854]a. Descriptions of sixty-six new land shells, from the collection of H. Cuming, Esq. *Proceedings of the Zoological Society of London* **20**[1852]: 56–70. [Publication split: p. 56–64 (22 March); p. 65–70 (23 May)]

Pfeiffer, L. 1854b. Skizze einer Monographie der Gattung *Achatinella* Swains. *Malakozoologische Blätter* **1**: 112–44. [June]

Pfeiffer, L. 1854c. Synopsis Auriculaceorum. *Malakozoologische Blätter* **1**: 145–56. [August]

Pfeiffer, L. [1854]d. Descriptions of nineteen new species of Helicea, from the collection of H. Cuming, Esq. *Proceedings of the Zoological Society of London* **21**[1853]: 124–28. [14 November]

Pfeiffer, L. [1855]a. Descriptions of forty-two new species of *Helix*, from the collection of H. Cuming, Esq. *Proceedings of the Zoological Society of London* **22**[1854]: 49–57. [10 January]

Pfeiffer, L. 1855b. Weitere Beobachtungen über die Gattung *Achatinella*. *Malakozoologische Blätter* **2**(8): 1–7. [January]

> We consider the signature footer on p. 1 ("Malak. Bl. VII. Dec, 54") to be a typographical error.

Pfeiffer, L. [1855]c. Descriptions of twenty-seven new species of *Achatinella*, from the collection of H. Cuming, Esq., collected by Dr. Newcomb and by Mons. D. Frick, late Consul-general of France at the Sandwich Islands. *Proceedings of the Zoological Society of London* **23**[1854]: 1–7, pl. 30. [27 March]

Pfeiffer, L. 1855d. Descriptions of sixteen new species of Helicea, from the collection of H. Cuming, Esq. *Proceedings of the Zoological Society of London* **22** [1854]: 122–26. [7 April]

Pfeiffer, L. [1855]e. Descriptions of fifty-seven new species of Helicea, from Mr. Cuming's collection. *Proceedings of the Zoological Society of London* **22** [1854]: 286–98. [8 May]

Pfeiffer, L. 1855f. Descriptions of forty-seven new species of Helicea, from the collection of H. Cuming, Esq. *Proceedings of the Zoological Society of London* **23**: 91–101, pl. 31. [23 July]

Pfeiffer, L. 1855g. Weitere Beobachtungen über die Gattung *Achatinella*. *Malakozoologische Blätter* **2**: 64–70. [July]

Pfeiffer, L. 1855h. Descriptions of nine new species of Helicea, from Mr. Cuming's collection. *Proceedings of the Zoological Society of London* **23**: 106–08, pl. 32. [13 August]

Pfeiffer, L. 1856a. Versuch einer Anordnung der Heliceen nach natürlichen Gruppen [concl.]. *Malakozoologische Blätter* **2**: 145–85. [January]

> This is the concluding part of this article. The first part was published, 1855, *Malakozoologische Blätter* 2: 112–44.

Pfeiffer, L. [1856]b. Descriptions of twenty-three new species of *Achatinella*, collected by Mr. D. Frick in the Sandwich Islands; from Mr. Cuming's collection. *Proceedings of the Zoological Society of London* **23**[1855]: 202–06. [5 February]

Pfeiffer, L. [1856]c. Descriptions of sixteen new species of *Achatinella*, from Mr. Cuming's collection, collected by Dr. Newcomb in the Sandwich Islands. *Proceedings of the Zoological Society of London* **23**[1855]: 207–10. [5 February]

Pfeiffer, L. [1856]d. Descriptions of five new species of terrestrial Mollusca, from the collection of H. Cuming, Esq. *Proceedings of the Zoological Society of London* **23** [1855]: 210–11. [5 February]

Pfeiffer, L. 1856e. Descriptions of twenty-five new species of land-shells, from the collection of H. Cuming, Esq. *Proceedings of the Zoological Society of London* **24**: 32–36. [16 June]

Pfeiffer, L. 1856f. *Monographia auriculaceorum viventium. Sistens descriptiones systematicas et criticas omnium hujus familiae generum et specierum hodie cognitarum, nec non fossilium enumeratione. Accedente proserpinaceorum nec non generis truncatellae historia.* T. Fischer, Casselis [= Cassel]. xiii + 209 p. [after May]

Pfeiffer, L. [1857]a. Descriptions of fifty-eight new species of Helicea from the collection of H, Cuming, Esq. *Proceedings of the Zoological Society of London* **24**[1856]: 324–36. [10 March]

Pfeiffer, L. 1857b. Diagnosen neuer Heliceen. *Malakozoologische Blätter* **4**: 229–32. [December]

Pfeiffer, L. 1858. Descriptions of eleven new species of land-shells, from the collection of H. Cuming, Esq. *Proceedings of the Zoological Society of London* **26**: 20–23, pl. 40. [9 March]

Pfeiffer, L. 1859a. Descriptions of twenty-seven new species of land-shells, from the collection of H. Cuming, Esq. *Proceedings of the Zoological Society of London* **27**: 23–29. [1 February–29 June]

Pfeiffer. L. 1859b. Descriptions of two new species of *Melampus*, from Mr. Cuming's collection. *Proceedings of the Zoological Society of London* **27**: 29–30. [1 February–29 June]

Pfeiffer. L. 1859c. Descriptions of eight new species of *Achatinella,* from Mr. Cuming's collection. *Proceedings of the Zoological Society of London* **27**: 30–32. [30 June]

Pfeiffer, L. 1859d. *Monographia heliceorum viventium. Sistens descriptiones systematicas et criticas omnium huius familiae generum et specierum hodie cognitarum. Volumen quartum. Supplementum secundum. Sistens enumerationem auctam omnium huius familiae generum et specierum hodie cognitarum, accedentibus descriptionibus novarum specierum.* F.A. Brockhaus, Lipsiae [= Leipzig]. ix + 920 p.

Pfeiffer, L. 1861. Descriptions of forty-seven new species of land-shells, from the collection of H. Cuming, Esq. *Proceedings of the Zoological Society of London* **1861**(1): 20–29. [May]

Pfeiffer, L. 1868. *Monographia heliceorum viventium. Sistens descriptiones systematicas et criticas omnium huius familiae generum et specierum hodie cognitarum. Volumen sextum. Supplementum tertium. Sistens enumerationem auctam omnium huius familiae generum et specierum hodie cognitarum. Accedentibus descriptionibus novarum specierum. Volumen secundum.* F.A. Brockhaus, Lipsiae [= Leipzig]. 598 p. [December]

Pfeiffer, L. 1870–1876. *Novitates Conchologicae. Series prima. Mollusca extramarina. Tome IV. Descriptions et figures de coquilles extramarines nouvelles ou peu connues.* T. Fischer, Cassel. 171 p., pls. 109–37.

Slightly different subtitle in German on page facing the title page. Published in parts. Collation per year is as follows:

Pages	Plates	Date of publication
1–8	109–111	1870
9–40	112–117	1871
41–88	118–125	1872
89–112	126–130	1874
113–144	131–134	1875a
145–171	135–137	1876a

Pfeiffer, L. 1875–1877. *Monographia heliceorum viventium. Sistens descriptiones systematicas et criticas omnium huius familiae generum et specierum hodie cognitarum. Volumen octavum. Supplementum quartum. Sistens enumerationem auctam omnium huius familiae generum et specierum hodie cognitarum. Accedentibus descriptionibus novarum specierum. Volumen secundum.* F.A. Brockhaus, Lipsiae [= Leipzig]. 729 p.

This is the second volume of the fourth supplement and was published in 4 parts from 1875-1877 as follows:

Pages	Date of publication
1–160	1875b
161–320	1876b
321–480	1876c
481–729	1877

Pfeiffer, L. & Clessin, S. [1878]–1881. *Nomenclator heliceorum viventium quo continetur nomina omnium hujus familiae generum et specierum hodie cognitarum disposita ex affinitate naturali. Opus postumum Ludovici Pfeiffer Dr.* T. Fischer, Casselis [= Cassel]. [1] + 617 p.

The title page has the date "1881". However, it was published in parts from 1878 to 1881 as follows.

Pages	Date of publication
1–152	1878
153–384	1879
385–617	after July 1881

Philippi, R.A. 1845. Diagnosen einiger neuen Conchylien. *Archiv für Naturgeschichte* **11**(1): 50–71.

The author's initials are given variously in the malacological literature as "A.", "R.A.", and "A.R."

Philippi, R.A. 1846. Diagnoses testaceorum quorundam novorum. *Zeitschrift für Malakozoologie* **1846**: 97–106. [July]

Philippi, R.A. 1847. Testaceorum novorum centuria. *Zeitschrift für Malakozoologie* **4**: 113–27. [August]

Pilsbry, H.A. [1893–1895]. *Manual of conchology: structural and systematic. With illustrations of the species. By George W. Tryon, Jr. Second series: Pulmonata.* Vol. IX. (Helicidae, vol. 7.). Guide to the study of helices. Academy of Natural Sciences, Philadelphia. xlviii + 366 p., [1] + 71 pls.

> The series was begun by Tryon. Pilsbry continued it after Tryon's death and is the sole author of the material contained in this volume. Published in parts as follows:
>
> | Part | Pages | Date of publication |
> |------|-------|---------------------|
> | 33 | 1–48 | 16 November 1893 |
> | 34 | 49–112 | 19 March 1894a |
> | 35 | 113–160 | 27 July 1894b |
> | 36 | 161–336 | 2 February 1895a |
> | 33a | i–xlviii | 2 February 1895b |
>
> An "Index to the Helices" (126 p.) was published in April 1895

Pilsbry, H.A. 1900. Note on Polynesian and East Indian Pupidae. *Proceedings of the Academy of Natural Sciences of Philadelphia* **52**: 431–33. [Publication split: p. 431–32 (9 August); p. 433 (14 August)]

Pilsbry, H.A. 1904. A new Hawaiian *Limnaea*. *Proceedings of the Academy of Natural Sciences of Philadelphia* **55**: 790–91. [1 February]

Pilsbry, H.A. 1906–1907. *Manual of Conchology. Structural and systematic. With illustrations of the species. Founded by George W. Tryon, Jr. Second series: Pulmonata.* Vol. XVIII. Achatinidae: Stenogyrinae and Coeliaxinae. Academy of Natural Sciences, Philadelphia. xii + 357 p., 51 pls.

> Published in parts as follows:
>
> | Part | Pages | Plates | Date of publication |
> |------|-------|--------|---------------------|
> | 69 | 1–64 | 1–10 | 20 January 1906a |
> | 70 | 65–160 | 11–20 | 10 April 1906b |
> | 71 | 161–272 | 21–34 | 2 October 1906c |
> | 72 | 273–357, i–xii | 35–51 | January 1907 |

Pilsbry, H.A. 1910. Notes on the classification of the Tornatellinidae. *The Nautilus* **23**(10): 122–24. [8 March]

Pilsbry, H.A. 1913. Two new Achatinellidae of Oahu. *The Nautilus* **27**(4): 39–40. [7 August]

Pilsbry, H.A. 1914. A new species of *Leptachatina*. *The Nautilus* **28**(5): 61–62. [22 September]

Pilsbry, H.A. 1916–1918. *Manual of Conchology. Second series: Pulmonata.* Vol. XXIV. Pupillidae (Gastrocoptinae). Academy of Natural Sciences, Philadelphia. xii + 380 p., 49 pls.

> Published in parts as follows:
>
> | Part | Pages | Plates | Date of publication |
> |------|-------|--------|---------------------|
> | 93 | 1–112 | 1–13 | 18 December 1916 |
> | 94 | 113–176 | 14–29 | 18 July 1917a |
> | 95 | 177–256 | 30–38 | 9 November 1917b |
> | 96 | 257–380, i–xii | 39–49 | March 1918 |

Pilsbry, H.A. 1920–1921. *Manual of Conchology. Second series: Pulmonata.* Vol. XXVI. Pupillidae (Vertigininae, Pupillinae). Academy of Natural Sciences, Philadelphia. iv + 254 p., 24 pls.

Published in parts as follows:

Part	Pages	Plates	Date of publication
101	1–64	1–8	23 December 1920
102	65–128	9–13	13 May 1921a
103	129–192	14–18	4 August 1921b
104	193–254, i–iv	19–24	November 1921c

Pilsbry, H.A. 1922a. Observations on the nomenclature of slugs. *The Nautilus* 35(3): 77–80. [23 January]

Pilsbry, H.A. 1922b. Observations on the nomenclature of slugs. II. *The Nautilus* 35(4): 117–18. [24 April]

Pilsbry, H.A. 1922–1926. *Manual of Conchology. Second series: Pulmonata.* Vol. XXVII. Pupillidae (Orculinae, Pagodulinae, Acanthinulinae, etc). Academy of Natural Sciences, Philadelphia. v + 369 p., 32 pls.

Pilsbry is the sole author of the material contained in this volume except as noted below. Published in parts as follows:

Part	Pages	Plates	Date of publication
105	1–80	1–5	29 August 1922c
106	81–128	6–11	13 June 1923
107	129–176	12–18	16 July 1924
108	177–369, i–v	19–32	1 April 1926*

* A handwritten note (on p. 177) in a copy in the Academy of Natural Sciences stating date of issue.

The article "Abidas and Chondrinas of the Pyrenees and the Iberian Peninsula", p. 267–315, is by F. Haas.

Pilsbry, H.A. 1927–1935. *Manual of Conchology. Second series: Pulmonata.* Vol. XXVIII. Geographic distribution of Pupillidae; Strobilopsidae, Valloniidae and Pleurodiscidae. Academy of Natural Sciences, Philadelphia. xii + 226 p., 31 pls.

Pilsbry is the sole author of the material contained in this volume except as noted below. Published in parts as follows (stamped dates are those on which each part was mailed from the Academy):

Part	Pages	Plates	Printed Date	Stamped Date
109	1–48	1–8	November 1927	11 November 1927
110	49–96	9–12	April 1931	2 April 1931
111	97–160	13–23	14 June 1934	14 June 1934
112	161–226, i–xii	24–31	November 1935	7 November 1935

The article, "Review of the anatomy of Pupillidae and related groups", p. 191–209, is by H. B. Baker.

Pilsbry, H.A. 1928. *Georissa*, a land snail genus new to the Hawaiian Islands. *Bishop Museum Bulletin* 47: 3–4. [8 March]

Pilsbry, H.A. 1946. *Land Mollusca of North America (north of Mexico)*. Volume II. Part 1. Academy of Natural Sciences, Philadelphia. vi + [ii] + 520 p., 1 pl. [6 December]

Pilsbry, H.A. & Baker, H.B. 1958. Type of *Paludestrina*. *The Nautilus* 71(3): 116. [4 March]

Pilsbry, H.A. & Cooke, C.M., Jr. 1906. On Hawaiian species of *Sphyradium*. *Proceedings of the Academy of Natural Sciences of Philadelphia* 58: 215–16. [24 July]

Pilsbry, H.A. & Cooke, C.M., Jr. 1908. Hawaiian species of *Helicina*. *Occasional Papers of the Bernice P. Bishop Museum* 3(2): 199–210, 1 pl. [24 July]

Pilsbry, H.A. & Cooke, C.M., Jr. 1912–1914. *Manual of Conchology. Structural and systematic. With illustrations of the species. Second series: Pulmonata.* Vol. XXII. Achatinellidae. Academy of Natural Sciences, Philadelphia. lviii + 428 p., 63 pls.

Pilsbry and Cooke are authors of the material contained in this volume except as noted below. Published in parts as follows:

Part	Pages	Plates	Date of publication
85	1–64	1–12	21 November 1912
86	65–112	13–26	9 October 1913a
87	113–176	27–39	10 December 1913b
88	177–428, i–lviii	40–63	May 1914a

The article "Genealogy and migrations of the Achatinellidae", p. 370–99, consists of a manuscript edited after Hyatt's death by A.G. Mayer. Authorship is attributed to Hyatt. However, bracketed statements and footnotes provided by other authors are indicated by their initials; in instances of footnotes where other authors' initials are not provided, authorship is to be attributed to Mayer.

Pilsbry, H.A. & Cooke, C.M., Jr. [1914]–1916. *Manual of Conchology. Second series: Pulmonata.* Vol. XXIII. Appendix to Amastridae. Tornatellinidae. Index, vols. XXI–XXIII. Academy of Natural Sciences, Philadelphia. xi + 302 p., 55 pls.

Published in parts as follows:

Part	Pages	Plates	Date of publication
89	1–48	1–13	23 October 1914b
90	49–128	14–23	4 August 1915a
91	129–256	24–38	1 December 1915b
92	257–302, i–xi	39–55	February 1916

Pilsbry, H.A. & Cooke, C.M., Jr. 1918–1920. *Manual of Conchology. Second series: Pulmonata.* Vol. XXV. Pupillidae (Gastrocoptinae, Vertigininae). Academy of Natural Sciences, Philadelphia. ix + 401 p., 34 pls.

Published in parts as follows:

Part	Pages	Plates	Date of publication
97	1–64	1–5	5 November 1918
98	65–144	6–10	20 February 1919a
99	145–224	11–18	30 June 1919b
100	225–401, i–ix	19–34	April 1920

Pilsbry, H.A. & Cooke, C.M., Jr. 1921. A new *Achatinella* from Oahu. *The Nautilus* **34**(4): 109–10. [5 May]

Pilsbry, H.A. & Cooke, C.M., Jr. 1922. The identity of *Helix depressiformis* and *H. prostrata* Pease. *The Nautilus* **36**(1): 17. [24 July]

Pilsbry, H.A. & Cooke, C.M., Jr. 1933. Notes on the land snail family Tornatellinidae. *The Nautilus* **47**(2): 59–62. [1 November]

Pilsbry, H.A. & Cooke, C.M., Jr. 1934. Notes on the nomenclature of Hawaiian Helicinidae. *The Nautilus* **48**(2): 53–54. [15 October]

Pilsbry, H.A. & Vanatta, E.G. [1905]. Notes on some Hawaiian Achatinellidae and Endodontidae. *Proceedings of the Academy of Natural Sciences of Philadelphia* **57**: 570–75. [12 September]

Pilsbry, H.A. & Vanatta, E.G. 1906. Hawaiian species of *Endodonta* and *Opeas*. *Proceedings of the Academy of Natural Sciences of Philadelphia* **57**[1905]: 783–86. [26 January]

Pilsbry, H.A. & Vanatta, E.G. 1908. A new Hawaiian *Kaliella*. *The Nautilus* **22** (8): 73. [11 December]

Quick, H.E. 1939. Some particulars of four Indo-Pacific *Succinea*. *Proceedings of the Malacological Society of London* **23**(5): 298–302. [15 July]

Quick, H.E. 1960. British slugs (Pulmonata; Testacellidae, Arionidae, Limacidae). *Bulletin of the British Museum (Natural History)* (Zoology) **6**(3): 103–226, pl. 12. [11 March]

Quoy, J.R.C. & Gaimard, L.P. 1824–[1826]. Zoologie *In:* Freycinet, L.C.D de, *Voyage autour du monde, entrepris par ordre du Roi sous le ministère et conformément aux instructions de S. Exc. M. le Vicomte du Bouchage, Secrétaire d'État au Département de la Marine, exécuté sur le corvettes de S.M. l'Uranie et la Physicienne, pendant les années 1817, 1818, 1819 et 1820; publié sous les auspices de S. E. M. le*

Comte Corbière, Secrétaire d'État de l'Intérieur, pouv la partie historique et les sciences naturelles, et de S. E. M. le Marquis de Clermont-Tonnerre, Secrétaire d'État de la Marine et des Colonies, pouv la partie nautique. "1824." Pillet Aîné, Imprimeur-Libraire, Paris. vii + 712 p., 96 pls.

Published in livraisons as follows:

Livraison	Pages	Date of publication
1	1–40	26 June 1824a
2	41–88	31 July 1824b
3	89–128	28 August 1824c
4	129–184	18 September 1824d
5	185–232	9 October 1824e
6	233–280	20 November 1824f
7	281–328	18 December 1824g
8	329–376	29 January 1825a
9	377–424	26 March 1825b
10	425–464	7 May 1825c
11	465–496	18 June 1825d
12	497–536	6 August 1825e
13	537–576	1 October 1825f
14	577–616	17 December 1825g
15	617–664	26 April 1826a
16	665–712	14 June 1826b

In the "Préface" to this work (unnumbered page 3), Quoy & Gaimard thank Férussac for the nomenclature of the terrestrial molluscs. In the introduction to the terrestrial molluscs, Quoy & Gaimard state (p. 463–64) that "Nous devons à M. de Férussac la description des espèces que nous avons rapportées, dont il a fait figurer plusiers dans son magnifique ouvrage sur les mollusques terrestres et fluviatiles." Thus, authorship of the descriptions of the terrestrial molluscs (p. 462–96) is Férussac *in* Quoy & Gaimard, though Quoy & Gaimard are the authors of the introductory text to that chapter (p. 462–64).

Rafinesque, C.S. 1820. *Annals of Nature: or annual synopsis of new genera and species of animals, plants, &c. discovered in North America.* First annual number for 1820. Lexington, Kentucky. 16 p. [1 March]

Récluz, C.A. 1850. Notice sur le genre *Nérita* et sur le s.-g. *Neritina*, avec le catalogue synonymique des Néritines. *Journal de Conchyliologie* 1(2): 131–64, pls. 3–7. [15 April]

Reeve, L.A. 1848–1850. *Conchologia Iconica: or, illustrations of the shells of molluscous animals. Volume V. Containing the monographs of the genera* Bulimus. Achatina. Dolium. Cassis. Turritella. Mesalia. Eglisia. Cassidaria. Oniscia. Eburnia. Reeve, Benham & Reeve, London.

This is a lambda book (a book in which plates were published separately, along with unnumbered pages of explantatory text, as the plates were ready). After all plates were completed, they were bound into volumes. Dates of publication as given on the bottom of the explanatory text for each plate are as follows for *Achatina*:

Plates	Date of publication
1–8	February 1849a
9–13	March 1849b
14	April 1849c
15–16	May 1849d
17–19	June 1849f
20–23	March 1850a

Reeve, L.A. 1849–1851. *Conchologia Iconica: or, illustrations of the shells of molluscous animals. Volume VI. Containing the monographs of the genera* Voluta. Fissurella. Partula. Achatinella. Artemis. Lucina. Hemipecten. Oliva. Strombus. Pterocera. Rostellaria. Struthiolaria. Reeve & Benham, London.

See Reeve (1848–1850) above for details on the publication methods of this series. Dates of publication as given on the bottom of the explanatory text for each plate are as follows for *Partula* and *Achatinella*:

Partula:

Plates	Date of publication
1	May 1849e
2–3	April 1850b
4	May 1850d

Achatinella:

Plates	Date of publication
1,6	May 1850e
2–5	April 1850c

Reeve, L.A. 1851–1854. *Conchologia Iconica: or, illustrations of the shells of molluscous animals. Volume VII. Containing a monograph of the genus* Helix. L. Reeve, London. 210 pls.

See Reeve (1848–1850) above for details on the publication methods of this series. Dates of publication as given on the bottom of the explanatory text for each plate are as follows for *Helix*:

Plates	Date of publication
1–6	March 1851a
7–14	April 1851b
15–22	May 1851c
23–30	June 1851d
31–38	July 1851e
39–46	September 1851f
47–54	October 1851g
55–62	December 1851h
63–70	January 1852a
71–78	February 1852b
79–86	March 1852c
87–94	April 1852d
95–102	May 1852e
103–110	June 1852f
111–126	August 1852g
127–134	October 1852h
135–142	November 1852i
143–146	December 1852j
147–150	February 1853a
151–154	April 1853b
155–162	May 1853c
163–166	June 1853d
167–170	July 1853e
171–174	October 1853f
175–176	April 1854a
177	May 1854b
178–185	June 1854c
186–189	July 1854d
190–193	August 1854e
194–195	September 1854f
196–210	December 1854g

Reeve, L.A. 1855–1856. *Conchologia Iconica: or, illustrations of the shells of molluscous animals. Volume IX. Containing monographs of the genera* Spondylus. Neritina. Natica. Navicella. Siphonaria. Nerita. Latia. L. Reeve, London.

See Reeve (1848–1850) above for details on the publication methods of this series. Dates of publication as given on the bottom of the explanatory text for each plate are as follows for *Neritina*:

Plates	Date of publication
2	August 1855a
3–10	October 1855b
11–26	November 1855c
27–30	January 1856a
31–34	February 1856b
1, 35–37	March 1856c

Reeve, L.A. & Sowerby, G.B. 1870–1872. *Conchologia Iconica: or, illustrations of the shells of molluscous animals. Volume XVIII. Containing monographs of the genera* Philine. Bullina. Nucula. Utriculus. Ostraea. Linteria. Scaphander, Pholas. Yoldia. Laeda. Placuna. Etheria. Mallera. Solenella. Neilo. Pholadomya. Succinea. Magilus. Clavagella. Limnaea. Lima. Dentalium. Corbis. L. Reeve & Co., London.

> See Reeve (1848–1850) above for details on the publication methods of this series. Reeve died in 1865. According to Woodward (1913: 1663), Sowerby continued the series beginning with the genus *Pyramidella* in volume 15, although his name only appears for the first time on the title page of volume 18. Dates of publication as given on the bottom of the explanatory text for each plate are as follows for *Succinea*:
>
> | Plates | Date of publication |
> | --- | --- |
> | 2–8 | August 1872a |
> | 1, 9–12 | November 1872b |

Reinhardt, O. 1883. Ueber einige von Hrn. Hungerford gesammelt japanische Hyalinen. *Sitzungberichte der Gesellschaft Naturforschender Freunde zu Berlin* **1883**: 82–86. [after 22 May]

Riedel, A. 1980. *Genera Zonitidarum. Diagnosen supraspezifischer Taxa der Familie Zonitidae (Gastropoda, Sylommatophora).* W. Backhuys, Rotterdam. 197 p.

Röding, P.F. 1798. *Museum Boltenianum.* Pars Secunda. J. C. Trapp, Hamburgi [= Hamburg]. viii + 199 p.

> Authorship of this work determined by ICZN, 1956, Direction 48.

Schaufuss, L.W. 1869. *Molluscorum systema et catalogus. System und Aufzählung sämmtlicher Conchylien der Sammlung von Fr. Paetel.* Oscar Weiske, Dresden. [iii] + xiv + 119 p.

Schmeltz, J.D.E. 1865. *Catalog II der zum Verkauf stehenden Doubletten aus den naturhistorischen Expeditionen der Herren Joh. Ces. Godeffroy & Sohn in Hamburg.* [No publisher given], Hamburg. iv + 35 p. [24–31 March]

Schmeltz, J.D.E. 1869. *Museum Godeffroy. Catalog IV, nebst einer Beilage, enthaltend: topographische Notizen; Beschreibung neuer Bryozoen von Senator Dr. Kirchenpauer zu Hamburg und einer neuen Asteriden-Gattung von Dr. Chr. Lütken zu Kopenhagen.* Wilhelm Mauke Söhne, vormals Perthes-Besser & Mauke, Hamburg. xxxix + [iii] + 139 + [ii] p. [after 18 May]

Shuttleworth, R.J. 1854. Beiträge zur näheren Kenntniss der Land- und Süsswasser-Mollusken der Insel Portorico. *Mitteilungen der Naturforschenden Gesellschaft in Bern* **56**: 33–56.

Smith, B.J. 1992. Non-marine Mollusca. In: Houston, W.W.K., ed., *Zoological Catalogue of Australia.* Vol. 8. Australian Government Publishing Service, Canberra. xi + 399 p.

Solem, A. 1959. Systematics of the land and fresh-water Mollusca of the New Hebrides. *Fieldiana Zoology* **43**: 1–359. [19 October]

Solem, A. [1974]. Island size and species diversity in Pacific island land snails. *Malacologia* **14**(1–2)[1973]: 397–400. [23 January]

Solem, A. 1976. *Endodontoid land snails from Pacific Islands (Mollusca: Pulmonata: Sigmurethra). Part I. Family Endodontidae.* Field Museum of Natural History, Chicago. xii + 508 p. [29 October]

Solem, A. 1977. Fossil endodontid land snails from Midway Atoll. *Journal of Paleontology* **51**(5): 902–11. [September]

Solem, A. 1978. Land snails from Mothe, Lakemba, and Karoni Islands, Lau Archipelago, Fiji. *Pacific Science* **32**(1): 39–45. [29 September]

Solem, A. 1979. Biogeographic significance of land snails, Paleozoic to Recent, p. 277–87. In: Gray, J. & Boucot, A.J., eds., *Historical biogeography, plate tectonics, and the changing environment.* Oregon State University Press, Corvallis.

Solem, A. 1981. Land-snail biogeography: a true snail's pace of change, p. 197–237. *In:* Nelson, G. & Rosen, D.E. (eds.), *Vicariance biogeography.* Columbia University Press, New York.

Solem, A. 1983. *Endodontoid land snails from Pacific Islands (Mollusca: Pulmonata: Sigmurethra). Part II. Families Punctidae and Charopidae, Zoogeography.* Field Museum of Natural History, Chicago. ix + 336 p. [7 January]

Solem, A. [1989]. Non-camaenid land snails of the Kimberley and Northern Territory, Australia. I. Systematics, affinities and ranges. *Invertebrate Taxonomy* 2(4)[1988]: 455–604. [22 March]

Solem, A. 1990. How many Hawaiian land snail species are left? and what we can do for them. *Bishop Museum Occasional Papers* 30: 27–40. [4 June]

Solem, A. 1991. Distribution and diversity patterns of Australian pupilloid land snails (Mollusca: Pulmonata: Pupillidae, s.l.). *The Veliger* 34(3): 233–52. [1 July]

Solem, A. & Yochelson, E.L. 1979. North American Paleozoic land snails, with a summary of other non-marine snails. *Professional Papers of the United States Geological Survey* 1072, 42 p., 10 pls.

Souleyet, F.L.A. 1842. Description de quelques coquilles terrestres appartenant aux genres *Cyclostome, Hélice,* etc. *Revue Zoologique, par la Société Cuvierienne* 1842: 101–02. [3 April]

Souleyet, F.L.A. 1852. *Voyage autour du monde exécuté pendent les années 1836 et 1837 sur la corvette* La Bonite, *commandée par M. Vaillant. Zoologie. Tome deuxième.* A. Bertrand, Paris. 664 p.

Sowerby, G.B. [1st of the name] 1825. *A catalogue of the shells contained in the collection of the late Earl of Tankerville, arranged according to the Lamarckian conchological system; together with an appendix, containing descriptions of many new species.* G.B. Sowerby, London. viii + 92 + xxxiv p., 9 pls.

Sowerby, G.B. [2nd of the name] 1849. Monograph of the genus *Neritina. Thesaurus Conchyliorum* 2(10): 507–46, pls. 109–16.

Starmühlner, F. 1976. Beiträge zur Kenntnis der Süßwasser-Gastropoden pazifischer Inseln. *Annalen des Naturhistorischen Museums in Wien* 80: 473–656, pls. 1–21. [November]

Steenberg, C.M. 1925. *Études sur l'anatomie et la systématique des maillots (Fam. Pupillidae s. lat.).* C.A. Reitzel, Copenhague [= Copenhagen]. viii + 211 + iii p., 34 pls.

Swainson, W. 1828. The characters of *Achatinella,* a new group of terrestrial shells, with descriptions of six species. *Quarterly Journal of Science, Literature, and Art* 1828: 81–86. [January–June]

Sykes, E.R. 1894. Note on the value of *Laimodonta. Journal of Malacology* 3(4): 73–74. [12 December]

Sykes, E.R. 1896. Preliminary diagnoses of new species of non-marine Mollusca from the Hawaiian Islands. Part I. *Proceedings of the Malacological Society of London* 2(3): 126–32. [31 October]

Sykes, E.R. 1897. Preliminary diagnoses of new species of non-marine Mollusca from the Hawaiian Islands. Part II. *Proceedings of the Malacological Society of London* 2(6): 298–99. [27 November]

Sykes, E.R. 1899. Illustrations of, with notes on, some Hawaiian non-marine Mollusca. *Proceedings of the Malacological Society of London* 3(5): 275–76, pls. 13, 14. [15 July]

Sykes, E.R. 1900. *Fauna Hawaiiensis.* Volume II, Part IV, Mollusca. p. 271–412, pls. 11, 12. University Press, Cambridge. [19 May]

Sykes, E.R. 1903. Malacological notes. *Journal of Malacology* 10(1): 1–3. [31 March]

Sykes, E.R. 1904a. The Hawaiian species of *Opeas. Proceedings of the Malacological Society of London* 6(2): 112–13. [23 June]

Sykes, E.R. 1904b. On a new species of *Amastra* from the Hawaiian Islands. *Annals and Magazine of Natural History* (7) **14**: 159–60. [August]

Sykes, E.R. 1909. *Carelia pilsbryi*, n. sp., from the Hawaiian Islands. *Proceedings of the Malacological Society of London* **8**(4): 204. [7 May]

Thiele, J. 1931. *Handbuch der systematischen Weichtierkunde.* Vol. 1, part 2. G. Fischer, Jena. P. 377–778. [31 October]

Thompson, F.G. & Huck, E.L. 1985. The land snail family Hydrocenidae in Vanuatu (New Hebrides Islands), and comments on other Pacific island species. *The Nautilus* **99**(2–3): 81–84. [29 April]

Thwing, E.W. 1907. Reprint of the original descriptions of the genus *Achatinella* with additional notes. *Occasional Papers of the Bernice P. Bishop Museum* **3**(1): 1–196, pls. 1–3.

Tillier, S. 1989. Comparative morphology, phylogeny and classification of land snails and slugs (Gastropoda: Pulmonata: Stylommatophora). *Malacologia* **30**(1–2): 1–303. [1 August]

Tillier, S. & Clarke, B.C. 1983. Lutte biologique et destruction du patrimoine génétique: le cas des mollusques gastéropodes pulmonés dans les territoires français du Pacifique. *Génétique Sélection Évolution* **15**: 559–566.

Tryon, G.W., Jr. 1865. [Reviews]. Observations on the terrestrial Pulmonifera of Maine, including a catalogue of all the species of terrestrial and fluviatile Mollusca, known to inhabit the State: by Ed. S. Morse. *American Journal of Conchology* **1**: 71–73. [15 April]

Tryon, G.W., Jr. 1886. *Manual of Conchology; structural and systematic. With illustrations of the species. Second series: Pulmonata.* Vol. II. Zonitidae. Academy of Natural Sciences, Philadelphia. 265 p., 64 pls.

Published in parts as follows:

Part	Pages	Date of publication
5	1–64	23 January 1886a
6	65–128	3 May 1886b
7	129–192	28 July 1886c
8	193–265	24 November 1886d

Tryon, G.W., Jr. 1888. *Manual of Conchology; structural and systematic. With illustrations of the species.* [First series]. Vol. X. Neritidae, Adeboriidae, Cyclostrematidae, Liotidae, by Geo. W. Tryon Jr. Phasianellinae, Turbinidae, Delphinulinae. By Henry A. Pilsbry. Academy of Natural Sciences, Philadelphia. 323 p., 69 pls.

The series was begun by Tryon. Pilsbry continued it after Tryon's death. Published in parts as follows:

Part	Pages	Plates	Date of publication
37	1–64	1–12	16 March 1888a
38	65–144	13–30	1 July 1888b
39	145–208	31–45	1 October 1888c
40	209–323	46–69	3 January 1889

Pages 3–160 are by Tryon. Pages 161–323 are by Pilsbry.

Turner, R.D. 1956a. The eastern Pacific marine mollusks described by C.B. Adams. *Occasional Papers on Mollusks* **2**(20): 21–135. [22 September]

Turner, R.D. 1956b. Additions to the western Atlantic marine mollusks described by C.B. Adams. *Occasional Papers on Mollusks* **2**(20): 136. [22 September]

United States Fish and Wildlife Service. 1993. Recovery Plan for the Oʻahu Tree Snails of the genus *Achatinella*. United States Fish and Wildlife Service, Portland, Oregon. 64 p. + appendices [p. A1–A64].

Vaught, K.C. 1989. *A classification of the living Mollusca.* American Malacologists Inc., Melbourne, Florida. xii + 195 p.

Wagner, A.J. 1905. Helicinenstudien. *Denkschriften der Kaiserlichen Akademie der Wissenschaften, Wien. Mathematisch-Naturwissenschaftliche Klasse* **77**: 357–450, pls. 1–9. [before 25 May]

Wagner, A.J. 1907–1911. Die Familie der Helicinidae. *In*: Küster, H.C., ed., *Systematisches Conchylien-Cabinet von Martini und Chemnitz. Neu herausgegeben und vervollständigt.* Band I. Abtheilung XVIII. Theil 1. Neue Folge. Bauer & Raspe, Nürnberg [= Nuremberg]. 391 p., 70 pls.

Published in Lieferungen. The collation for this section is as follows:

Lieferung	Pages	Plates	Date of publication
518	1–32	1–6	1907a
522	33–72	7–12	1907b
526	73–104	13–18	1908a
530	105–136	19–24	1908b
534	137–160	25–30	1908c
535	161–184	31–36	1909a
538	185–216	37–42	1909b
541	217–248	43–48	February 1910a
542	249–272	49–54	May 1910b
543	273–296	55–60	August 1910c
544	297–328	61–66	October 1910d
547	329–391	67–70	May 1911

Walker, B. 1903. Notes on eastern American Ancyli. *The Nautilus* **17**(2): 13–19, pl. 1. [11 June]

Warén, A. & Gittenberger, E. 1993. Case 2820. *Turbo politus* Linnaeus, 1758 (currently *Melanella polita*; Mollusca, Gastropoda): proposed conservation of usage of the specific name, so conserving the specific name of *Buccinum acicula* Müller, 1774 (currently *Cecilioides acicula*). *Bulletin of Zoological Nomenclature* **50**(2): 107–11. [30 June]

Waterhouse, F.H. 1937. List of the dates of delivery of the sheets of the 'Proceedings' of the Zoological Society of London, from commencement in 1830 to 1859 inclusive. *Proceedings of the Zoological Society of London* (A) **1937**: 78–83. [15 April]

Welch, d'A.A. 1938. Distribution and variation of *Achatinella mustelina* Mighels in the Waianae Mountains, Oahu. *Bernice P. Bishop Museum Bulletin* **152**, 164 p., 13 pls. [1 September]

Welch, d'A.A. 1942. Distribution and variation of the Hawaiian tree snail *Achatinella apexfulva* Dixon in the Koolau Range, Oahu. *Smithsonian Miscellaneous Collections* **103**(1), 236 p., 12 pls. [16 December]

Welch, d'A.A. 1954. Distribution and variation of the Hawaiian tree snail *Achatinella bulimoides* Swainson on the leeward and northern slopes of the Koolau Range, Oahu. *Proceedings of the Academy of Natural Sciences of Philadelphia* **106**: 63–107, pls. 1–2. [20 August]

Welch, d'A.A. 1958. Distribution and variation of the Hawaiian tree snail *Achatinella bulimoides* Swainson on the windward slope of the Koolau Range, Oahu. *Proceedings of the Academy of Natural Sciences of Philadelphia* **110**: 123–211, pls. 10–14. [22 August]

Wenz, W. 1938–1944. *Handbuch der Paläozoologie.* Band 6 Gastropoda. Teil 1: Allgemeiner Teil und Prosobranchia. Gebrüder Borntraeger, Berlin.

Published in Lieferungen as follows:

Lieferung	Pages	Date of publication
1	1–240	March 1938a
2	241–480	October 1938b
3	481–720	July 1939
4	721–960	Augusu 1940
5	961–1200	October 1941
6	1201–1506	October 1943
7	1507–1639, i–xii	November 1944

Westerlund, C.A. 1878. *Fauna Europaea Molluscorum Extramarinorum Prodromus. Fasciculus II.* Berlingiana, Lundae [= Lund]. P. 161–320.
Whittemore, T.J. 1859. [Description of a new species of *Helix*, from Maine]. *Proceedings of the Boston Society for Natural History* **7**: 28–29. [April]
Wiktor, A. 1987. Milacidae (Gastropoda, Pulmonata) — systematic monograph. *Annales Zoologici* (Warsaw) **41**(3): 153–319. [31 December]
Wollaston, T.V. 1878. *Testacea Atlantica.* L. Reeve & Co., London. xi + [i] + 588 p. [14 October]
Wood, W. 1828. *Supplement to the Index Testaceologicus; or a catalogue of shells, British and foreign.* W. Wood, London. iv + [i] + 59 p., 8 pls. [March]
Wright, S. 1978. *Evolution and the genetics of populations. A treatise in four volumes. Volume 4. Variability within and among natural populations.* University of Chicago Press, Chicago and London. [x] + 580 p.
Zilch, A. 1959–1960. *Handbuch der Paläozoologie.* Band 6 Gastropoda. Teil 2: Euthyneura. Gebrüder Borntraeger, Berlin.

    Published in Lieferungen as follows:

Lieferung	Pages	Date of publication
1	1–200	17 July 1959a
2	201–400	25 November 1959b
3	401–600	30 March 1960a
4	601–835, i–xii	15 August 1960b

Zimmerman, E.C. 1948. *Insects of Hawaii.* Volume 1. Introduction. University of Hawaii Press, Honolulu. xx + 206 p. [27 July]

INDEX OF LISTED TAXA

Taxa treated in the catalog are listed here in alphabetical order by name, author, and (for species-group names) current generic combination. Original generic combination for species-group names, if different from the current combination, is listed in parentheses. Family-group names are in **BOLDFACE** capitalized letters. Genus-group names are in all CAPITALIZED letters. Unavailable names, nomina nuda, and misidentifications are listed in *italics*. Pagination in boldface refers to entries in the catalog proper; those in plain Roman type refer to listings of taxa in the checklist.

AA Baker, **157**, 202
abavus Hyatt & Pilsbry, Amastra, **96**, 188
abberans Hyatt & Pilsbry, Amastra, **90**, 186
abbreviata Ancey, Tornatellaria (Tornatellina), **81**, 184
abbreviata Reeve, Achatinella, **55**, 178
abeillei Ancey, Philonesia (Microcystis), **157**, 202
acanthinula Ancey, Pronesopupa (Pupa), **137**, 199
accincta Gould, Leptachatina (Achatina), **120**, 194
accineta Mighels, Leptachatina (Achatina), **120**, 194
achatinaceum Pfeiffer, Paropeas (Bulimus), **142**, 199
ACHATINELLA Swainson, **37**, **38**, 174
ACHATINELLASTRUM Pfeiffer, **48**, 176
ACHATINELLIDAE, **35**, 174
ACHATINELLINAE, **37**, 174
acicula Pilsbry & Cooke, Tornatellides, **83**, 185
acuminata Cooke, Amastra, **105**, 190
acuminata Gould, Leptachatina (Achatinella), **120**, 194
acuta Newcomb, Amastra (Achatinella), **103**, 189
acuta Swainson, Amastra (Achatinella), **108**, 191
ada Pilsbry & Cooke, Tornatellides, **83**, 185
adamsi Newcomb, Partulina (Achatinella), **68**, 181
adelinae Pilsbry & Cooke, Tornatellaria, **81**, 184
admodesta Mighels, Pronesopupa (Pupa), **137**, 199
adusta Gould, Carelia (Achatina), **111**, 192
adusta Reeve, Achatinella, **49**, 178
AEDITUANS Cooke & Kondo, **83**, 184
aemulator Hyatt & Pilsbry, Amastra, **105**, 190
affinis Newcomb, Amastra (Achatinella), **90**, 186
affinis Souleyet, Lymnaea, **32**, 173
agatha Schaufuss, Achatinella, **61**, 179
agglutinans Newcomb, Amastra (Achatinella), **100**, 189
AGRIOLIMAX Mörch, **167**, 204
aiea Hyatt & Pilsbry, Amastra, **108**, 191
alata Pfeiffer, Tropidoptera (Helix), **117**, 193
alba Jay, Achatinella, **38**, 175
alba Sykes, Achatinella, **38**, 174
albalabia Welch, Achatinella, **55**, 178
albescens Gulick, Achatinella, **49**, 177
albida Pfeiffer, Amastra (Achatinella), **108**, 191
albina Ancey, Nesophila (Charopa), **146**, 200
albipraetexta Welch, Achatinella, **38**, 174
albocincta Pilsbry & Cooke, Amastra, **90**, 186
albofasciata Smith, Achatinella (Apex), **38**, 174
albolabris Newcomb, Amastra (Achatinella), **105**, 190
albospira Smith, Achatinella (Apex), **38**, 176
alexanderi Cooke & Pilsbry, Columella (Sphyradium), **139**, 199
alexandri Cooke, Gulickia, **78**, 184
alexandri Newcomb, Laminella (Achatinella), **114**, 192
ALLOCHROA Ancey, **27**, 172
alloia Cooke & Pilsbry, Nesopupa, **135**, 198
ALLOPEAS Baker, **141**, 199
aloha Pilsbry & Cooke, Achatinella, **38**, 174
alpha Pilsbry & Cooke, Pleuropoma (Helicina), **20**, 171
altiformis Welch, Achatinella, **38**, 176
AMALIA Moquin-Tandon, **166**, 204
AMASTRA Adams & Adams, **90**, 186
AMASTRELLA Sykes, **96**, 188
AMASTRIDAE, **89**, 186
AMASTRINAE, **90**, 186

ambigua Pease, Lymnaea, **32**, 173
ambusta Pease, Auriculella, **75**, 183
amicta Smith, Amastra, **110**, 191
amoena Pfeiffer, Auriculella (Achatinella), **75**, 183
ampla Newcomb, Achatinella, **49**, 177
ampulla Gulick, Partulina (Achatinella), **68**, 182
anacardiensis Paetel, Achatinella, **61**, 179
anaglypta Cooke, Amastra, **105**, 190
analoga Gulick, Achatinella, **49**, 178
anceophila Cooke, Carelia, **111**, 191
anceyana Baldwin, Partulina (Achatinella), **66**, 180
anceyana Baldwin, Partulina (Achatinella), **66**, 180
anceyana Cooke, Leptachatina, **118**, 194
anceyana Cooke & Pilsbry, Lyropupa, **132**, 197
anceyana Cooke & Pilsbry, Nesopupa, **134**, 198
anceyana Cooke & Pilsbry, Tornatellaria, **82**, 184
anceyanum Pilsby & Cooke, Elasmias, **87**, 186
ANCYLIDAE, **34**, 173
angulata Pease, Carelia, **111**, 191
ANGULIDENS Pilsbry & Cooke, **118**, 194
angusta Cooke & Pilsbry, Nesopupa, **135**, 198
angusta Paetel, Achatinella, **49**, 177
anisoplax Pilsbry & Cooke, Tornatellides, **83**, 185
annosa Cooke, Amastra, **96**, 188
anthonii Newcomb, Amastra (Achatinella), **96**, 188
antiqua Baldwin, Amastra (Achatinella), **100**, 189
antiqua Baldwin, Amastra (Achatinella), **100**, 189
antiqua Cooke & Pilsbry, Lyropupa, **132**, 197
antiqua Pease, Leptachatina, **120**, 194
antiqua Solem, Cookeconcha, **143**, 199
antiquata, Leptachatina, **120**
antoni Pfeiffer, Helicina, **23**, 172
aperta Lea, Succinea, **152**, 201
APEX Martens, **37**, **38**, 174
apexfulva Dixon, Achatinella (Helix), **38**, 174
APHANOCONIA Wagner, **19**, 171
apicalis Ancey, Succinea, **150**, 201
apicalis Baldwin, Succinea, **150**, 201
apicata Pfeiffer, Achatinella, **38**, 174
apiculata Ancey, Endodonta, **145**, 200
aplustre Newcomb, Achatinella, **49**, 177
approximans Ancey, Leptachatina, **120**, 194
approximata Sowerby, Succinea, **150**, 201
aptycha Pfeiffer, Partulina (Achatinella), **68**, 181
arborea Sykes, Leptachatina, **120**, 194
arenarum Pilsbry & Cooke, Amastra, **100**, 189
arenorum Baker, Philonesia, **158**, 203
ARMIELLA Hyatt & Pilsbry, **100**, 189
armillata Cooke, Amastra, **100**, 189
ARMSIA Pilsbry, **118**, 194
arnemani Welch, Achatinella, **55**, 178
arnemanni Cooke & Kondo, Partulina, **68**, 181
artata Cooke, Pauahia (Leptachatina), **129**, 197
ascendens Baker, Philonesia, **159**, 203
aspera Baldwin, Laminella, **114**, 192
assimilis Newcomb, Amastra (Achatinella), **90**, 186
atra Jay, Neritina, **15**, 170
atroflava Hyatt & Pilsbry, Amastra, **90**, 187
attenuata Cooke, Leptachatina, **120**, 194
attenuata Pfeiffer, Partulina (Achatinella), **68**, 182
attenuatus Cooke & Pilsbry, Tornatellides, **83**, 185

INDEX [237]

augusta Smith, Achatinella, **49**, 177
aulacospira Ancey, Erinna (Limnea), **31**, 173
aurantium Pilsbry & Cooke, Laminella, **114**, 193
aureola Welch, Achatinella, **39**, 174
AURICULA Lamarck, **28**, 172
auricula Férussac, Auriculella (Partula), **74**, 183
AURICULASTRA Martens, **28**, 172
AURICULELLA Pfeiffer, **74**, 183
AURICULELLINAE, **74**, 183
aurora Pilsbry & Cooke, Amastra, **100**, 189
aurostoma Baldwin, Amastra, **90**, 186
aurulenta Ancey, Succinea, **150**, 201
auwahiensis Pilsbry & Cooke, Amastra, **103**, 190
avus Pilsbry & Cooke, Leptachatina, **120**, 195
azona Ancey, Carelia, **111**, 192

babori Collinge, Milax (Amalia), **166**, 204
bacca Pease, Nesopupa (Vertigo), **136**, 198
bacca Reeve, Achatinella, **55**, 178
badia Baldwin, Amastra (Achatinella), **105**, 190
baetica Jay, Amastra (Achatinella), **108**, 191
baileyana Gulick, Partulina (Achatinella), **68**, 182
bakeri Welch, Achatinella, **39**, 174
baldwini Ancey, Catinella (Succinea), **148**, 200
baldwini Ancey, Cecilioides (Caecilianella), **140**, 199
baldwini Ancey, Nesophila (Charopa), **146**, 200
baldwini Ancey, Nesopupa, **137**, 198
baldwini Ancey, Nesovitrea (Hyalinia), **165**, 204
baldwini Ancey, Orobophana (Helicina), **17**, 170
baldwini Ancey, Pacificella (Tornatellina), **81**, 184
baldwini Ancey, Philonesia (Microcystis), **159**, 203
baldwini Ancey, Thiara (Melania), **26**, 172
baldwini Cooke, Carelia, **111**, 192
baldwini Cooke, Leptachatina, **120**, 194
BALDWINIA Ancey, **64**, 180
baldwiniana Cooke, Lyropupa, **131**, 197
baldwiniana Cooke & Pilsbry, Tornatellaria, **82**, 184
baldwiniana Hyatt & Pilsbry, Amastra, **90**, 186
baldwini Newcomb, Amastra (Achatinella), **90**, 187
balteata Hyatt & Pilsbry, Amastra, **91**, 187
balteata Pease, Leptachatina, **120**, 184
balteata Pilsbry & Cooke, Perdicella (Partulina), **73**, 182
beata Pilsbry & Cooke, Achatinella, **39**, 174
bella Reeve, Partulina (Achatinella), **66**, 181
bellula Smith, Achatinella, **49**, 176
bellus Cooke & Pilsbry, Tornatellides, **83**, 185
bembicodes Cooke, Amastra, **101**, 189
bensonia Schaufuss, Achatinella, **61**, 179
berniceia Pilsbry & Cooke, Pleuropoma (Helicina), **20**, 171
beta Pilsbry & Cooke, Orobophana (Helicina), **17**, 171
bevenoti Collinge, Deroceras (Agriolimax), **167**, 204
bicolor Jay, Carelia (Achatina), **111**, 192
bicolor Pfeiffer, Achatinella, **39**, 176
bicolorata Ancey, Succinea, **150**, 201
bigener Hyatt, Amastra, **91**, 186
bilineata Reeve, Achatinella, **49**, 177
binaria Pilsbry, Endodonta (Helix), **145**, 200
binominis Sykes, Lymnaea, **32**, 173
biplicata Newcomb, Amastra (Achatinella), **91**, 186
bishopi Cooke & Pilsbry, Nesopupa, **134**, 198
BLAUNERIA Shuttleworth, **30**, 173
boettgeri Cooke & Pilsbry, Pronesopupa, **137**, 199
boettgeriana Ancey, Philonesia (Microcystis), **159**, 203
borcherdingi Hyatt & Pilsbry, Amastra, **91**, 186
borealis Neal, Orobophana, **17**, 171
breviana Baldwin, Amastra (Achatinella), **110**, 191
breviata Baldwin, Amastra, **105**, 190
brevicula Pease, Leptachatina (Helicter), **120**, 194
brevis Cooke, Leptachatina, **121**, 196
brevis Pfeiffer, Amastra (Achatinella), **96**, 188
bronniana Philippi, Pleuropoma (Helicina), **20**, 171

bronnii Philippi, Allochroa (Auricula), **27**, 172
brownii Adams & Adams, Allochroa (Ellobium), **27**, 172
bruneola Welch, Achatinella, **39**, 174
brunibasis Welch, Achatinella, **39**, 176
brunicolor Welch, Achatinella, **39**, 176
brunnea Smith, Auriculella, **75**, 183
brunneus Cooke & Pilsbry, Tornatellides, **83**, 185
brunosa Welch, Achatinella, **39**, 174
bryani Cooke & Pilsbry, Tornatellides, **83**, 185
bryani Neal, Orobophana, **17**, 171
bryani Pilsbry & Cooke, Amastra, **105**, 191
buddii Newcomb, Achatinella, **49**, 176
buena Welch, Achatinella, **39**, 174
bulbosa Gulick, Laminella (Achatinella), **114**, 193
BULIMELLA Pfeiffer, **55**, 178
bulimoides Swainson, Achatinella, **55**, 178
byronensis Beck, Achatinella (Helicteres), **55**, 178
byronii Wood, Achatinella (Helix), **55**, 178

cacuminis Pilsbry & Cooke, Auriculella, **75**, 183
caduca Mighels, Succinea, **150**, 201
CAECILIANELLA Bourguignat, **140**, 199
caesia Gulick, Achatinella, **49**, 176
caesiapicta Welch, Achatinella, **55**, 178
callosa Pfeiffer, Leptachatina (Achatinella), **119**, 194
canaliculata Baldwin, Newcombia (Achatinella), **62**, 180
canaliculata Baldwin, Newcombia (Achatinella), **62**, 180
canalifera Ancey, Auriculella, **75**, 183
candida Pfeiffer, Achatinella, **55**, 178
candida Pilsbry & Cooke, Partulina, **64**, 180
canella Gould, Succinea, **150**, 201
canyonensis Neal, Pleuropoma, **20**, 171
capax Pilsbry & Cooke, Achatinella, **55**, 179
caperata Gould, Godwinia (Vitrina), **164**, 204
capillata Pease, Nesophila (Helix), **146**, 200
captiosa Cooke, Leptachatina, **121**, 194
caputadamantis Hyatt & Pilsbry, Amastra, **105**, 190
carbonaria Ancey, Lyropupa, **131**, 197
CARELIA Adams & Adams, **111**, 191
carinata Gulick, Amastra, **101**, 189
CARINELLA Pfeiffer, **104**, 190
carinella Baldwin, Perdicella (Newcombia), **73**, 182
carnicolor Baldwin, Partulina, **68**, 181
CASSIDULINAE, **27**, 172
casta Ancey, Succinea, **150**, 201
casta Newcomb, Achatinella, **49**, 176
castanea Hyatt & Pilsbry, Amastra, **96**, 188
castanea Pfeiffer, Auriculella (Tornatellina), **75**, 183
castanea Reeve, Achatinella, **49**, 178
castaneus Megerle von Mühlfeld, Melampus (Voluta), **29**, 172
CATINELLA Pease, **148**, 200
CATINELLINAE, **148**, 200
CECILIOIDES Férussac, **140**, 199
centralis Ancey, Nesopupa, **137**, 198
cepulla Gould, Succinea, **150**, 201
cerea Pfeiffer, Auriculella (Achatinella), **75**, 183
cerealis Gould, Leptachatina (Achatinella), **121**, 194
cervina Gulick, Achatinella, **49**, 176
cervixnivea Pilsbry & Cooke, Achatinella, **39**, 174
cestus Newcomb, Achatinella, **39**, 174
chamissoi Pfeiffer, Auriculella (Achatinella), **75**, 183
chamissoi Pfeiffer, Philonesia (Helix), **158**, 203
CHETOSYNA Baker, **154**, 202
chlorotica Pfeiffer, Amastra (Achatinella), **108**, 191
christopherseni Welch, Achatinella, **39**, 174
chromatacme Pilsbry & Cooke, Achatinella, **39**, 174
chrysallis Pfeiffer, Leptachatina (Achatina), **129**, 197
cicercula Gould, Philonesia (Helix), **159**, 203
cincta Ancey, Tornatellaria (Tornatellina), **82**, 184
cinderella Hyatt, Amastra, **91**, 186
cinerea Sykes, Achatinella, **39**, 174

cinerosa Pfeiffer, Achatinella, **40**, 174
cingula Mighels, Leptachatina (Achatinella), **121**, 194
cingulata Paetel, Achatinella, **61**, 179
cinnamomea Ancey, Succinea, **150**, 201
cinnamomea Pfeiffer, Newcombia (Achatinella), **62**, 180
circulata Paetel, Achatinella, **61**, 179
circulospadix Welch, Achatinella, **56**, 178
circumcincta Hyatt & Pilsbry, Laminella, **114**, 193
citrea Sykes, Amastra, **91**, 187
citrina Pfeiffer, Laminella (Achatinella), **114**, 193
clara Pfeiffer, Leptachatina (Achatinella), **121**, 196
clathratula Ancey, Lyropupa, **131**, 197
clausa Adams & Adams, Laemodonta (Plecotrema), **29**, 173
clausinus Mighels, Bulimus, **87**, 186
clementina Pfeiffer, Achatinella, **56**, 178
coccoglypta Ancey, Succinea, **150**, 201
cochlea Reeve, Carelia (Achatina), **111**, 192
cognata Gulick, Achatinella, **49**, 176
collaris Welch, Achatinella, **40**, 176
colorata Reeve, Achatinella, **50**, 177
COLUMELLA Westerlund, **139**, 199
columna Ancey, Pauahia (Leptachatina), **129**, 197
comes Pilsbry & Cooke, Tornatellides, **83**, 185
compacta Pease, Leptachatina (Helicter), **121**, 194
compacta Pease, Lymnaea, **32**, 173
compactus Sykes, Tornatellides (Tornatellina), **84**, 185
compressa Paetel, Achatinella, **61**, 179
compta Pease, Partulina, **68**, 181
concavospira Pfeiffer, Achatinella, **40**, 175
concentrata Pilsbry & Vanatta, Endodonta, **145**, 200
concidens Gulick, Achatinella, **50**, 176
concinna Newcomb, Laminella (Achatinella), **114**, 193
concolor Cooke, Leptachatina, **121**, 194
concolor Martens, Laminella (Achatinella), **114**, 193
concolor Smith, Achatinella, **50**, 176
concomitans Hyatt, Partulina, **68**, 181
confusa Sykes, Partulina (Achatinella), **64**, 180
confusus Sykes, Tornatellides (Tornatellina), **84**, 185
conica Baldwin, Amastra, **96**, 188
conica Pease, Allochroa (Laimodonta), **27**, 172
conicoides Sykes, Leptachatina, **121**, 194
conicospira Smith, Amastra, **108**, 191
conifera Smith, Amastra, **91**, 186
coniformis Gulick, Achatinella (Apex), **40**, 174
consanguinea Smith, Achatinella, **50**, 177
conspersa Pfeiffer, Amastra (Achatinella), **105**, 190
conspicienda Cooke, Leptachatina, **121**, 194
constricta Pfeiffer, Helicina, **23**, 172
contigua Pease, Thiara (Melania), **26**, 172
contorta Férussac, Cookeconcha (Helix), **143**, 199
contracta Gulick, Achatinella, **50**, 177
CONULUS Fitzinger, **153**
convexior Pilsbry & Cooke, Tornatellaria, **82**, 184
convexiuscula Sykes, Leptachatina, **121**, 194
COOKECONCHA Solem, **143**, 199
cookei Baldwin, Achatinella, **40**, 174
cookei Cockerell, Cookeconcha (Endodonta), **143**, 199
cookei Hyatt & Pilsbry, Amastra, **105**, 191
cookei Hyatt & Pilsbry, Tropidoptera (Pterodiscus), **117**, 193
cookei Neal, Orobophana, **17**, 171
cookei Pilsbry, Georissa, **16**, 170
cookei Pilsbry, Leptachatina, **118**, 194
cookii Baldwin, Achatinella, **40**, 174
cooperi Baldwin, Partulina, **66**, 181
cornea Newcomb, Amastra (Achatinella), **105**, 190
corneiformis Hyatt & Pilsbry, Amastra, **96**, 188
corneola Pfeiffer, Leptachatina (Achatinella), **121**, 194
corrugata Gulick, Achatinella, **56**, 179
corticicola Cooke & Pilsbry, Pronesopupa, **138**, 199
corusca Pfeiffer, Partulina (Achatinella), **69**, 182
corusca Pilsbry & Cooke, Partulina, **68**, 182
coruscans Hartman, Leptachatina (Achatinella), **121**, 194

costata Borcherding, Newcombia, **62**, 180
costata Pease, Lyropupa (Vertigo), **133**, 198
costulata Gulick, Leptachatina (Achatinella), **121**, 194
costulosa Pease, Leptachatina, **121**, 194
costulosa Pease, Nesopupa (Vertigo), **135**, 198
crassa Ancey, Succinea, **150**, 201
crassa Newcomb, Partulina (Achatinella), **69**, 181
crassidentata Pfeiffer, Achatinella, **50**, 177
crassilabris Philippi, Helicina, **23**, 172
crassilabrum Newcomb, Amastra (Achatinella), **105**, 190
crassula Smith, Auriculella, **75**, 183
crocea Gulick, Partulina (Achatinella), **69**, 181
cryptoportica Gould, Philonesia (Helix), **159**, 203
crystallina Gulick, Leptachatina (Achatinella), **121**, 194
cubana Dall, Lyropupa (Vertigo), **133**, 198
cucumis Gulick, Achatinella, **50**, 178
cumingi Newcomb, Newcombia (Achatinella), **62**, 180
cumingiana Pfeiffer, Carelia (Spiraxis), **111**, 192
cuneata Cooke, Leptachatina, **121**, 195
cuneus Pfeiffer, Achatinella, **50**, 176
curta Newcomb, Achatinella, **50**, 177
CYCLAMASTRA Pilsbry & Vanatta, **100**, 189
cyclostoma Baldwin, Amastra (Achatinella), **101**, 189
cyclostoma Baldwin, Amastra (Achatinella), **101**, 189
cylindrata Pease, Leptachatina (Helicter), **121**, 195
cylindrata Pilsbry & Cooke, Lyropupa, **133**, 198
cylindrella Cooke, Leptachatina, **121**, 196
cylindrica Newcomb, Amastra (Achatinella), **108**, 191
cylindrica Sykes, Lamellidea (Tornatellina), **79**, 184
cyphostyla Ancey, Tornatellides (Tornatellina), **84**, 185
cyrta Cooke & Pilsbry, Lyropupa, **133**, 198

dautzenbergi Welch, Achatinella, **40**, 176
davisiana Cooke, Amastra, **105**, 190
debilis Pilsbry & Cooke, Amastra, **101**, 189
decepta Adams, Amastra (Achatinella), **108**, 191
decepta Baker, Philonesia, **159**, 203
deceptor Cockerell, Leptachatina, **121**, 195
decipiens Newcomb, Achatinella, **56**, 179
decolor Welch, Achatinella, **40**, 176
decora Férussac, Achatinella (Helix), **40**, 175
decorata Pilsbry & Cooke, Newcombia, **62**, 180
decorticata Gulick, Amastra, **96**, 188
decussatula Pease, Cookeconcha (Helix), **143**, 199
defuncta Cooke, Leptachatina, **121**, 195
delicata Ancey, Succinea, **150**, 201
delicata Cooke, Amastra, **101**, 189
delta Gulick, Achatinella, **50**, 177
delta Pilsbry & Cooke, Pleuropoma (Helicina), **21**, 171
densilineata Reeve, Partulina (Partula), **69**, 182
dentata Pease, Lamellidea (Tornatellina), **79**, 184
dentata Pfeiffer, Leptachatina (Achatinella), **119**, 194
depicta Baldwin, Laminella (Achatinella), **114**, 193
depicta Baldwin, Laminella (Achatinella), **115**, 193
depressula Ancey, Philonesia (Microcystis), **159**, 203
DEROCERAS Rafinesque, **167**, 204
deshaysii Morelet, Amastra (Achatinella), **91**, 186
dextra Schmeltz, Partulina (Achatinella), **69**, 182
dextroversa Pilsbry & Cooke, Achatinella, **56**, 179
diaphana Smith, Auriculella, **75**, 183
dichroma Cooke, Amastra, **106**, 190
diducta Baker, Philonesia, **157**, 202
diffusa Welch, Achatinella, **40**, 176
digonophora Ancey, Planamastra (Patula), **117**, 193
diluta Smith, Achatinella, **50**, 178
dimidiata Pfeiffer, Leptachatina (Achatinella), **121**, 195
dimissa Hyatt & Pilsbry, Amastra, **91**, 187
dimondi Adams, Laminella (Achatinella), **115**, 193
dimorpha Gulick, Achatinella, **50**, 177
diptyx Pilsbry & Cooke, Tornatellides, **84**, 185
disculus Pfeiffer, Hiona (Helix), **156**, 202

INDEX [239]

discus Baker, Striatura, **162**, 203
discus Pilsbry & Vanatta, Tropidoptera (Amastra), **117**, 193
disjuncta Pilsbry & Cooke, Nesopupa, **135**, 198
dispersa Cooke & Pilsbry, Nesopupa, **137**, 198
dispersa Hyatt & Pilsbry, Amastra, **106**, 190
dissimiliceps Hyatt & Pilsbry, Amastra, **91**, 187
dissimilis Cooke, Leptachatina, **121**, 194
dissotropis Ancey, Pleuropoma (Helicina), **21**, 172
distans Pease, Nesophila (Helix), **146**, 200
diversa Gulick, Achatinella, **51**, 177
dixoni Borcherding, Partulina (Achatinellastrum), **66**, 181
dolei Ancey, Carelia, **111**, 192
dolei Baldwin, Partulina (Achatinella), **69**, 181
dolium Pfeiffer, Achatinella, **40**, 175
dormitor Pilsbry & Cooke, Leptachatina, **121**, 195
drepanophorus Cooke & Pilsbry, Tornatellides, **84**, 185
dubia Newcomb, Partulina (Achatinella), **64**, 180
dubiosa Adams, Partulina (Achatinella), **69**, 182
dubitabilis Cooke & Pilsbry, Nesopupa, **134**, 198
dulcis Cooke, Leptachatina, **128**, 197
dumartroyii Souleyet, Auriculella (Partula), **75**, 183
dunkeri Pfeiffer, Achatinella, **51**, 177
duoplicata Baldwin, Laminella, **115** 192
duplocincta Pilsbry & Cooke, Achatinella, **40**, 174
durandi Ancey, Amastra, **91**, 186
dwighti Pfeiffer, Partulina (Achatinella), **69**, 181
dwightii Cooke, Amastra, **103**, 189
dwightii Newcomb, Partulina (Achatinella), **69**, 181

eburnea Gulick, Partulina (Achatinella), **69**, 182
EBURNELLA Pease, **66**, 180
EDENTULOPUPA Cooke & Pilsbry, **137**, 199
ekahanuiensis Solem, Endodonta, **145**, 200
ELAMELLIDEA Cooke & Kondo, **79**, 184
ELASMIAS Pilsbry, **87**, 186
elegans Newcomb, Achatinella, **56**, 178
elegantula Hyatt & Pilsbry, Amastra, **91**, 186
elephantina Cooke, Amastra, **101**, 189
elevata Pfeiffer, Leptachatina (Achatinella), **121**, 195
elisae Ancey, Cookeconcha (Pitys), **144**, 199
ellipsoidea Gould, Amastra (Achatinella), **106**, 191
elliptica Gulick, Amastra, **97**, 188
ELLOBIIDAE, 27, 172
ELLOBIINAE, 28, 172
ELLOBIUM Röding, **28**, 172
elongata Küster, Auriculastra (Auricula), **28**, 172
elongata Newcomb, Amastra (Achatinella), **103**, 189
elongata Pease, Succinea, **152**, 201
elongata Pfeiffer, Achatinella, **41**, 175
emerita Sykes, Leptachatina, **121**, 195
emmersonii Newcomb, Achatinella, **51**, 177
emortua Cooke, Amastra, **97**, 188
ENDODONTA Albers, **145**, 200
ENDODONTIDAE, 143, 199
ENTERODONTA Sykes, **29**, 173
eos Pilsbry & Cooke, Amastra, **106**, 190
erecta Pease, Amastra (Helicter), **108**, 191
ERINNA Adams & Adams, **31**, 173
ernestina Baldwin, Achatinella, **51**, 178
ernestina Baldwin, Achatinella, **51**, 178
errans Hyatt & Pilsbry, Amastra, **106**, 190
errans Pilsbry & Cooke, Partulina, **64**, 180
EUCONULINAE, 153, 202
EUCONULUS Reinhardt, **153**, 202
euryomphala Ancey, Tornatellides (Tornatellina), **84**, 185
evelynensis Cooke & Kondo, Carelia, **111**, 192
ewaensis Pilsbry, Tropidoptera (Pterodiscus), **118**, 193
ewaensis Welch, Achatinella, **41**, 174
exaequata Gould, Hiona (Helix), **156**, 202
exanima Neal, Orobophana, **18**, 171
exilis Gulick, Leptachatina (Achatinella), **121**, 195

exoptabilis Cooke, Leptachatina, **121**, 195
expansa Pease, Auriculella, **75**, 183
explanata Gould, Catinella (Succinea), **148**, 200
exserta Pfeiffer, Helix, **165**, 204
extensa Pease, Leptachatina, **121**, 195
extincta Ancey, Lamellidea (Tornatellina), **80**, 184
extincta Hyatt & Pilsbry, Carelia, **111**, 192
extincta Pfeiffer, Amastra (Achatinella), **101**, 189

faba Pfeiffer, Achatinella, **56**, 179
fallax Baker, Philonesia, **159**, 203
farcimen Pfeiffer, Amastra (Achatinella), **110**, 191
fasciata Gulick, Partulina (Achatinella), **69**, 182
fastigata Cooke, Amastra, **97**, 188
FERRISSIA Walker, **34**, 173
ferruginea Baldwin, Amastra (Achatinella), **110**, 191
ferruginea Neal, Pleuropoma, **21**, 171
ferussaci Pfeiffer, Laminella (Achatinella), **115**, 193
FERUSSACIIDAE, 140, 199
filicostata Cooke & Pilsbry, Lyropupa, **133**, 198
filicostata Pease, Cookeconcha (Pitys), **144**, 200
flavescens Newcomb, Amastra (Achatinella), **97**, 188
flavida Clessin, Lymnaea (Physa), **32**, 173
flavida Cooke, Auriculella, **75**, 183
flavida Gulick, Achatinella (Apex), **41**, 174
flavitincta Welch, Achatinella, **41**, 174
flemingi Baldwin, Partulina, **66**, 181
flemingi Cooke, Amastra, **103**, 189
forbesi Cooke, Amastra, **106**, 190
forbesi Cooke & Pilsbry, Nesopupa, **135**, 198
forbesi Cooke & Pilsbry, Tornatellides, **84**, 185
forbesiana Pfeiffer, Achatinella, **41**, 174
formosa Gulick, Achatinella, **51**, 176
fornicata Gould, Helix, **169**, 204
fossilis Baldwin, Amastra, **97**, 188
fossilis Cooke, Leptachatina, **118**, 194
fossilis Cooke & Pilsbry, Lyropupa, **131**, 197
fragilis Gulick, Leptachatina (Achatinella), **121**, 195
fragilis Pilsbry & Cooke, Amastra, **101**, 189
fragilis Souleyet, Succinea, **150**, 201
fragosa Cooke, Amastra, **97**, 188
fraterna Cooke, Leptachatina, **119**, 194
fraterna Sykes, Amastra, **103**, 189
FRICKELLA Pfeiffer, **74**, 183
fricki Pfeiffer, Achatinella, **56**, 178
fricki Pfeiffer, Endodonta (Helix), **145**, 200
fricki Pfeiffer, Melampus, **29**, 173
frickii, Endodonta (Helix), **145**
frit Pilsbry & Cooke, Tornatellides, **84**, 185
frondicola Cooke & Pilsbry, Pronesopupa, **138**, 199
frosti Ancey, Amastra, **108**, 191
fucosa Lyons, Partulina (Achatinella), **65**, 180
fulgens Newcomb, Achatinella, **51**, 177
fulgida Cooke, Leptachatina, **121**, 195
fulgora Gould, Helicina, **23**, 172
fulgurans Sykes, Perdicella, **73**, 182
fuliginea Pfeiffer, Carelia (Achatina), **111**, 192
fuliginosa Gould, Amastra (Achatinella), **97**, 189
fulva Pfeiffer, Partulina (Achatinella), **66**, 181
fulvicans Baldwin, Partulina, **66**, 181
fulvula Welch, Achatinella, **56**, 178
fumida Gulick, Leptachatina (Achatinella), **121**, 195
fumosa Newcomb, Leptachatina (Achatinella), **121**, 194
fumositincta Welch, Achatinella, **41**, 174
fusca Newcomb, Leptachatina (Achatinella), **119**, 194
fuscobasis Smith, Achatinella (Bulimella), **56**, 179
fuscolineata Smith, Achatinella, **51**, 177
fuscospira Pilsbry & Cooke, Partulina, **65**, 180
fuscostriata Welch, Achatinella, **41**, 174
fuscozona Smith, Achatinella, **51**, 176
fuscozonata Pilsbry & Cooke, Partulina, **65**, 180
fuscula Gulick, Leptachatina (Achatinella), **128**, 197
fuscum Ancey, Elasmias (Tornatellina), **87**, 186
fuscum Küster, Ellobium (Auricula), **28**, 172
fusiformis Pfeiffer, Amastra (Achatinella), **91**, 187

fusoidea Newcomb, Partulina (Achatinella), **69**, 181

gaetanoi Pilsbry & Vanatta, Euconulus (Kaliella), **154**, 202
gagates Draparnaud, Milax (Limax), **166**, 204
gamma Pilsbry & Cooke, Pleuropoma (Helicina), **21**, 171
garrettiana Ancey, Succinea, **150**, 201
GASTROCOPTA Wollaston, **130**, 197
GASTROCOPTINAE, **130**, 197
GASTRODONTINAE, **162**, 203
gayi Cooke, Leptachatina, **121**, 195
gayi Cooke & Pilsbry, Lamellidea (Tornatellina), **80**, 184
gemina Neal, Pleuropoma, **21**, 172
gemma Pfeiffer, Newcombia (Achatinella), **62**, 180
gentilis Cooke, Amastra, **97**, 188
georgii Pilsbry & Cooke, Amastra, **91**, 187
GEORISSA Blanford, **16**, 170
germana Newcomb, Partulina (Achatinella), **66**, 180
gibba Henshaw, Succinea, **150**, 201
gigantea Newcomb, Amastra (Achatinella), **91**, 187
gigas Lesson, Neritina, **14**, 170
glabra Newcomb, Achatinella, **56**, 178
glauca Gulick, Achatinella, **51**, 177
glaucopicta Welch, Achatinella, **41**, 174
globosa Cooke, Amastra, **101**, 189
globosa Hyatt & Pilsbry, Amastra (Achatinella), **92**, 187
globosa Pfeiffer, Achatinella, **41**, 174
globosum Collinge, Deroceras (Agriolimax), **167**, 204
globuloidea Neal, Pleuropoma, **21**, 171
glossema Cooke, Carelia, **112**, 192
glutinosa Pfeiffer, Leptachatina (Achatinella), **123**, 195
glypha Baker, Philonesia, **159**, 203
gnampta Cooke & Pilsbry, Nesopupa, **135**, 198
GODWINIA Sykes, **164**, 204
goniops Pilsbry & Cooke, Amastra, **92**, 186
goniostoma Pfeiffer, Amastra (Achatinella), **92**, 186
gouldi Newcomb, Partulina (Achatinella), **69**, 182
gouldi Pfeiffer, Partulina (Bulimus), **69**, 182
gouldi Pfeiffer, Partulina (Bulimus), **69**, 182
gouldi Pilsbry & Cooke, Lyropupa, **131**, 197
gouveiae Cooke & Pilsbry, Nesopupa, **136**, 198
gouveiana Baker, Philonesia, **157**, 202
gouveii Cooke, Amastra, **101**, 189
gracile Hutton, Allopeas (Bulimus), **141**, 199
gracilior Hyatt & Pilsbry, Laminella, **115**, 193
gracilis Férussac, Laminella (Helix), **115**, 193
gracilis Pease, Blauneria, **30**, 173
gracilis Pease, Lamellidea (Tornatellina), **80**, 184
gracilis Pfeiffer, Leptachatina (Achatinella), **123**, 195
grana Newcomb, Leptachatina (Achatinella), **123**, 195
grandis Baker, Philonesia, **160**, 203
granifera Gulick, Leptachatina (Achatinella), **123**, 194
granosa Sowerby, Neritina, **14**, 170
gravida Férussac, Laminella (Helix), **115**, 193
gravis Paetel, Achatinella, **61**, 179
grayana Pfeiffer, Amastra (Achatinella), **92**, 186
gregoryi Cooke, Amastra, **101**, 189
grisea Newcomb, Partulina (Achatinella), **65**, 180
griseibasis Welch, Achatinella, **41**, 174
griseipicta Welch, Achatinella, **41**, 176
griseitincta Welch, Achatinella, **41**, 176
griseizona Pilsbry & Cooke, Achatinella, **41**, 175
grossa Pfeiffer, Amastra (Achatinella), **109**, 191
guavarum Baker, Philonesia, **157**, 202
GULICKIA Cooke, **78**, 184
gulickiana Hyatt & Pilsbry, Amastra, **106**, 190
gulickiana Pilsbry & Cooke, Achatinella, **51**, 177
gulickii Smith, Achatinella (Apex), **42**, 174
gummea Gulick, Leptachatina (Achatinella), **123**, 195
guttula Gould, Leptachatina (Achatinella), **123**, 195
gyrans Hyatt, Amastra, **97**, 188

haena Cooke, Laxisuccinea, **149**, 200
haenensis Cockerell, Leptachatina, **123**, 195
hahakeae Baker, Philonesia, **158**, 202
halawaensis Borcherding, Partulina (Achatinella), **70**, 182
HALEAKALA Baker, **157**, 202
hanleyana Pfeiffer, Achatinella, **42**, 174
hartmani Newcomb, Amastra (Leptachatina), **101**, 189
hartmanii, Amastra (Leptachatina), **102**
hartmanni Ancey, Philonesia (Microcystis), **159**, 203
hartmanni Clessin, Lymnaea (Physa), **32**, 173
haupuensis Cooke, Godwinia, **164**, 204
havaiana Schaufuss, Achatinella, **61**, 179
hawaiana Paetel, Achatinella, **61**, 179
hawaiensis Pilsbry, Erinna (Limnaea), **31**, 173
hawaiensis Pilsbry & Cooke, Pupoidopsis, **138**, 199
HAWAIIA Gude, **163**, 203
hawaiiense Sykes, Allopeas (Opeas), **141**, 199
hawaiiensis Ancey, Lyropupa, **132**, 197
hawaiiensis Ancey, Nesovitrea (Vitrea), **165**, 204
hawaiiensis Baldwin, Partulina (Achatinella), **65**, 180
hawaiiensis Baldwin, Partulina (Achatinella), **65**, 180
hawaiiformis Cooke, Tornatellina, **87**, 186
hawaiiensis Hyatt & Pilsbry, Amastra, **97**, 188
hawaiiensis Pilsbry & Cooke, Pleuropoma (Helicina), **19**, 171
hawaiiensis Pilsbry & Cooke, Tornatellaria, **82**, 184
hayseldeni Baldwin, Partulina, **66**, 181
helena Newcomb, Perdicella (Achatinella), **74**, 182
HELICAMASTRA Pilsbry & Vanatta, **117**, 193
HELICARIONIDAE, **153**, 202
heliciformis Ancey, Tropidoptera (Amastra), **118**, 193
HELICINA Lamarck, **23**, 172
HELICINIDAE, **17**, 170
HELICTER, **48**
HELICTERELLA Gulick, **48**, 176
HELICTERES Beck, **48**, 176
HELICTERES Férussac, **48**, 176
helvina Baldwin, Laminella (Achatinella), **115**, 193
helvina Baldwin, Laminella (Achatinella), **115**, 193
henshawi Ancey, Cookeconcha (Endodonta), **144**, 199
henshawi Ancey, Succinea, **150**, 201
henshawi Ancey, Tornatellaria (Tornatellina), **82**, 184
henshawi Baldwin, Amastra, **97**, 188
henshawi Sykes, Leptachatina, **128**, 197
henshawi Sykes, Paropeas (Opeas), **142**, 199
hepatica Borcherding, Partulina (Achatinellastrum), **67**, 181
herbacea Gulick, Achatinella, **51**, 177
hesperia Pilsbry & Cooke, Leptachatina, **123**, 199
HETERAMASTRA Pilsbry, **103**, 189
hibrida Neal, Orobophana, **18**, 171
HILOAA Baker, **158**, 203
hiloi Baker, Philonesia, **158**, 203
HIONA Baker, **154**, **155**, 202
HIONARION Baker, **155**, 202
HIONELLA Baker, **155**, 202
hitchcocki Cooke, Amastra, **92**, 186
honokowaiensis Neal, Pleuropoma, **21**, 171
honomuniensis Pilsbry & Cooke, Newcombia, **62**, 180
horneri Ancey, Punctum, **147**, 200
horneri Baldwin, Partulina (Achatinella), **65**, 180
horneri Baldwin, Partulina (Achatinella), **65**, 180
humilis Newcomb, Amastra (Achatinella), **92**, 186
humilis Pfeiffer, Amastra (Achatinella), **92**, 186
hutchinsonii Pease, Amastra (Helicter), **103**, 189
hyattiana Pilsbry, Carelia, **112**, 192
hybrida Newcomb, Achatinella, **51**, 177
HYDROBIIDAE, **24**, 172
HYDROCENIDAE, **16**, 170
hyperleuca Hyatt & Pilsbry, Carelia, **112**, 192
hyperodon Pilsbry & Cooke, Leptachatina, **119**, 194
hystricella Cooke & Pilsbry, Pronesopupa, **138**, 199
hystricella Pfeiffer, Cookeconcha (Helix), **144**, 199
hystrix Pfeiffer, Cookeconcha (Helix), **144**, 200

INDEX [241]

idae Borcherding, Partulina, **70**, 182
idae Cooke & Pilsbry, Tornatellides, **84**, 185
ignominiosus Paetel, Achatinella, **61**, 179
ihiihiensis Welch, Achatinella, **42**, 174
ILIKALA Cooke, **119**, 194
illibata Cooke & Pilsbry, Tornatellaria, **82**, 184
illimis Cooke, Leptachatina, **123**, 195
imitatrix Sykes, Leptachatina, **123**, 195
implicata Cooke, Amastra, **103**, 189
impressa Paetel, Achatinella, **61**, 179
impressa Sykes, Leptachatina, **123**, 195
inaequalis Adams & Adams, Laemodonta (Plecotrema), **30**, 173
incerta Cooke & Pilsbry, Pronesopupa, **138**, 199
inconspicua Ancey, Succinea, **150**, 201
indefinita Ancey, Philonesia (Microcystis), **158**, 202
indefinita Lea & Lea, Thiara (Melania), **26**, 172
induta Gulick, Partulina (Achatinella), **70**, 181
inelegans Pilsbry & Cooke, Achatinella, **56**, 178
infelix Hyatt & Pilsbry, Amastra, **97**, 188
inflata Pfeiffer, Amastra (Achatinella), **97**, 188
INFRANESOPUPA Cooke & Pilsbry, **134**, 198
infrequens Cooke, Carelia, **112**, 192
infrequens Cooke & Pilsbry, Nesopupa, **135**, 198
innotabilis Smith, Achatinella (Apex), **42**, 174
inopinata Cooke, Amastra, **92**, 186
inornata Mighels, Amastra (Achatinella), **109**, 191
inornatus Pilsbry & Cooke, Tornatellides, **84**, 185
insignis Pilsbry & Cooke, Tornatellides, **84**, 185
insignis Reeve, Partulina (Bulimus), **70**, 182
intercarinata Mighels, Cookeconcha (Helix), **144**, 199
interjecta Baker, Philonesia, **158**, 202
interjecta Hyatt & Pilsbry, Amastra, **103**, 190
intermedia Newcomb, Amastra (Achatinella), **109**, 191
interrupta Cooke & Pilsbry, Nesopupa, **135**, 198
interrupta Pilsbry & Cooke, Lyropupa, **133**, 198
irregularis Cooke & Pilsbry, Tornatellides, **84**, 185
irregularis Pfeiffer, Leptachatina (Achatinella), **119**, 194
irregularis Pfeiffer, Leptachatina (Achatinella), **84**, 194
irwini Pilsbry & Cooke, Achatinella, **42**, 174
irwiniana Cooke, Amastra, **106**, 190
isenbergi Cooke, Carelia, **112**, 192
isthmica Ancey, Leptachatina, **123**, 195

janeae Cooke, Amastra, **97**, 188
johnsoni Hyatt & Pilsbry, Amastra, **92**, 186
johnsoni Newcomb, Achatinella, **51**, 177
jucunda Bland & Binney, Auriculella, **76**, 184
jucunda Pilsbry & Cooke, Auriculella, **76**, 184
juddii Baldwin, Achatinella, **51**, 177
juddii Baldwin, Achatinella, **52**, 177
juddii Cooke, Amastra, **102**, 189
juddii Pilsbry & Cooke, Orobophana (Helicina), **18**, 171
jugosa Mighels, Cookeconcha (Helix), **144**, 200
juncea Gulick, Achatinella, **52**, 177
junceus Gould, Allopeas (Bulimus), **141**, 199

kaaeana Baldwin, Partulina, **70**, 181
kaaensis Neal, Pleuropoma, **21**, 171
KAALA Baker, **157**, 202
kaalaensis Cooke & Pilsbry, Nesopupa, **135**, 198
kaalaensis Welch, Achatinella, **42**, 176
kahakuloensis Pilsbry & Cooke, Amastra, **92**, 186
kahana Hyatt & Pilsbry, Amastra, **98**, 188
kahoolavensis Cooke & Pilsbry, Tornatellides, **84**, 185
kahoolavensis Pilsbry & Cooke, Lyropupa, **133**, 198
kahoolawensis Neal, Pleuropoma, **21**, 171
kahukuensis Pilsbry & Cooke, Achatinella, **42**, 174
kahukuensis Pilsbry & Cooke, Tornatellides, **84**, 185
kailuana Pilsbry, Gastrocopta, **130**, 197
kailuanus Pilsbry & Cooke, Tornatellides, **84**, 185
kaipapauensis Welch, Achatinella, **56**, 178
kaipaupauensis Hyatt & Pilsbry, Amastra, **106**, 190
kalaeloana Christensen, Endodonta, **145**, 200
kalalauensis Cooke, Carelia, **112**, 192
kalamaulensis Pilsbry & Cooke, Amastra, **92**, 186
kalekukei Welch, Achatinella, **57**, 178
kaliella Baker, Philonesia, **161**, 203
kalihiensis Pilsbry & Cooke, Laminella, **115**, 193
kaliuwaaensis Pilsbry & Cooke, Achatinella, **57**, 179
kaluaahacola Pilsbry & Cooke, Partulina, **70**, 182
kamaloensis Cooke & Pilsbry, Nesopupa, **136**, 198
kamaloensis Pilsbry & Cooke, Lamellidea (Tornatellina), **80**, 184
kamaloensis Pilsbry & Cooke, Laminella, **115**, 193
kamaloensis Pilsbry & Cooke, Partulina, **70**, 182
kamaloensis Pilsbry & Cooke, Tornatellides, **85**, 185
kamehameha Pilsbry & Vanatta, Endodonta, **145**, 200
kanaiensis Pfeiffer, Bulimus, **87**, 186
kapuana Gouveia & Gouveia, Partulina, **65**, 180
kapuensis Welch, Achatinella, **42**, 176
KAUAIA Sykes, **104**, 190
kauaiensis Ancey, Nesopupa, **135**, 198
kauaiensis Baker, Philonesia, **159**, 203
kauaiensis Newcomb, Amastra (Achatinella), **104**, 190
kauaiensis Pease, Thiara (Melania), **26**, 172
kauaiensis Pfeiffer, Bulimus, **87**, 186
kauaiensis Pilsbry, Georissa, **16**, 170
kauaiensis Pilsbry & Cooke, Pleuropoma (Helicina), **19**, 171
kauensis Pilsbry & Cooke, Amastra, **98**, 188
kaunakakai Baker, Euconulus, **154**, 202
kaunakakaiensis Pilsbry & Cooke, Amastra, **92**, 186
kaupakaluana Hyatt & Pilsbry, Amastra, **92**, 186
kaupakaluana Pilsbry & Cooke, Partulina, **70**, 181
kawaiensis Pfeiffer, Hawaiia (Helix), **163**, 203
kawaiensis Reeve, Hawaiia (Helix), **163**, 203
kawaihapaiensis Pilsbry & Cooke, Amastra, **102**, 189
kawaiiki Welch, Achatinella, **42**, 174
kiekieensis Neal, Pleuropoma, **21**, 171
kilauea Pilsbry & Cooke, Tornatellides, **85**, 185
kilohanana Pilsbry & Cooke, Lamellidea (Tornatellina), **80**, 184
KIPUA Baker, **158**, 203
kipui Baker, Hiona, **156**, 202
knudseni Cooke, Carelia, **112**, 192
knudseni Cooke, Leptachatina, **123**, 195
knudseni Pilsbry & Cooke, Pleuropoma (Helicina), **19**, 171
knudsenii Baldwin, Amastra (Achatinella), **100**, 189
kobelti Borcherding, Carelia, **112**, 192
kohalensis Hyatt & Pilsbry, Amastra, **98**, 188
kona Pilsbry & Cooke, Lyropupa, **133**, 198
konaensis Neal, Pleuropoma, **21**, 171
konaensis Sykes, Euconulus (Kaliella), **154**, 202
konaensis Sykes, Leptachatina, **123**, 195
konaensis Sykes, Succinea, **150**, 201
konahuanui Baker, Philonesia, **160**, 203
konana Pilsbry & Cooke, Partulina, **65**, 180
koolauensis Cooke, Planamastra, **117**, 193
kualii Baker, Philonesia, **160**, 203
kuesteri Pfeiffer, Auriculella (Tornatellina), **76**, 183
kuhnsi Ancey, Succinea, **150**, 201
kuhnsi Cooke, Laminella (Amastra), **115**, 193
kuhnsi Cooke, Leptachatina, **123**, 195
kuhnsi Pilsbry, Perdicella (Partulina), **74**, 182
kulaensis Neal, Pleuropoma, **21**, 171

labiata Newcomb, Leptachatina (Achatinella), **119**, 194
LABIELLA Pfeiffer, **119**, 194
laciniosa Mighels, Pleuropoma (Helicina), **21**, 171
lacrima Gulick, Leptachatina (Achatinella), **123**, 195
lactea Gulick, Partulina (Achatinella), **67**, 181
laddi Solem, Protoendodonta, **146**, 200
LAEMODONTA Philippi, **29**, 173
laeva Baldwin, Amastra, **103**, 190

laeve Müller, Deroceras (Limax), **167**, 204
laevigata Cooke, Leptachatina, **124**, 195
laevis Pease, Leptachatina, **124**, 195
lagena Gulick, Leptachatina (Achatinella), **120**, 194
lahainana Pilsbry & Cooke, Amastra, **92**, 187
laiensis Pilsbry & Cooke, Achatinella, **57**, 179
LAMELLARIA Liardet, **79**, 184
LAMELLIDEA Pilsbry, **79**, 184
LAMELLINA Pease, **79**, 184
lamellosa Férussac, Endodonta (Helix), **145**, 200
lamellosa Férussac, Endodonta (Helix), **145**, 200
laminata Pease, Endodonta (Helix), **145**, 200
LAMINELLA Pfeiffer, **114**, 192
lanaiensis Cooke, Auriculella, **76**, 183
lanaiensis Cooke, Leptachatina, **124**, 195
lanaiensis Cooke, Lyropupa, **131**, 197
lanaiensis Pilsbry & Cooke, Nesopupa, **137**, 198
lanaiensis Sykes, Cookeconcha (Endodonta), **144**, 200
lanaiensis Sykes, Nesovitrea (Vitrea), **165**, 204
lanceolata Cooke, Leptachatina, **124**, 195
lanceolata Cooke & Pilsbry, Lamellidea (Tornatellina), **80**, 184
lata Adams, Laminella (Achatinella), **115**, 123
lathropae Welch, Achatinella, **42**, 176
laticeps Hyatt & Pilsbry, Amastra, **103**, 190
latizona Borcherding, Partulina (Achatinellastrum), **67**, 181
laula Neal, Pleuropoma, **21**, 171
laurani Welch, Achatinella, **42**, 174
LAXISUCCINEA Cooke, **149**, 200
lehuiensis Smith, Achatinella, **52**, 177
leiahiensis Cooke, Leptachatina, **124**, 195
lemkei Welch, Achatinella, **42**, 174
lemmoni Baldwin, Partulina, **70**, 181
lenta Cooke, Leptachatina, **119**, 194
lepida Cooke, Leptachatina, **124**, 195
LEPTACHATINA Gould, **118**, **120**, 194
LEPTACHATININAE, **118**, 194
leptospira Cooke & Pilsbry, Tornatellides, **85**, 185
leucochila Gulick, Leptachatina (Achatinella), **124**, 195
leucoderma Pilsbry & Cooke, Laminella, **115**, 193
leucophaea Gulick, Achatinella (Apex), **42**, 176
leucorrhaphe Gulick, Achatinella (Apex), **42**, 174
leucozona Gulick, Achatinella (Apex), **43**, 174
leucozonalis Beck, Helicteres, **87**, 186
libera Cooke, Laxisuccinea, **149**, 200
ligata Smith, Achatinella, **52**, 177
lignaria Gulick, Partulina (Achatinella), **70**, 182
lihueensis Neal, Orobophana, **18**, 170
lila Pilsbry, Achatinella, **57**, 179
lilacea Gulick, Achatinella (Apex), **43**, 174
lilae Cooke & Pilsbry, Tornatellaria, **82**, 184
liliacea Pfeiffer, Achatinella, **52**, 177
LIMACIDAE, **167**, 204
limatula Cooke & Pilsbry, Nesopupa, **135**, 198
LIMAX Linnaeus, **167**, 204
limbata Gulick, Achatinella, **57**, 178
LIMBATIPUPA Cooke & Pilsbry, **135**, 198
lineipicta Welch, Achatinella, **43**, 174
lineolata Newcomb, Amastra (Achatinella), **92**, 187
lirata Cooke, Carelia, **112**, 192
lirata Pfeiffer, Newcombia (Bulimus), **62**, 180
LIRATOR Beck, **29**, 173
lita Pilsbry, Tropidoptera (Pterodiscus), **118**, 193
litoralis Cooke & Pilsbry, Nesopupa, **137**, 199
littoralis Pilsbry & Cooke, Achatinella, **52**, 176
livida Swainson, Achatinella, **52**, 177
longa Sykes, Amastra, **93**, 187
longior Pilsbry & Cooke, Partulina, **70**, 182
longispira Smith, Achatinella, **52**, 178
longiuscula Cooke, Leptachatina, **124**, 195
lorata Férussac, Achatinella (Helix), **43**, 175
lorata Férussac, Achatinella (Helix), **43**, 175
luakahaense Pilsbry & Cooke, Elasmias, **87**, 186
lubricellus Ancey, Euconulus (Kaliella), **154**, 202
lucida Ancey, Succinea, **151**, 201
lucida Pease, Leptachatina, **124**, 195

lucidus Pease, Melampus, **29**, 173
luctifera Pilsbry & Vanatta, Cookeconcha (Endodonta), **144**, 200
luctuosa Pfeiffer, Amastra (Achatinella), **98**, 188
lugubris Chemnitz, Achatinella (Turbo), **43**, 174
lugubris Férussac, Achatinella (Helix), **43**, 174
lugubris Férussac, Achatinella (Helix), **43**, 174
lumbalis Gould, Succinea, **151**, 201
lurida Pfeiffer, Auriculella (Achatinella), **76**, 183
lutea Pfeiffer, Partulina (Achatinella), **70**, 182
luteola Férussac, Amastra (Helix), **110**, 191
luteostoma Baldwin, Achatinella, **57**, 179
luteostoma Baldwin, Achatinella, **57**, 179
lutulenta Ancey, Succinea, **151**, 201
lymani Cooke, Carelia, **112**, 192
lymaniana Baker, Hiona, **156**, 202
lymaniana Baldwin, Achatinella, **43**, 176
lymaniana Baldwin, Achatinella, **43**, 176
lymaniana Cooke & Pilsbry, Pronesopupa, **138**, 199
lymaniana Pilsbry & Cooke, Orobophana (Helicina), **18**, 171
lymanniana Ancey, Hiona (Microcystis), **156**, 202
LYMNAEA Lamarck, **31**, 173
LYMNAEIDAE, 31, 173
lyonsiana Ancey, Gastrocopta (Pupa), **130**, 197
lyonsiana Baldwin, Achatinella, **57**, 179
lyonsiana Baldwin, Achatinella, **57**, 179
lyrata Gould, Lyropupa (Pupa), **131**, 197
LYROPUPA Pilsbry, **131**, 197
LYROPUPILLA Pilsbry & Cooke, **132**, 197

macerata Hyatt & Pilsbry, Amastra, **93**, 187
macrodon Borcherding, Partulina, **70**, 181
macromphala Ancey, Tornatellides (Tornatellina), **85**, 185
macroptychia Ancey, Tornatellides (Tornatellina), **85**, 185
macrostoma Pfeiffer, Achatinella, **57**, 179
magdalenae Ancey, Lyropupa (Pupa), **132**, 197
magdalenae Ancey, Orobophana (Helicina), **18**, 171
magna Adams, Amastra (Achatinella), **93**, 187
magnapustulata Cooke & Kondo, Carelia, **112**, 192
magnifica Schaufuss, Achatinella, **61**, 179
mahogani Gulick, Achatinella, **57**, 179
mailiensis Welch, Achatinella, **43**, 176
makahaensis Pilsbry & Cooke, Achatinella, **43**, 176
makalii Neal, Pleuropoma, **20**, 171
makawaoensis Hyatt & Pilsbry, Amastra, **93**, 187
makuaensis Neal, Orobophana, **18**, 171
malleata Ancey, Auriculella, **76**, 183
malleata Smith, Amastra, **93**, 187
mamillaris Ancey, Succinea, **151**, 201
manana Pilsbry & Cooke, Leptachatina, **124**, 195
maniensis Pfeiffer, Leptachatina (Achatina), **124**, 195
maniensis Pfeiffer, Perdicella (Achatinella), **74**, 182
manoaensis Pfeiffer, Leptachatina (Achatinella), **124**, 196
manoensis Pease, Perdicella, **74**, 183
mapulehuae Baker, Philonesia, **157**, 202
margaretae Pilsbry & Cooke, Achatinella, **52**, 177
margarita Pfeiffer, Leptachatina (Achatina), **124**, 194
marginata Gulick, Leptachatina (Achatinella), **124**, 195
marmorata Gould, Partulina (Achatinella), **70**, 181
marsupialis Pilsbry & Vanatta, Endodonta, **146**, 200
martensi Borcherding, Partulina (Achatinellastrum), **67**, 181
mastersi Newcomb, Amastra (Achatinella), **93**, 187
matutina Neal, Pleuropoma, **21**, 171
mauiensis Ancey, Succinea, **151**, 201
mauiensis Lea, Tarebia (Melania), **25**, 172
mauiensis Neal, Pleuropoma, **20**, 171
mauiensis Pfeiffer, Perdicella (Partulina), **74**, 182
MAUKA Baker, **158**, 203
maunahoomae Baker, Hiona, **155**, 201
maunalei Baker, Philonesia, **160**, 203
maunaloae Pilsbry & Cooke, Lyropupa, **133**, 198

INDEX [243]

maura Ancey & Sykes, Amastra, **93**, 187
maxima Henshaw, Succinea, **151**, 201
maxima Welch, Achatinella, **44**, 176
mcgregori Pilsbry & Cooke, Leptachatina, **124**, 195
mcgregori Pilsbry & Cooke, Pacificella (Tornatellina), **81**, 184
meadowsi Welch, Achatinella, **44**, 174
media Hyatt & Pilsbry, Amastra, **106**, 190
megodonta Baker, Hiona, **155**, 202
meineckei Baker, Hiona, **156**, 202
meineckei Cooke, Amastra, **98**, 188
meineckei Cooke, Carelia, **112**, 192
meineckei Neal, Orobophana, **18**, 171
meineckei Pilsbry & Cooke, Achatinella, **52**, 177
meinickei, Achatinella, **52**
MELAMPODINAE, **28**, 172
melampoides Pfeiffer, Leptachatina (Achatinella), **125**, 196
MELAMPUS Montfort, **28**, 172
MELANIA Lamarck, **26**, 172
melanogama Pilsbry & Cooke, Achatinella, **44**, 175
melanosis Newcomb, Amastra (Achatinella), **98**, 188
melanostoma Newcomb, Achatinella, **57**, 179
meniscus Ancey, Striatura (Pseudohyalinia), **163**, 203
METAMASTRA Hyatt & Pilsbry, **104**, 190
metamorpha Pilsbry & Cooke, Amastra, **102**, 189
meyeri Borcherding, Partulina, **70**, 182
micans Pfeiffer, Amastra (Achatinella), **109**, 191
micra Cooke, Leptachatina, **125**, 195
micra Cooke & Pilsbry, Lyropupa, **133**, 198
MICROCYSTINAE, **154**, 202
microdon Pilsbry & Cooke, Leptachatina, **119**, 194
micromphala Pilsbry & Cooke, Tornatellides, **85**, 185
microstoma Gould, Amastra (Achatinella), **106**, 191
microthauma Ancey, Lyropupa, **132**, 197
mighelsiana Pfeiffer, Partulina (Achatinella), **67**, 181
MILACIDAE, **166**, 204
MILAX Gray, **166**, 204
milolii Baker, Hiona, **156**, 202
minor Borcherding, Carelia, **112**, 192
minuscula Binney, Hawaiia (Helix), **163**, 203
minuscula Pfeiffer, Perdicella (Achatinella), **74**, 182
minuta Cooke & Pilsbry, Auriculella, **76**, 183
mirabilis Ancey, Lyropupa (Pupa), **132**, 197
mirabilis Cooke, Amastra, **93**, 187
mirabilis Cooke, Carelia, **112**, 192
mirabilis Henshaw, Succinea, **151**, 201
MIRAPUPA Cooke & Pilsbry, **133**, 198
mistura Pilsbry & Cooke, Achatinella, **57**, 178
mixta Welch, Achatinella, **44**, 176
modesta Adams, Amastra (Achatinella), **93**, 187
modicella Cooke, Amastra, **102**, 189
moesta Newcomb, Amastra (Achatinella), **93**, 187
mokuleiae Baker, Philonesia, **160**, 203
moloaaensis Cooke & Kondo, Carelia, **112**, 192
molokaiensis Cooke, Leptachatina, **125**, 195
molokaiensis Cooke & Pilsbry, Pronesopupa, **138**, 199
molokaiensis Neal, Pleuropoma, **21**, 171
molokaiensis Sykes, Nesovitrea (Vitrea), **165**, 204
monacha Pfeiffer, Achatinella, **58**, 178
montagui Pilsbry, Amastra, **106**, 190
montagui Pilsbry, Partulina, **71**, 181
montana Baldwin, Amastra, **93**, 187
montana Cooke, Auriculella, **76**, 183
montivaga Cooke, Amastra, **106**, 190
moomomiensis Neal, Pleuropoma, **21**, 171
moomomiensis Pilsbry & Cooke, Amastra, **93**, 186
moomomiensis Pilsbry & Cooke, Lyropupa, **134**, 198
moomomiensis Pilsbry & Cooke, Tornatellides, **85**, 185
morbida Cooke, Leptachatina, **128**, 197
morbida Pfeiffer, Partulina (Achatinella), **65**, 180
moreletiana Clessin, Lymnaea (Physa), **32**, 173
morticina Hyatt & Pilsbry, Amastra, **102**, 189
mucida Baldwin, Partulina (Achatinella), **71**, 181
mucida Baldwin, Partulina (Achatinella), **71**, 181
mucronata Newcomb, Amastra (Achatinella), **93**, 187
multicolor Pfeiffer, Achatinella, **58**, 179

multidentata Cooke & Pilsbry, Nesopupa, **135**, 198
multilineata Newcomb, Achatinella, **44**, 176
multistrigata Pilsbry & Cooke, Partulina, **71**, 181
multisulcatus Beck, Laemodonta (Lirator), **30**, 173
multizonata Baldwin, Achatinella, **52**, 176
multizonata Baldwin, Achatinella, **52**, 176
muricolor Welch, Achatinella, **44**, 174
muscaria Hyatt & Pilsbry, Laminella, **116**, 193
mustelina Mighels, Achatinella, **44**, 176
mutabilis Baldwin, Partulina, **67**, 181
myrrhea Pfeiffer, Partulina (Achatinella), **71**, 182

nacca Gould, Gastrocopta (Vertigo), **131**, 197
nana Baldwin, Amastra (Achatinella), **93**, 187
nana Baldwin, Amastra (Achatinella), **94**, 187
nannodes Cooke, Amastra, **103**, 189
nanus Cooke & Pilsbry, Tornatellides, **85**, 185
napus Pfeiffer, Achatinella, **44**, 175
naticoides Clessin, Lymnaea (Physa), **32**, 173
nattii Hartman, Partulina (Achatinella), **67**, 181
neali Pilsbry, Georissa, **16**, 170
neckeri Cooke & Kondo, Tornatellides, **83**, 184
necra Cooke, Carelia, **113**, 192
neglecta Pilsbry & Cooke, Amastra, **94**, 187
neglectus Smith, Achatinella (Apex), **44**, 175
nematoglypta Pilsbry & Cooke, Leptachatina, **119**, 194
NERIPTERON Lesson, **14**, 170
NERITIDAE, **13**, 170
NERITINA Lamarck, **13**, 170
NERITINAE, **13**, 170
NERITONA Martens, **14**, 170
NESOCONULUS Baker, **154**, 202
NESOCYCLUS Baker, **156**, 202
NESODAGYS Cooke & Pilsbry, **136**, 198
NESOPHILA Pilsbry, **146**, 200
NESOPUPA Pilsbry, **134**, 198
NESOPUPILLA Pilsbry & Cooke, **136**, 198
NESOPUPINAE, **131**, 197
NESOVITREA Cooke, **164**, 204
NEUTRA Baker, **156**, 202
newcombi Adams & Adams, Erinna, **31**, 173
newcombi Pfeiffer, Auriculella (Balea), **76**, 183
newcombi Pfeiffer, Carelia (Achatina), **113**, 192
newcombi Pfeiffer, Catinella (Succinea), **148**, 200
newcombi Pfeiffer, Godwinia (Helix), **164**, 204
newcombi Pfeiffer, Nesopupa (Pupa), **136**, 198
newcombi Pfeiffer, Tornatellaria (Tornatellina), **82**, 184
newcombi Reeve, Godwinia (Helix), **164**, 204
NEWCOMBIA Pfeiffer, **62**, 180
NEWCOMBIANA, **62**
newcombiana Garrett, Succinea, **151**, 201
newcombianus Pfeiffer, Newcombia (Bulimus), **62**, 180
newcombii Lea, Thiara (Melania), **26**, 172
nigra Newcomb, Amastra (Achatinella), **94**, 187
nigra Pfeiffer, Amastra (Achatinella), **94**, 187
nigricans Pilsbry & Cooke, Achatinella, **58**, 178
nigripicta Welch, Achatinella, **44**, 175
nigrolabris Smith, Amastra, **109**, 191
niihauensis Neal, Pleuropoma, **20**, 171
nitida Newcomb, Leptachatina (Achatinella), **125**, 195
nivea Baldwin, Partulina (Achatinella), **71**, 181
nivea Baldwin, Partulina (Achatinella), **71**, 181
nivosa Newcomb, Achatinella, **58**, 178
nobilis Pfeiffer, Achatinella, **44**, 175
nocturna Welch, Achatinella, **44**, 176
nonouensis Neal, Pleuropoma, **22**, 172
normalis Hyatt & Pilsbry, Amastra, **98**, 188
nubifera Hyatt & Pilsbry, Amastra, **94**, 187
nubigena Pilsbry & Cooke, Amastra, **104**, 190
nubilosa Mighels, Amastra (Achatinella), **94**, 187
nucleola Gould, Amastra (Achatinella), **98**, 188
nucula Smith, Amastra, **94**, 187
nuda Ancey, Cookeconcha (Endodonta), **144**, 200

nuuanuensis Pilsbry & Cooke, Orobophana (Helicina), **18**, 171
nympha Gulick, Achatinella, **58**, 179
oahouensis Souleyet, Lymnaea, **32**, 173
oahuensis Ancey, Philonesia (Microcystis), **160**, 200
oahuensis Brot, Thiara (Melania), **26**, 172
oahuensis Cooke & Pilsbry, Nesopupa, **136**, 198
oahuensis Cooke & Pilsbry, Tornatellides, **85**, 185
oahuensis Green, Amastra (Achatina), **109**, 191
oahuensis Pilsbry & Cooke, Pleuropoma (Helicina), **22**, 172
obclavata Pfeiffer, Leptachatina (Achatinella), **125**, 196
obeliscus Pfeiffer, Auriculella (Achatinella), **76**, 183
obeliscus Reeve, Carelia (Achatina), **113**, 192
obesa Newcomb, Amastra (Achatinella), **102**, 189
obesiformis Welch, Achatinella, **44**, 176
obesula Ancey, Succinea, **151**, 201
obliqua Ancey, Auriculella, **76**, 183
obliqua Gulick, Achatinella, **58**, 178
oblonga Pease, Lamellidea (Tornatellina), **80**, 184
obscura Newcomb, Amastra (Achatinella), **94**, 187
obsoleta Pfeiffer, Leptachatina (Spiraxis), **125**, 195
obtusa Pfeiffer, Leptachatina (Achatinella), **125**, 195
obtusangula Pfeiffer, Hiona (Helix), **156**, 202
obtusum Pilsbry & Cooke, Elasmias, **87**, 186
occidentalis Cooke, Leptachatina, **125**, 195
occidentalis Pilsbry & Cooke, Partulina, **71**, 181
occidentalis Pilsbry & Cooke, Tornatellaria, **82**, 184
octanfracta Jonas, Laemodonta (Pedipes), **30**, 173
octavula Paetel, Leptachatina (Achatinella), **125**, 196
octogyrata Gulick, Leptachatina (Achatinella), **125**, 195
oioensis Welch, Achatinella, **44**, 175
olaaensis Cooke, Leptachatina, **125**, 195
olaaensis Pilsbry, Columella, **139**, 199
olesonii Baldwin, Achatinella, **53**, 178
olivacea Cooke, Auriculella, **76**, 183
olivacea Pease, Carelia, **113**, 192
olivacea Reeve, Achatinella, **53**, 178
oliveri Welch, Achatinella, **44**, 175
omphalodes Ancey, Leptachatina (Thaanumia), **128**, 197
OMPHALOPS Baker, **164**, 204
oncospira Cooke & Pilsbry, Tornatellides, **85**, 185
oomorpha Gulick, Achatinella, **58**, 178
OPEAS Albers, **142**, 199
opella Pilsbry & Vanatta, Opeas, **142**, 199
opipara Cooke, Leptachatina, **125**, 195
optabilis Cooke, Leptachatina, **128**, 197
oregonensis Lea, Succinea, **152**, 201
orientalis Hyatt & Pilsbry, Amastra, **106**, 190
orientalis Hyatt & Pilsbry, Laminella, **116**, 193
orientalis Neal, Pleuropoma, **20**, 171
ornata Newcomb, Perdicella (Achatinella), **74**, 182
OROBOPHANA Wagner, **17**, 170
orophila Ancey, Succinea, **151**, 201
orycta Cooke & Pilsbry, Pronesopupa, **138**, 199
oryza Pfeiffer, Leptachatina (Achatinella), **125**, 195
oswaldi Cooke, Amastra, **106**, 190
oswaldi Cooke & Kondo, Tornatellides, **85**, 185
oswaldi Welch, Achatinella, **58**, 178
ovata Cooke, Leptachatina, **125**, 196
ovata Newcomb, Achatinella, **58**, 178
ovatula Cooke, Amastra, **98**, 188
ovatula Cooke & Pilsbry, Lyropupa, **134**, 198
oviformis Newcomb, Achatinella, **58**, 179
oviformis Pfeiffer, Achatinella, **58**, 179
ovum Pfeiffer, Achatinella, **45**, 175
owaihiensis Chamisso, Auriculella (Auricula), **76**, 183

paalaensis Welch, Achatinella, **45**, 175
pachystoma Pease, Leptachatina (Helicter), **125**, 196
PACIFICELLA Odhner, **80**, 184
PACIFICELLINAE, **78**, 184
pagodula Cooke, Amastra, **98**, 188
palawai Baker, Philonesia, **160**, 203
palehuae Baker, Philonesia, **160**, 203

pallida Jay, Achatinella, **45**, 175
pallida Reeve, Achatinella, **45**, 175
papakokoensis Welch, Achatinella, **58**, 178
papillosa Jay, Neritina, **14**, 170
papyracea Gulick, Achatinella, **53**, 177
paradoxa Pfeiffer, Carelia (Spiraxis), **113**, 192
PARAMASTRA Hyatt & Pilsbry, **108**, 191
PAROPEAS Pilsbry, **142**, 199
paropsis Cooke, Catinella, **149**, 200
PARTULINA Pfeiffer, **63**, **68**, 180
PARTULINELLA Hyatt, **63**, **68**, 181
parva Baker, Philonesia, **161**, 203
parvicolor Welch, Achatinella, **45**, 175
parvula Gulick, Leptachatina (Achatinella), **126**, 196
parvulus Pfeiffer, Melampus, **29**, 173
parvulus Pfeiffer, Melampus, **29**, 173
patula Mighels, Catinella (Succinea), **149**, 200
patula Smith, Auriculella, **76**, 183
PAUAHIA Cooke, **129**, 197
paucicostata Pease, Cookeconcha (Pithys), **144**, 200
paucilamellata Ancey, Cookeconcha (Endodonta), **144**, 200
paulla Brot, Thiara (Melania), **26**, 172
paulula Cooke, Amastra, **107**, 190
paumaluensis Welch, Achatinella, **45**, 175
pauxilla Gould, Nesovitrea (Helix), **165**, 204
peaseana Pilsbry, Planamastra, **117**, 193
peasei Clessin, Lymnaea (Physa), **33**, 173
peasei Smith, Amastra, **110**, 191
PEDIPEDINAE, **29**, 173
PEDIPES Férussac, **30**, 173
pellucida Baldwin, Amastra (Achatinella), **107**, 190
pellucida Baldwin, Amastra (Achatinella), **107**, 190
pellucida Gulick, Auriculella, **77**, 183
pellucida Pilsbry & Cooke, Auriculella, **76**, 183
PELLUCIDOMUS Baker, **154**, 202
peponum Gould, Lamellidea (Pupa), **80**, 184
perantiqua Cooke & Kondo, Partulina, **65**, 180
percitrea Neal, Orobophana, **18**, 171
percostata Pilsbry & Cooke, Lyropupa, **134**, 198
PERDICELLA Pease, **73**, 182
perdix Reeve, Partulina (Achatinella), **71**, 181
perfecta Pilsbry, Partulina, **71**, 182
perforata Cooke, Leptachatina, **128**, 197
perforata Pfeiffer, Partulina (Achatinella), **71**, 182
periscelis Cooke, Carelia, **113**, 192
perkinsi Collinge, Deroceras (Agriolimax), **167**, 204
perkinsi Sykes, Auriculella, **77**, 183
perkinsi Sykes, Hiona (Macrochlamys), **155**, 202
perkinsi Sykes, Leptachatina, **126**, 196
perkinsi Sykes, Newcombia, **62**, 180
perkinsi Sykes, Tornatellides (Tornatellina), **85**, 185
perlonga Pease, Lyropupa (Vertigo), **134**, 198
perlucens Ancey, Philonesia (Microcystis), **160**, 203
perparva Neal, Pleuropoma, **22**, 172
perplexa Pilsbry & Cooke, Achatinella, **45**, 175
perpusilla Smith, Auriculella, **77**, 183
persubtilis Cooke, Leptachatina, **126**, 196
perversa Cooke, Auriculella, **77**, 183
perversa Hyatt & Pilsbry, Amastra, **104**, 190
perversa Swainson, Achatinella, **45**, 175
petasus Ancey, Armsia (Pterodiscus), **118**, 194
petila Gulick, Leptachatina (Achatinella), **119**, 194
petitiana Pfeiffer, Auriculella (Tornatellina), **77**, 183
petricola Newcomb, Amastra (Achatinella), **98**, 188
petricola Pfeiffer, Amastra (Achatinella), **98**, 188
PETTANCYLUS Iredale, **34**, 173
pexa Gulick, Partulina (Achatinella), **65**, 180
pfeifferi Newcomb, Newcombia (Achatinella), **63**, 180
phaeostoma Ancey, Partulina, **65**, 180
phaeozona Gulick, Achatinella, **53**, 177
philippiana Pfeiffer, Newcombia (Achatinella), **63**, 180
PHILONESIA Sykes, **157**, **159**, 203
PHILOPOA Cooke & Kondo, **81**, 184
physa Newcomb, Partulina (Achatinella), **66**, 180
pica Swainson, Achatinella, **45**, 174
picta Mighels, Laminella (Achatinella), **116**, 193

INDEX [245]

picta Pfeiffer, Laminella (Achatinella), **116**, 193
PIENA Baker, **160**, 203
piihonuae Baker, Philonesia, **158**, 203
piliformis Neal, Pleuropoma, **22**, 172
pilsbryi Baker, Hiona, **155**, 202
pilsbryi Cooke, Amastra, **104**, 190
pilsbryi Cooke, Leptachatina, **126**, 196
pilsbryi Cooke, Tornatellides, **85**, 185
pilsbryi Sykes, Carelia, **113**, 192
pilsbryi Welch, Achatinella, **45**, 175
PIRA Adams & Adams, **28**, 172
pisum Philippi, Helicina, **23**, 172
plagioptyx Pilsbry & Cooke, Lyropupa, **134**, 198
plagioptyx Pilsbry & Cooke, Tornatellides, **85**, 185
PLANAMASTRA Pilsbry, **117**, 193
planospira Pfeiffer, Achatinella, **59**, 179
platyla Ancey, Hiona (Microcystis), **155**, 202
platystyla Gulick, Partulina (Achatinella), **66**, 180
PLECOTREMA Adams & Adams, **29**, 173
PLEUROPOMA Möllendorff, **19**, **20**, 171
plicata Pfeiffer, Newcombia (Achatinella), **63**, 180
plicifera Ancey, Nesopupa, **137**, 199
plicosa Ancey, Philonesia (Microcystis), **160**, 203
plumata Gulick, Achatinella, **53**, 177
plumbea Gulick, Partulina (Achatinella), **71**, 181
pluris Pilsbry & Cooke, Lyropupa, **132**, 197
pluscula Cooke, Amastra, **102**, 189
poamohoensis Welch, Achatinella, **45**, 175
poholuae Baker, Hiona, **155**, 202
polita Baker, Philonesia, **158**, 203
polita Newcomb, Partulina (Achatinella), **67**, 181
polygnampta Pilsbry & Cooke, Lamellidea (Tornatellina), **80**, 184
polymorpha Gulick, Achatinella (Apex), **45**, 175
ponderosa Ancey, Auriculella, **77**, 183
popouelensis Pilsbry & Cooke, Tornatellides, **85**, 185
popouwelae Baker, Philonesia, **160**, 203
popouwelensis Pilsbry & Cooke, Leptachatina, **126**, 196
popouwelensis Welch, Achatinella, **45**, 176
porcellana Ancey, Auriculella, **77**, 183
porcellana Newcomb, Partulina (Achatinella), **67**, 181
porcus Hyatt & Pilsbry, Amastra, **98**, 188
porphyrea Newcomb, Amastra (Achatinella), **109**, 191
porphyrostoma Pease, Amastra (Helicter), **109**, 191
porrecta Mighels, Paludina, **24**, 172
praemagna Neal, Orobophana, **18**, 171
praeopima Cooke, Amastra, **107**, 190
praeparva Neal, Pleuropoma, **22**, 172
praestabilis Cooke, Leptachatina, **126**, 196
prasinus Reeve, Achatinella, **53**, 178
priggei, Cooke, Carelia, **113**, 192
prionoptychia Cooke & Pilsbry, Tornatellides, **86**, 185
prisca Ancey, Lyropupa, **132**, 197
pristina Henshaw, Succinea, **151**, 201
problematica Cooke, Amastra, **102**, 189
procera Pilsbry & Cooke, Partulina (Achatinella), **66**, 180
procerulus Ancey, Tornatellides (Tornatellina), **86**, 185
producta Mighels, Lymnaea (Physa), **33**, 173
producta Reeve, Achatinella, **53**, 177
productus Ancey, Tornatellides (Tornatellina), **86**, 185
PRONESOPUPA Iredale, **137**, 199
propinquella Cooke, Carelia, **113**, 192
PROTOENDODONTA Solem, **146**, 200
protracta Sykes, Succinea, **151**, 201
proxima Pease, Partulina (Helicter), **71**, 181
PSEUDISIDORA Thiele, **31**, 173
pseudocommodus Schmeltz, Melampus, **29**, 173
PSEUDOHYALINA Morse, **162**, 203
PSEUDOVITREA Baker, **163**, 203
PTERODISCUS Pilsbry, **117**, 193
PTYCHOCHILUS Boettger, **134**
PTYCHOCHYLUS, **134**
pudorina Gould, Succinea, **152**, 201
puella Pilsbry & Cooke, Amastra, **94**, 187
pugetensis Dall, Striatura (Patulastra), **163**, 203
pulchella Pfeiffer, Achatinella, **45**, 175

pulcherrima Swainson, Achatinella, **59**, 179
pulchra Cooke, Leptachatina, **126**, 196
pulchra Pease, Auriculella, **77**, 183
pulla Pfeiffer, Amastra (Achatinella), **94**, 187
pullata Baldwin, Amastra (Achatinella), **94**, 187
pullata Baldwin, Amastra (Achatinella), **94**, 187
pumicata Mighels, Leptachatina (Bulimus), **126**, 196
pumila Pfeiffer & Clessin, Amastra (Achatinella), **94**, 187
punctata Pfeiffer, Succinea, **151**, 201
PUNCTIDAE, **147**, 200
PUNCTUM Morse, **147**, 200
punicea Welch, Achatinella, **46**, 175
PUPILLIDAE, **130**, 197
PUPILLINAE, **138**, 199
pupoidea Cooke, Leptachatina, **126**, 196
pupoidea Newcomb, Amastra (Achatinella), **94**, 186
PUPOIDOPSIS Pilsbry & Cooke, **138**, 199
pupukanioe Pilsbry & Cooke, Achatinella, **59**, 179
pusilla Baker, Philonesia, **158**, 202
pusilla Neal, Pleuropoma, **22**, 172
pusilla Newcomb, Amastra (Achatinella), **94**, 187
pusillus Gould, Nesovitrea (Helix), **165**, 204
puukolekolensis Pilsbry & Cooke, Lyropupa, **134**, 198
puukolekolensis Pilsbry & Cooke, Tornatellides, **86**, 185
pygmaea Smith, Achatinella, **53**, 177
pyramidalis Gulick, Partulina (Achatinella), **71**, 181
pyramidatus Ancey, Tornatellides (Tornatellina), **86**, 185
pyramis Pfeiffer, Leptachatina (Achatinella), **126**, 196
pyrgiscus Pfeiffer, Allopeas (Opeas), **142**, 199
PYTHIINAE, **30**, 173

quadrata Ancey, Succinea, **151**, 201
quernea Pilsbry & Cooke, Achatinella, **46**, 175

radiata Gould, Partulina (Achatinella), **72**, 182
radiata Pfeiffer, Achatinella, **59**, 179
recta Newcomb, Achatinella, **53**, 177
redfieldi Newcomb, Partulina (Achatinella), **72**, 182
reevei Adams, Achatinella, **53**, 177
remota Cooke, Amastra, **98**, 188
remyi Newcomb, Laminella (Achatinella), **116**, 193
remyi Pfeiffer, Laminella (Achatinella), **116**, 193
resinula Gulick, Leptachatina (Achatinella), **126**, 196
reticulata Gould, Lymnaea (Physa), **33**, 173
reticulata Newcomb, Amastra (Achatinella), **107**, 190
rex Sykes, Tropidoptera (Amastra), **118**, 193
rhabdota Cooke & Pilsbry, Lyropupa, **132**, 197
rhadina Cooke & Pilsbry, Nesopupa, **136**, 198
rhodorhaphe Smith, Achatinella, **53**, 177
rhodostoma Pfeiffer, Orobophana (Helicina), **18**, 171
ricei Cooke, Amastra, **100**, 189
rigida Hyatt, Carelia, **113**, 192
ringens Sykes, Cookeconcha (Endodonta), **144**, 200
rohri Pfeiffer, Partulina (Achatinella), **72**, 182
ronaldi Cooke & Pilsbry, Tornatellides, **86**, 185
rosea Swainson, Achatinella, **59**, 178
rosealimbata Welch, Achatinella, **59**, 178
roseata Welch, Achatinella, **46**, 175
roseipicta Welch, Achatinella, **46**, 175
roseoplica Pilsbry & Cooke, Achatinella, **59**, 179
roseotincta Sykes, Amastra, **94**, 187
rotelloidea Mighels, Pleuropoma (Helicina), **20**, 171
rotunda Gulick, Achatinella, **59**, 178
rotundata Gould, Catinella (Succinea), **149**, 200
rubella Lea, Lymnaea, **33**, 173
rubella Pease, Succinea, **152**, 201
rubens Gould, Amastra (Achatinella), **99**, 188
rubicunda Baldwin, Amastra (Achatinella), **99**, 188
rubicunda Baldwin, Amastra (Achatinella), **99**, 188
rubida Gulick, Amastra, **99**, 188
rubida Pease, Catinella, **149**, 200
rubidilinea Welch, Achatinella, **46**, 175
rubidipicta Welch, Achatinella, **46**, 175

rubiginosa Gould, Cookeconcha (Helix), **144**, 200
rubiginosa Newcomb, Achatinella, **59**, 179
rubinia Hyatt & Pilsbry, Amastra, **99**, 188
rubristoma Baldwin, Amastra, **95**, 187
rudicostatus Ancey, Tornatellides (Tornatellina), **86**, 186
rudis Pfeiffer, Amastra (Achatinella), **109**, 191
rufa Newcomb, Partulina (Achatinella), **72**, 182
rufapicta Welch, Achatinella, **59**, 178
rufobrunnea Ancey, Hiona (Microcystis), **155**, 202
rugata Pease, Endodonta (Helix), **146**, 200
rugosa Newcomb, Achatinella, **59**, 179
RUGOSELLA Coen, **64**, 180
RUGOSELLA Mienis, **64**, 180
rugulosa Pease, Amastra, **99**, 188
russi Welch, Achatinella, **46**, 176
rustica Gulick, Amastra, **95**, 186
rutila Newcomb, Achatinella, **59**, 179

saccata Pfeiffer, Partulina (Achatinella), **67**, 181
saccula Hartman, Leptachatina (Achatinella), **126**, 196
sagittata Pilsbry & Cooke, Leptachatina, **126**, 196
sandvicensis Pfeiffer, Helix, **169**, 204
sandvicensis Pease, Pedipes, **30**, 173
sandvicensis Pfeiffer, Leptachatina (Achatina), **126**, 196
sandvicensis Philippi, Lymnaea, **33**, 173
sandwichensis Clessin, Lymnaea (Physa), **33**, 173
sandwichensis Pfeiffer, Spiraxis, **88**, 186
sandwichiensis Souleyet, Allochroa (Auricula), **27**, 172
sandwichiensis Souleyet, Limax, **168**, 204
sandwichiensis Souleyet, Pleuropoma (Helicina), **22**, 172
sanguinea Newcomb, Laminella (Achatinella), **116**, 193
saxatilis Gulick, Leptachatina (Achatinella), **126**, 196
saxicola Baldwin, Amastra, **99**, 188
scabra Pilsbry & Cooke, Lyropupa, **132**, 197
scamnata Schaufuss, Achatinella, **61**, 179
schauinslandi Borcherding, Partulina, **72**, 181
scitula Gulick, Achatinella, **53**, 176
sculpta Pilsbry & Cooke, Leptachatina (Achatina), **126**, 196
sculptus Pfeiffer, Melampus, **29**, 173
scutilus Mighels, Leptachatina (Bulimus), **126**, 196
semicarinata Newcomb, Partulina (Achatinella), **67**, 181
semicarnea Ancey & Sykes, Amastra, **95**, 187
semicostata Pfeiffer, Achatinella, **129**, 197
seminigra Hyatt & Pilsbry, Amastra, **99**, 188
seminigra Lamarck, Achatinella (Monodonta), **46**, 174
seminuda Baldwin, Amastra (Achatinella), **95**, 187
seminuda Baldwin, Amastra, **95**, 187
seminulum Boettger, Nesopupa (Pupa), **136**, 198
semipicta Sykes, Leptachatina, **127**, 196
semiplicatum Adams & Adams, Melampus (Ellobium), **29**, 173
semitecta Paetel, Achatinella, **61**, 180
semivenulata Borcherding, Laminella, **116**, 193
semivestita Hyatt & Pilsbry, Laminella, **116**, 193
senilis Baldwin, Amastra, **99**, 188
senistra Schmeltz, Achatinella, **53**, 178
sepulta Pilsbry & Cooke, Amastra, **95**, 186
sericans Ancey, Philonesia (Microcystis), **157**, 202
sericata Cooke & Pilsbry, Pronesopupa, **138**, 199
sericea Pfeiffer, Amastra (Achatinella), **107**, 190
SERICIPUPA Cooke & Pilsbry, **138**, 199
serrarius Pilsbry & Cooke, Tornatellides, **86**, 185
serrula Cooke, Auriculella, **77**, 183
servilis Gould, Gastrocopta (Pupa), **131**, 197
setigera Gould, Cookeconcha (Helix), **144**, 200
sharpi Pilsbry & Cooke, Columella (Sphyradium), **139**, 199
sharpi Pilsbry & Cooke, Tornatellaria, **82**, 184
sharpi Sykes, Ferrissia (Ancylus), **34**, 173
signata Neal, Pleuropoma, **22**, 172
similaris Baker, Philonesia, **159**, 203
similaris Pease, Amastra, **102**, 189
simplex Pease, Leptachatina (Helicter), **127**, 196

simulacrum Pilsbry & Cooke, Achatinella, **46**, 175
simulans Reeve, Achatinella, **46**, 175
simularis Hartman, Amastra, **95**, 187
simulator Pilsbry & Cooke, Achatinella, **46**, 175
SINALBINULA Pilsbry, **131**, 197
sinclairi Ancey, Carelia, **113**, 192
singularis Cooke & Kondo, Philopoa, **81**, 184
singularis Cooke & Pilsbry, Nesopupa, **136**, 198
sinistra Schaufuss, Achatinella, **61**, 180
sinistrorsa Baldwin, Amastra, **104**, 190
sinistrorsa Chamisso, Auriculella (Auricula), **77**, 183
sinistrorsus Chemnitz, Achatinella (Turbo), **46**, 175
sinistrorsus Martens, Lymnaea, **33**, 173
sinulifera Pilsbry & Cooke, Lyropupa, **133**, 197
smithi Cooke & Pilsbry, Tornatellaria, **82**, 184
smithi Sykes, Leptachatina, **127**, 196
sola Hyatt & Pilsbry, Amastra, **102**, 189
sola Neal, Pleuropoma, **20**, 171
solida Ancey, Auriculella, **77**, 183
solida Pease, Amastra (Helicter), **110**, 188, 190
solida Pfeiffer, Partulina (Achatinella), **72**, 182
solidissima Bland & Binney, Auriculella, **77**, 183
solidissima Sowerby, Neritina, **14**, 170
solitaria Newcomb, Achatinella, **53**, 177
somniator Pilsbry & Cooke, Leptachatina, **127**, 196
sordida Newcomb, Achatinella, **46**, 176
soror Newcomb, Amastra (Achatinella), **104**, 190
sororem Pfeiffer, Amastra (Achatinella), **104**, 189
souleyeti Ancey, Succinea, **151**, 201
sowerbyana Pfeiffer, Achatinella, **60**, 179
spadicea Gulick, Achatinella, **60**, 178
spaldingi Cooke, Amastra, **107**, 190
spaldingi Cooke, Carelia, **113**, 192
spaldingi Cooke, Planamastra, **117**, 193
spaldingi Cooke & Pilsbry, Tornatellides, **86**, 185
spaldingi Neal, Pleuropoma, **22**, 172
spaldingi Pilsbry & Cooke, Achatinella, **54**, 177
spaldingi Pilsbry & Cooke, Lyropupa, **133**, 197
sparna Cooke & Pilsbry, Lyropupa, **133**, 197
sphaerica Pease, Amastra, **102**, 189
SPHAEROCONIA Wagner, **19**, 171
SPHYRADIUM, **139**
spicula Cooke, Amastra, **99**, 188
spinigera Cooke & Pilsbry, Pronesopupa, **138**, 199
spirizona Férussac, Amastra (Helix), **109**, 191
spirizona Férussac, Amastra (Helix), **109**, 191
splendida Newcomb, Partulina (Achatinella), **72**, 182
steeli Welch, Achatinella, **46**, 175
stellula Gould, Cookeconcha (Helix), **145**, 200
stewartii Green, Achatinella (Achatina), **54**, 177
stiria Gulick, Leptachatina (Achatinella), **127**, 196
stokesi Pilsbry & Cooke, Tornatellides (Tornatellaria), **86**, 185
stokesii Neal, Orobophana, **18**, 171
straminea Cooke, Auriculella, **77**, 183
straminea Reeve, Laminella (Achatinella), **116**, 193
striata Baker, Philonesia, **160**, 203
striata Newcomb, Leptachatina (Tornatellina), **127**, 196
striata Philippi, Laemodonta (Auricula), **30**, 173
striatella Gulick, Leptachatina (Achatinella), **119**, 194
striatula Gould, Leptachatina (Achatinella), **127**, 196
striatula Pease, Lyropupa (Vertigo), **132**, 197
STRIATURA Morse, **162**, 202
STRIATURA Morse, **162**
STURANYA Wagner, **19**, **20**, 171
STURANYELLA Pilsbry & Cooke, **19**, **20**, 171
STURYANELLA, **19**, **20**
subangulatus Ancey, Tornatellides (Tornatellina), **86**, 185
subassimilis Hyatt & Pilsbry, Amastra, **95**, 186
subcentralis Cooke & Pilsbry, Nesopupa, **135**, 198
subcornea Hyatt & Pilsbry, Amastra, **107**, 190
subcostata Pilsbry & Cooke, Nesopupa, **137**, 198
subcrassilabris Hyatt & Pilsbry, Amastra, **95**, 187
subcylindracea Cooke, Leptachatina, **119**, 194
subdola Baker, Hiona, **155**, 202
subglobosa Cooke, Amastra, **99**, 188

INDEX [247]

subnigra Hyatt & Pilsbry, Amastra, **95**, 187
subobscura Hyatt & Pilsbry, Amastra, **95**, 187
subovata Cooke, Leptachatina, **127**, 196
subovata Schaufuss, Achatinella, **61**, 180
subpolita Pilsbry & Cooke, Partulina, **72**, 182
subpulla Hyatt & Pilsbry, Amastra, **95**, 186
subrostrata Pfeiffer, Amastra (Achatinella), **107**, 190
subrugosa Pilsbry & Cooke, Pacificella (Tornatellina), **81**, 184
subrutila Mighels, Kaala (Helix), **157**, 202
subsculpta Neal, Pleuropoma, **22**, 172
subsoror Hyatt & Pilsbry, Amastra, **104**, 190
subtenuis Neal, Orobophana, **18**, 171
subtilissimus Gould, Euconulus (Helix), **154**, 202
subula Gulick, Leptachatina (Achatinella), **127**, 196
SUBULINIDAE, **141**, 199
SUBULININAE, **141**, 199
subvirens Newcomb, Achatinella, **60**, 179
succincta Newcomb, Leptachatina (Achatinella), **127**, 196
succincta Pfeiffer, Leptachatina (Achatinella), **127**, 196
SUCCINEA Draparnaud, **149**, 201
SUCCINEIDAE, **148**, 200
SUCCINEINAE, **149**, 201
suffusa Reeve, Laminella (Achatinella), **116**, 193
sulcata Pfeiffer, Newcombia (Achatinella), **63**, 180
sulculosa Ancey, Pleuropoma (Helicina), **20**, 171
sulphuratus Beck, Helicteres, **88**, 186
sulphurea Ancey, Amastra, **99**, 188
supracostata Sykes, Leptachatina, **127**, 196
suturafusca Welch, Achatinella, **47**, 175
suturalba Welch, Achatinella, **47**, 175
suturalis Ancey, Carelia, **113**, 192
suturalis Pilsbry & Cooke, Achatinella, **54**, 178
swainsoni Pfeiffer, Achatinella, **60**, 179
swiftii Newcomb, Achatinella, **47**, 176
sykesi Hyatt & Pilsbry, Amastra, **95**, 187
sykesii Cooke & Pilsbry, Tornatellaria, **82**, 184

taeniolata Pfeiffer, Achatinella, **60**, 179
tahitensis Brot, Tarebia (Melania), **25**, 172
tahitensis Pfeiffer, Succinea, **151**, 201
talpina Gulick, Partulina (Achatinella), **72**, 182
tantalus Pilsbry & Cooke, Auriculella, **77**, 183
tantalus Pilsbry & Cooke, Lamellidea (Tornatellina), **79**, 184
tantilla Cooke, Pauahia (Leptachatina), **129**, 197
tappaniana Adams, Partulina (Achatinella), **72**, 182
TAREBIA Adams & Adams, **25**, 172
tenebrosa Cooke, Carelia, **113**, 192
tenebrosa Pease, Leptachatina, **127**, 196
tenella Ancey, Auriculella, **77**, 183
tenella Gould, Vitrina, **164**, 204
tenerrima Ancey, Succinea, **151**, 201
tenerrima Baldwin, Succinea, **151**, 201
tenuicostata Pease, Leptachatina (Helicter), **127**, 196
tenuilabris Gulick, Amastra, **99**, 189
tenuis Smith, Auriculella, **78**, 184
tenuispira Baldwin, Amastra (Achatinella), **109**, 191
tenuispira Baldwin, Amastra (Achatinella), **109**, 191
terebra Ancey, Tornatellides (Tornatellina), **86**, 185
terebra Newcomb, Partulina (Achatinella), **73**, 182
terebralis Gulick, Leptachatina (Achatinella), **127**, 196
teres Pfeiffer, Leptachatina (Achatinella), **127**, 196
tessellata Newcomb, Partulina (Achatinella), **73**, 182
testudinea Baldwin, Amastra (Achatinella), **110**, 191
tetragona Ancey, Succinea, **151**, 201
tetrao Newcomb, Laminella (Achatinella), **116**, 193
tetrao Pfeiffer, Laminella (Achatinella), **116**, 193
textilis Ancey, Amastra (Helix), **107**, 191
textilis Férussac, Amastra (Helix), **107**, 191
thaamuni, Achatinella, 54
thaanumi Ancey, Euconulus (Kaliella), **154**, 202
thaanumi Ancey, Nesopupa, **136**, 198
thaanumi Ancey, Succinea, **152**, 201
thaanumi Cooke, Carelia, **113**, 192

thaanumi Cooke, Leptachatina, **128**, 197
thaanumi Cooke & Pilsbry, Lyropupa, **132**, 197
thaanumi Cooke & Pilsbry, Tornatellaria (Tornatellides), **82**, 184
thaanumi Hyatt & Pilsbry, Amastra, **107**, 191
thaanumi Pilsbry, Tropidoptera (Pterodiscus), **118**, 193
thaanumi Pilsbry & Cooke, Achatinella, **54**, 177
thaanumi Pilsbry & Vanatta, Cookeconcha (Endodonta), **145**, 200
THAANUMIA Ancey, **128**, 197
thaanumiana Pilsbry, Partulina, **66**, 180
thaumasia Cooke & Pilsbry, Lyropupa, **134**, 198
theodorei Baldwin, Partulina (Achatinella), **73**, 182
theodorii Baldwin, Partulina (Achatinella), **73**, 182
THIARA Röding, **26**, 172
THIARIDAE, **25**, 172
thurstoni Baker, Euconulus, **154**, 202
thurstoni Cooke, Amastra, **102**, 189
thurstoni Pilsbry & Cooke, Achatinella, **60**, 179
thwingi Ancey, Cookeconcha (Endodonta), **145**, 200
thwingi Pilsbry & Cooke, Perdicella (Partulina), **74**, 183
tiara Mighels, Nesophila (Helix), **146**, 200
TORNATELLARIA Pilsbry, **81**, 184
TORNATELLIDES Pilsbry, **83**, 184
TORNATELLIDINAE, **81**, 184
TORNATELLININAE, **87**, 186
TORNATELLINOPS, **80**
torquata Schaufuss, Achatinella, **61**, 180
torrida Gulick, Achatinella, **60**, 179
transversalis Pfeiffer, Amastra (Achatinella), **107**, 191
tricincta Hyatt & Pilsbry, Amastra, **95**, 187
tricolor Smith, Achatinella, **54**, 177
trilineata Gulick, Achatinella, **54**, 177
triplicata Pease, Auriculella, **78**, 183
tristis Férussac, Amastra, **99**, 189
tristis Férussac, Amastra (Helix), **99**, 189
triticea Gulick, Leptachatina (Achatinella), **127**, 196
trochoides Sykes, Tornatellaria (Tornatellina), **83**, 184
TROPIDOPTERA Ancey, **117**, 193
TRUELLA Pease, **152**, 201
truncata Cooke, Lyropupa, **132**, 197
tryphera Cooke & Pilsbry, Nesopupa, **136**, 198
tsunami Cooke & Kondo, Carelia, **113**, 192
tuba Pfeiffer, Partulina (Achatinella), **73**, 182
tuberans Gulick, Achatinella (Apex), **47**, 175
tuberculata Cooke, Catinella, **149**, 200
tumefactus Gulick, Achatinella (Apex), **47**, 175
turbinata Jay, Achatinella, **61**, 180
turbiniformis Gulick, Achatinella (Apex), **47**, 175
turgida Ancey, Philonesia (Microcystis), **158**, 202
turgida Newcomb, Achatinella, **47**, 176
turgidula Pease, Leptachatina, **128**, 196
turgidula Pease, Lymnaea, **33**, 173
turricula Mighels, Carelia (Achatina), **113**, 192
turrita Gulick, Leptachatina, **128**, 196
turritella Cooke, Auriculella, **78**, 184
turritella Férussac, Amastra (Helix), **110**, 191

ualapuensis Pilsbry & Cooke, Newcombia, **63**, 180
uberta Gould, Orobophana (Helicina), **18**, 171
ultima Pilsbry & Cooke, Amastra, **102**, 189
umbilicata Ancey, Tornatellaria (Auriculella), **83**, 184
umbilicata Mighels, Physa, **33**, 173
umbilicata Pfeiffer, Amastra, **102**, 189
umbrosa Baldwin, Amastra (Achatinella), **96**, 187
umbrosa Baldwin, Amastra (Achatinella), **96**, 187
uncifera Cooke & Pilsbry, Lyropupa, **132**, 197
undata Baldwin, Amastra (Achatinella), **107**, 191
undata Baldwin, Amastra (Achatinella), **107**, 191
undosa Gulick, Partulina (Achatinella), **73**, 181
undulata Newcomb, Achatinella, **54**, 177
undulata Pfeiffer, Achatinella, **54**, 177
unicolor Ancey, Amastra, **110**, 191
uniplicata Hartman, Amastra (Achatinella), **96**, 187
uniplicata Pease, Auriculella, **78**, 184

ustulata Gulick, Partulina (Achatinella), **73**, 182
ustulata Pfeiffer, Achatinella, **54**, 177

valida Pfeiffer, Achatinella, **47**, 176
vana Sykes, Leptachatina, **128**, 196
varia Cooke, Leptachatina, **128**, 196
varia Gulick, Achatinella, **54**, 177
variabilis Newcomb, Partulina (Achatinella), **67**, 181
variabilis Pease, Carelia, **113**, 192
variegata Pfeiffer, Amastra (Achatinella), **110**, 191
ventrosa Pfeiffer, Achatinella, **47**, 175
ventulus Férussac, Leptachatina (Helix), **128**, 196
ventulus Férussac, Leptachatina (Helix), **128**, 196
venulata Newcomb, Achatinella, **54**, 177
venusta Gould, Succinea, **152**, 201
venusta Mighels, Laminella (Achatinella), **116**, 193
verrauiana Lea, Thiara (Melania), **26**, 172
versicolor Gulick, Achatinella (Apex), **47**, 175
versipellis Gulick, Achatinella, **54**, 177
VERTIGININAE, 139, 199
vesicalis Gould, Succinea, **152**, 201
vespertina Baldwin, Achatinella, **47**, 175
vespertina Baldwin, Achatinella, **47**, 175
vespertina Jay, Neritina, **14**, 170
vespertina Pilsbry & Cooke, Amastra, **107**, 190
vespertina Sowerby, Neritina, **14**, 170
vestita Mighels, Achatinella, **48**, 176
vetuscula Cooke, Amastra, **96**, 188
vetusta Baldwin, Amastra (Achatinella), **107**, 191
vetusta Baldwin, Amastra (Achatinella), **108**, 191
vidua Pfeiffer, Achatinella, **60**, 178
villosa Sykes, Amastra, **104**, 189
violacea Newcomb, Amastra (Achatinella), **96**, 188
virens Gulick, Achatinella, **55**, 178
virgatifulva Welch, Achatinella, **48**, 175
virgula Cooke, Tornatellides, **86**, 185
virgula Cooke & Pilsbry, Tornatellides, **86**, 185
virgulata Mighels, Partulina (Partula), **73**, 182
viridans Mighels, Achatinella, **60**, 179
viridans Pease, Carelia, **113**, 192
viridis Pease, Carelia, **113**, 192
viriosa Cooke, Amastra, **100**, 189
vitrea Newcomb, Leptachatina (Achatinella), **128**, 194
VITREINAE, 163, 203
vitreola Gulick, Leptachatina (Achatinella), **128**, 196
vitreus Dohrn, Tornatellides (Tornatellina), **86**, 185
VITRINA Draparnaud, **164**, 204
VITRININAE, 164, 204
VITRINUS Montfort, **164**, 204

vittata Reeve, Achatinella, **48**, 175
vivens Baker, Euconulus, **154**, 202
volutata Gould, Lymnaea, **33**, 173
vulpina Férussac, Achatinella (Helix), **55**, 178
vulpina Férussac, Achatinella (Helix), **55**, 178

wahiawa Welch, Achatinella, **48**, 175
wahiawae Baker, Hiona, **155**, 202
waialaeensis Welch, Achatinella, **48**, 175
waianaeensis Welch, Achatinella, **48**, 176
waianaensis Ancey, Succinea, **152**, 201
waianaensis Cooke & Pilsbry, Nesopupa, **137**, 199
waianaensis Pilsbry & Cooke, Laminella, **116**, 193
waianaensis Pilsbry & Cooke, Tornatellides, **86**, 186
waiawa Hyatt & Pilsbry, Amastra, **110**, 191
waiheensis Baker, Philonesia, **157**, 202
WAIHOUA Baker, **161**, 203
wailauensis Hyatt & Pilsbry, Amastra, **96**, 188
wailauensis Pilsbry & Cooke, Newcombia, **63**, 180
wailelensis Welch, Achatinella, **48**, 175
wailuaensis Sykes, Partulina (Achatinella), **67**, 181
waimaluensis Welch, Achatinella, **48**, 175
waimanoensis Pilsbry & Cooke, Achatinella, **60**, 179
waimanoi Baker, Hiona, **155**, 202
WAIMEA Cooke & Pilsbry, **86**, 186
waimeae Baker, Hiona, **156**, 202
waipouliensis Cooke & Kondo, Carelia, **113**, 192
welchi Baker, Philonesia, **159**, 203
wesleyana Ancey, Nesopupa, **136**, 198
wesleyi Sykes, Tropidoptera (Endodonta), **118**, 193
westerlundiana Ancey, Auriculella, **78**, 183
wheatleyana Pilsbry & Cooke, Achatinella, **60**, 178
wheatleyi Newcomb, Achatinella, **60**, 179
whitei Cooke, Amastra, **100**, 189
wilderi Neal, Orobophana, **18**, 171
wilderi Pilsbry, Achatinella, **60**, 179
winniei Baldwin, Partulina, **73**, 182

zebra Newcomb, Perdicella (Achatinella), **74**, 183
zebrina Pfeiffer, Perdicella (Achatinella), **74**, 183
zonata Borcherding, Carelia, **113**, 192
zonata Borcherding, Carelia, **113**, 192
zonata Gulick, Achatinella, **55**, 177
ZONITIDAE, 162, 203
ZONITINAE, 164, 204